Information Fusion and Data Science

Series Editor
Henry Leung, University of Calgary, Calgary, AB, Canada

This book series provides a forum to systematically summarize recent developments, discoveries and progress on multi-sensor, multi-source/multi-level data and information fusion along with its connection to data-enabled science. Emphasis is also placed on fundamental theories, algorithms and real-world applications of massive data as well as information processing, analysis, fusion and knowledge generation.

The aim of this book series is to provide the most up-to-date research results and tutorial materials on current topics in this growing field as well as to stimulate further research interest by transmitting the knowledge to the next generation of scientists and engineers in the corresponding fields. The target audiences are graduate students, academic scientists as well as researchers in industry and government, related to computational sciences and engineering, complex systems and artificial intelligence. Formats suitable for the series are contributed volumes, monographs and lecture notes.

More information about this series at http://www.springer.com/series/15462

Qilong Xue

Data Analytics for Drilling Engineering

Theory, Algorithms, Experiments, Software

Springer

Qilong Xue
China University of Geosciences
Beijing, China

ISSN 2510-1528 ISSN 2510-1536 (electronic)
Information Fusion and Data Science
ISBN 978-3-030-34034-6 ISBN 978-3-030-34035-3 (eBook)
https://doi.org/10.1007/978-3-030-34035-3

This Springer imprint is published by the registered company Springer Nature Switzerland AG
The registered company address is: Gewerbestrasse 11, 6330 Cham, Switzerland

Preface

Digitalization reshaping the oil industry has become the industry consensus, and data have become a torrent flowing into every area of the global economy. According to IDC's estimation, the data increases by 50% annually all the time, and much information is contained in plenty of data. The new data revolutionary has emerged, and big data and data mining technologies are pushing science into a new era of big science. In line with this, this book focuses on drilling. The digital application of drilling engineering has developed too fast in recent years. I don't want to discuss the digital technologies purely, because the oil and gas industry is too vast. I hope this book can open a window and let more people pay attention to some aspects of drilling application in data analysis and processing. Some of the results in the book are our own research results, and we have published related papers. In other chapters, we try our best to explain the data model and technical difficulties according to our understanding. I am very pleased that Professor Henry has approved this field and invited me to write this volume. After more than a year of hard work, this book is finally available.

This book is mainly for data processing and mining in drilling engineering and is committed to build a bridge between drilling engineers and signal processing scientists. In drilling engineering, a lot of signal processing technologies are required to solve practical problems, such as downhole information transmission, spatial attitude of drillstring, drillstring dynamics, seismic while drilling, and so on. The current main problem is that signal processing experts do not understand the actual project and drilling engineers lack knowledge of signal processing, so there is an urgent need for a document which summarizes the signal processing issues in drilling engineering as a mathematical problem understandable to the average drilling scientist. This book presents the difficulties and challenges in signal processing encountered in drilling engineering, such as continuous wave downhole information transmission and dynamic measurement of spatial attitude at the bottom rotating drillstring, in which the signal processing algorithm is one of the core technologies. In conclusion, this book will show the importance of signal processing to drilling engineers and open up a new area of application for signal processing scientists.

So, hopefully, the book will be easily understood. It is intended for IT people, engineers, managers, business users, data analysts, and everyone involved in the drilling business who are looking for ways to improve drilling processes by using data as one of their pillars. Thanks to Dr. Wang Lu from China University of Geosciences (Beijing) and Associate Professor Wang ZiZhen of China University of Petroleum (East) for contributing to Chapters 4 and 7, respectively, to my wife and family for their support, and to my classmates and students for helping me organize the materials. My apologies if there are inappropriateness or mistakes found in the book, as these are inevitable. I hope that the interested readers can send me an email if they have any questions. I welcome any in-depth discussions and exchanges with any of the same occupation.

Beijing, China
September 2019

Qilong Xue

Acknowledgments

The author(s) disclosed receipt of the following financial support for the research, authorship, and/or publication of this article: This work was supported by the Natural Science Foundation of China (51704264) and the Fundamental Research Funds for the Central Universities (2652018096).

Thanks to Associate Professor Wang ZiZhen of China University of Petroleum (East) for contributing to Chapter 7; Dr. Wang Lu and Dr. Jin Wang from China University of Geosciences (Beijing) for contributing to Chapters 4 and 5, respectively; to Fangtao Li for contributing to Chapters 6 and 8; to my wife and family for their support; and to my classmates and students for helping me organize the materials. In addition, the authors express their appreciation to the Drilling Technology Research Institute, Shengli Petroleum Administration of Sinopec Corp., for providing data and materials.

Contents

Chapter 1
Application of Data Processing in Drilling Engineering: A Review

Abstract Digitalization will reshape the oil industry has become the industry consensus, data have become a torrent flowing into every area of the global economy. This book is mainly for data processing and mining in drilling engineering, and is committed to build a bridge between drilling engineers and signal processing scientists. The technical development of drilling engineering focuses on commercial operations, so the final presentation of any kind of data processing theory is software and some kind of instrument, Many oil companies only give the results of the test, more is used for advertising, however, we need to get to the bottom of its theoretical basis.

Keywords Data analytics · Drilling engineering · Theory · Algorithms · Experiments · Software

1.1 Theory

In drilling engineering, the following four aspects are extremely demanding on the level of signal processing, they are downhole information transmission, spatial attitude of drillstring, drillstring dynamics measurement and seismic while drilling respectively. The last three items are about downhole measurements, and the first is about how to transfer the measurement data to the ground. Various downhole parameters can be obtained through Measurement While Drilling (MWD) and Logging While Drilling (LWD) in real time. Drilling fluid pulse signal generator, as a type of data transmission mode, can be divided into positive pulse, negative pulse and continuous wave generator according to the signal type. We focus on the continuous wave generator because it is the fastest and most difficult to achieve. Currently, only a few companies have products for MWD systems [1–3] that transmit information in continuous wave mode. PowerPulser™ of Schlumberger, the continuous wave signal generator with continuous rotary valve, sends signals at a frequency of 24 Hz with the highest data transmission rate of 12bit/s. The transmission characteristics of continuous wave signal mainly include signal reflection, signal transmission and signal attenuation. At present, there are two main theories

Q. Xue, *Data Analytics for Drilling Engineering*, Information Fusion and Data Science, https://doi.org/10.1007/978-3-030-34035-3_1

on the attenuation of pressure waves in pipelines: one is the acoustic attenuation model based on Lamb law [4], and the other is the attenuation model based on wall friction which is commonly used in the pipeline water hammer theory. The acoustic attenuation model based on Lamb's law is simple and easy to use. Therefore, almost all oil companies use this formula for attenuation estimation. The attenuation model based on wall friction, in the course of its development, has undergone the stages of quasi-static wall shear model, frequency-related friction of laminar flow [5], and frequency-related friction of turbulent flow [6]. Although the water hammer theory is developing continuously, it is limited to Newtonian fluid.

Several other new transmission modes are also worthy of our attention. XACT company successively published the invention patents of related equipment and technology of acoustic remote transmission [7, 8]. The acoustic transmission system [9] uses sound waves to transmit signals through the drill pipe. Since the signal decays rapidly in the drill pipe string, relay stations should be installed every 400–500 m in the drill pipe. According to the attenuation characteristics of electromagnetic signals during channel transmission [10–12], the following conclusions can be drawn: (1) electromagnetic signals are seriously attenuated by formation, and the formation resistivity suitable for EM-MWD operation is usually 2–200; (2) if the transmission distance of EM-MWD signals reaches thousands of meters, the working frequency of electromagnetic signals should not be more than 100 Hz; (3) the working frequency, transmission distance, formation resistivity and drill pipe resistivity of electromagnetic signals can all affect the signal voltage and signal current between ground receiving electrodes. IntelliPipe [13] system, successfully developed by American IntelliServ, realizes the non-contact transmission of signals by using the electromagnetic induction coupling principle, and solved a series of problems such as line wear and poor contact in the wired transmission mode [14]. The attenuation of signal propagation can manifest as the decrease of signal output amplitude [15]. The signal output amplitude changes with the transmission distance.

As shown in the Fig. 1.1, data is transmitted to the ground through mud pulses, acoustic, electromagnetic waves or IntelliPipe. Although increasing the amount of transmitted data is the bottleneck in the development of drilling engineering technology, how to effectively apply data after it is transmitted to the ground is also an important topic. The solution platform will use 3 stages, first is data visualization and dashboarding, second is shallow analysis in which the linear regression analysis and cross plots are created and third, which is the most value add will be deep data analysis and machine learning.

In the oil & gas drilling engineering, measurement-while-drilling (MWD) system is usually used to provide real-time monitoring of the position and orientation of the bottom hole. Particularly in the rotary steerable drilling technology and application, that is a challenging task to measurement the spatial attitude of the bottom drillstring accurately in real time while the drillstring rotating. The regular wellbore position calculations are typically performed by measuring azimuth and inclination with the MWD system in a stationary mode (the drillstring non-rotating). However, the attitude of the bottom rotating drilling tool should be obtained at the real time in

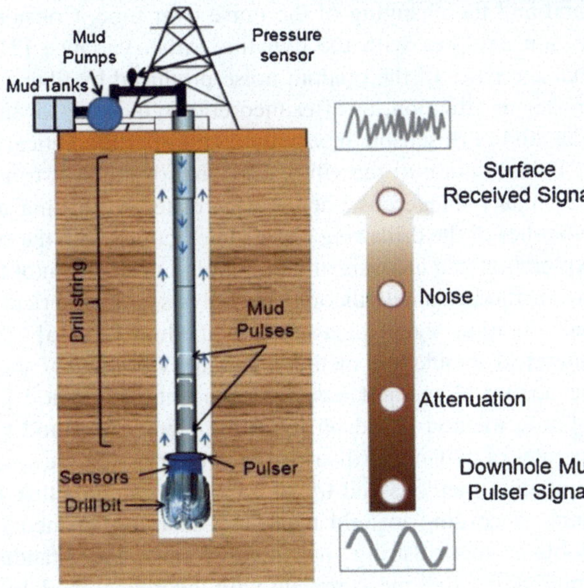

Fig. 1.1 Data transfer and application mode

both of the rotary steerable system (RSS) and automatic vertical drilling system [16–18]. Even with development of the drilling technology, continuous measurement of well trajectory becomes increasingly important.

A set of "strap-down" measurement system was developed in our research, the triaxial accelerometer and triaxial fluxgate were installed near the bit, and real-time inclination and azimuth can been measured while the drillstring rotating [19, 20]. Furthermore, established the mathematical model of continuous measurement while drilling. We proposed a dynamic solution approach to azimuth and inclination of the bottom rotating drillstring. The real-time signals of the accelerometer and the fluxgate sensors are processed and analyzed in a time window, and the movement patterns of the drilling bit will be observed, such as stationary, uniform rotation and stick-slip. The different signal processing methods will be used for different movement patterns. Additionally, put forward a scientific approach to improve the solver accuracy benefit from the use of stick-slip vibration phenomena. Based on the dynamic measurement theoretical models, we develop the Kalman filter to improve the solver accuracy as well. Although the stick-slip vibration should be avoided point of view the endurance of the drillstring, we can still take advantage of it because of it cannot be avoided entirely. Finally, develop continuous measurement while drilling algorithm processor using Kalman filtering. The actual measurement data through drilling process verify that the algorithm proposed is reliable and effective and the dynamic measurement errors of inclination and azimuth are effectively reduced. In addition, strong downhole vibration will generate greater random noise of fiber-optic gyroscope and accelerometer during MWD operation. The random noise in vibration features concerning time series mutation, slowness and periodicity in its varying. The results in Chap. 4 shows a wide internal noise band and the changing of the noise over time. Considering all these features, this section explores with the dynamic Allan variance [21, 22] method the dynamic characteristics of the random noise produced by fiber-optic gyroscope and accelerometer in vibration, to offer theoretical guidance for improving the environmental adaptability of sensors in vibration and offer theoretical support for noise modeling.

Interpretation of the vibration signal of the drillstring is another important issue [23], many dangerous phenomena related to drilling are normally caused by the dynamics of the drillstring and its interactions with the surroundings. At present, the exploration and analysis of the vibration mechanism of the drillstring mainly adopts two methods, namely theoretical analysis and numerical simulation method [24–29] and vibration signal measurement method [30–36]. The theoretical analysis and numerical simulation methods for the vibration of the drillstring mainly include the analytical method based on the energy method [26, 27] and the numerical solution method based on the finite element method [28, 29]. In order to clarify the rules of dynamic drillstring in the rotary steerable system, we fully utilize modern computing methods and try to establish a model close to the real drilling environment. According to field measured data correct theory model; complex dynamic problems of drillstring may have a clear understanding. Our innovation lies in combining actual measurement with theoretical modeling. Two evaluation methods are compared systematically, such as theoretical and measurement methods. In the Dynamic Modeling we should coupling the separate vibration form at the same time decomposition the coupled vibration form in the vibration measurement data. This is like forward and inversion algorithms in geophysics.

Seismic While Drilling (SWD) [37–40] is a well seismic method developed in recent years which is based on reverse vertical seismic logging. It is a newly-developed well seismic technology which combines the seismic exploration technology with petroleum drilling engineering technology. Compared to conventional VSP, SWD has its own characteristics and unique advantages in that it uses bit vibration in the process of drilling as the source for seismic measurement, without interfering with the drilling or occupying drilling time, and without any risk to the hole, especially the bit can be predicted in real-time structure details of the formation in front of the bit through the field seismic imaging processing, with the main purpose of reducing drilling risks. The key of this technology is how to collect and recover the weak bit reflection signal under strong disturbance noise and make it the equivalent formation impulse response.

Logging while drilling (LWD) [41, 42] is a technique of conveying well logging tools into the well borehole downhole as part of the bottom hole assembly (BHA). Although the terms Measurement while drilling (MWD) and LWD are related, within the context of this section, the term MWD refers to directional-drilling measurements, e.g., for decision support for the smooth operation of the drilling, while LWD refers to measurements concerning the geological formation made while drilling. LWD tools work with its measurement while drilling (MWD) system to transmit partial or complete measurement results to the surface via typically a drilling mud pulse or other improved techniques, while LWD tools are still in the borehole, which is called "real-time data". Complete measurement results can be downloaded from LWD tools after they are pulled out of hole, which is called "memory data".

Big data is a new area for drilling engineers, we don't care about big data theory itself, we only focus on how big data technology can be applied in drilling engineering. In Chap. 8, We have introduced the application status of big data technology in drilling engineering as comprehensively as possible. Of course, it is still far from enough. In recent years, big data technology has developed spurt and there are thousands of papers produced every year. The big data technology will generate disruptive influences on the whole oil & gas industry. Better utilize the big data, improve the petroleum and petrochemical industries and take it as the global leading industry to take the preemptive opportunities on the tide of economic development.

1.2 Experiments

Drilling engineering is science and technology that focuses on applications, the establishment of each scientific theory or the invention of new technologies requires the test of field tests. Just like the products of downhole information transmission, digital communication for drilling has to overcome unique challenges in comparison to modern wireless communication systems such as fourth generation long-term evolution and Wi-Fi.

PowerPulser™ of Schlumberger, the continuous wave signal generator with continuous rotary valve, sends signals at a frequency of 24 Hz with the highest data transmission rate of 12bit/s. Baker Hughes' continuous wave signal generator with oscillating shear valve has an experimental transmission rate of 40 bps at the well depth of 3000 ft. and of 15 bps at well depth of 24,000 ft. Halliburton is also working on developing MWD systems that transmit information in a continuous wave mode, aiming at achieving transmission rates of 20–30bit/s. XACT and Extreme Engineering conducted on-site testing of acoustic MWD tools with two diameter specification, one tool with the diameter of 121 mm, well depth of 1205 m and the transmission rate of 10bit/s, another tool with the diameter of 165 mm, well depth of 2489 m and the transmission rate of 20bit/s [43]. Up to 2011, XACT has offered service for more than 400 Wells, among which two relay amplifiers are used to achieve a maximum depth of 4000 meters. However, the use of relay amplifiers makes the reliability of the instrument under great restriction. Its system uses 640–680 Hz single passband and BPSK encoding [44]. In 2013 XACT used six relay amplifiers to offer service for wells at the greatest depth of 4000 m (13,200 ft) [45].

In Chap. 3, dynamic measurement algorithms developed were tested through laboratory bench and field measurements data respectively [46]. The field tests were carried out on wells using automatic vertical or rotary steerable drilling. Three and four ribs actuators were adopted. A lot of raw measurement data were accumulated. These data were input into our algorithm for authentication. Conclusions on bit movement rules were gained through analyzing and summarizing large amounts of data. Two wells in China (Xuanye1 and Anshun1) are selected here to evaluate the algorithm. Simulation and experimental results show that the dynamic solver methods can meet the engineering requirements. However, according to the drilling field tests, the storage data playback from the underground measurements indicated that drillstring vibration is the enormous noise, which completely submerged the desired signals.

In Chap. 5, We develop a strap down measurement while drilling surveying system that incorporates three axis magnetometers and three axis accelerometers arranged in three mutually orthogonal direction. E1~E9 in the Fig. 5.21 indicate the field experiments in China. Field observations based on downhole and surface vibration measurements have indicated that drillstrings exhibit severe vibrations. These vibrations are observed to become more severe at the BHA. The value of the angular speed fluctuates is between 0~120 rpm, which indicates that the system is in the state of stick slip. This is another new discovery about the dynamics of RSS Tool, it is observed he re that drilling vibration can lead to chaos [47, 48], since most researches in universities are limited to laboratory or theoretical studies due to the high cost of drilling, our findings reduce this gap by calculating physical nonlinear dynamical model with real drilling experiments. The existence of chaos in drilling may open a new concept of drilling chaos in the solid flow mechanics that will benefit to both the physicists and the drilling engineers.

"Drilling Change Requires Changing Drillers", drilling data is mainly obtained through on-site construction or laboratory experiments. How effective use of data poses new challenges for drilling engineers, as show the Fig. 1.2, Corva has grown

Fig. 1.2 Corva data analyst monitors many wells at a time, using programs tracking the status of wells and sending alerts on potential issues. The digital drilling advisory company regularly releases new tools to automate that job. (Source: Corva)

rapidly due to the strong demand for its real-time data and analysis system. In a year, it has gone from a couple of rigs equipped with its drilling advisory system to 250 rigs and 35 clients by September, said Ryan -Dawson, chief executive officer of Corva [49].

At present, most of data in the oil & gas industry is applied into the control and monitoring, not to optimize the assets performance. If the oil & gas industry can analyze and comprehend all data generated by itself, the operation efficiency can be enhanced by 20%. The research of Bain Capital shows that the bid data analysis can enhance the performances of oilfield and factor by 6–8% [50]. For now, the oil price and the economy are in downturn, and the profit gaining capability at the upstream will be challenged, which makes the completion among oil & gas companies more and more violent and cruel. How to reduce the operation cost of upstream enterprise, enhance production efficiency and avoid environmental risk? This is a subject that each oil & gas company must be confronted with.

The big data technology can be used to treat mass of isomeric data easily, thus providing valuable reference information for the oil & gas industry, so as to help the operation enterprise make a decision better. The digital technology can change the upstream business, and create extra profit from existing production capacity, thus making the industry more productive and flexible. Work on the application of Big Data and analytics in the oil & gas industry is in the experimental stage [53]. Only a handful companies have adopted Big Data in the field [54].

- Chevron proof-of-concept using Hadoop (IBM BigInsights) for seismic data processing;
- Shell piloting Hadoop in Amazon Virtual Private Cloud (Amazon VPC) for seismic sensor data;
- Cloudera Seismic Hadoop project combining Seismic Unix with Apache Hadoop;
- PointCross Seismic Data Server and Drilling Data Server using Hadoop and NoSQL;
- University of Stavanger data acquisition performance study using Hadoop.

1.3 Algorithms and Software

An algorithm is a procedure or formula for solving a problem, based on conducting a sequence of specified actions. In this book, we present the basic formula as the basis for algorithm programming. Of course, in drilling engineering, most algorithms are applied to underground embedded systems, which is fundamentally different from the traditional IT industry. Such as MWD or LWD tools, the measurement algorithm of the downhole attitude needs to complete the operation in the downhole MCU. But for wireless transmission systems, a perfect fit between downhole data encoding and ground data processing is required to achieve high quality signals.

For mature products, algorithms are certainly confidential in commercial companies. The software is also used as a commercial product supporting hardware or sold separately. In Chap. 3, we developed an algorithm to dynamic measurement of spatial attitude at the bottom rotating drillstring, although the effectiveness of the algorithm is proved by experimental data, it does not form a product for commercial application, so we cannot say that the algorithm must be successful [19, 20].

The future development direction is the effective combination of downhole algorithms and ground computer programs, of course, combined with theoretical analysis. As shown in Chap. 5, our innovation lies in combining actual measurement with theoretical modeling. Two evaluation methods are compared systematically, such as theoretical and measurement methods. Interpretation and development of downhole data processing algorithms is the future trend of development, the formation can effectively promote the development of related software engineering drilling technology.

The application of big data in the drilling industry is most likely to form large commercial software in the future. Big data is considered the enormous amounts of data sets, so complex that the traditional methods of analysis cannot even begin to crack it. It is thus where the big-data analytics come in to process and from it generate consistent patterns and correlations that could be used for discoveries and communication of essential and useful tips and trends. During the entire process, mathematical and statistical formulae and tools are used alongside algorithms [51]. What oil companies can do with such information generated from a successful analysis of the data is overwhelming. The only challenge is in the processing. Drilling and performance, as well as prediction of the market trends, could be taken to a whole new level. Subsurface geology, geographical locations, the type of soil, and the depth are a few of the factors the tools take into consideration. Much of the software innovation that's key to the digitization of big oil is happening at oil service contracting companies, such as Halliburton and Schlumberger, and big IT providers including Microsoft, IBM, Oracle and Open Source Projects [54].

With the rise in exploratory frontiers, the increase in rig day rates and costs, and the introduction of new regulatory policies, the demand for innovative technologies has never been higher. To reduce costs and mitigate risks, oil and gas operators need technologies that can enhance drilling efficiency and optimize drilling operations across all phases. With SAS Visual Analytics [55], drilling engineers, advisors and asset team members can use intuitive analytic tools to improve understanding of their wells and operations – without being statisticians. And with an easy-to-use interface atop a robust framework of data management and advanced analytic solutions, engineers can ask and get answers to unanticipated questions on the fly.

The eDrilling [56] is a new and innovative system for real time drilling simulation, 3D visualization and control from a remote drilling expert center. The concept uses all available real time drilling data (surface and downhole) in combination with real time modelling to monitor and optimize the drilling process. This information is used to visualize the wellbore in 3D in real time.

As show in the Fig. 1.3, iSolutions [52] is working to generate an advanced visualization and machine learning platform targeted at understanding drilling rig

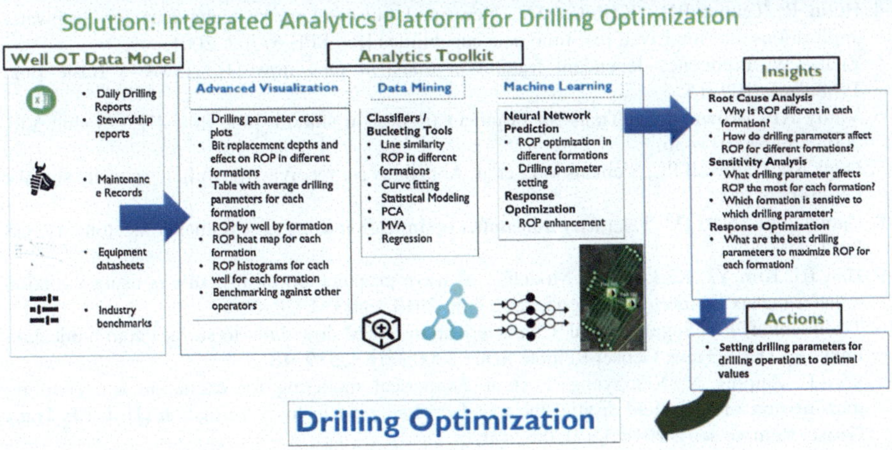

Fig. 1.3 Analytics platform for drilling optimization

parameter impacts on Rate of Penetration (ROP). The platform will allow for drilling rig parameters to be tuned to optimize ROP for wells on existing and nearby Pads in the intermediate section. Using drilling information's from historical wells, the model will be trained to estimate ROP's given various input settings for all formations intersected in the vertical portion of a drilling operation. The model will consider parameters such as weight on bit, mud rate, bit type, RPM etc.

An optimization function will be generated to provide drilling engineers with critical drilling parameter settings for each formation intersected in the drilling of the wells. These recommended settings can be used as a starting point to provide a basis for further optimization work. The optimization function will allow engineers to perform 'what-if' analysis of a particular set of parameters without having to alter the actual drilling plan of a well. Hereby the situation (challenge), complication and solution for the project above are illustrated: The Challenge Drilling interbedded formations is challenging … • Drilling parameters are important and must be optimized • Different formations have different geological and geo-mechanical characteristics • Bit life is vital in estimating the ROP and needs to be captured in the drilling optimization effort There is a gap in OT analytics tools that can be leveraged to facilitate drilling optimization.

References

1. Perry CA, Burgess DE, Turner WE. Rotary pulser for transmitting information to the surface from a drill string down hole in a well [P]. United States: US7327634B2, 2008.
2. Malone D, Johnson M. Logging while drilling tools, systems, and methods capable of transmitting data at a plurality of different frequencies [P]. United States: 5375098, 1994.
3. Hahn D, Peters V, Rouatbi C, Eggers H. Oscillating shear valve for mud pulse telemetry and associated methods of use [P]. United States: US6975244B2, 2005.

4. Hutin R, Tennet RW, Kashikar SV. New mud pulse telemetry techniques for deepwater applications and improved real-time data capabilities [R]. SPE 67762, 2001.
5. Zielke W. Frequency-dependent friction in transient pipe flow [J]. ASME J Basic Eng. 1968;90(1):109–15.
6. Vardy AE, Brown JMB. Transient turbulent friction in smooth pipe flows[J]. J Sound Vib. 2003;259(5):1011–36.
7. Dopf AR, Camwell PL, Siemens WL, et al. Apparatus for receiving downhole acoustic signals: US 2005.
8. Camwell PL, Neff JM. Telemetry transmitter optimization using time domain reflectometry: US 2011.
9. Han JH, Kim YJ, Karkoub M. Modeling of wave propagation in drill strings using vibration transfer matrix methods[J]. J Acoust Soc Am. 2013;134(3):1920–31.
10. Trofimenkoff FN, Segal M, et al. Characterization of EM downhole-to-surface communication links [J]. TEES Trans Geosci Remote Sens. 2000;38(6):2539–48.
11. Wei L, Zaiping N, Xiangyang S, et al. Numerical modeling for excitation and coupling transmission of near field around the metal drilling pipe in lossy formation [J]. IEEE Trans Geosci Remote sens. 2014;52(7):3862–71.
12. Poh Kheong V, Rodger D, Marshall A. Modeling an electromagnetic telemetry system for signal transmission in oil fields [J]. IEEE Trans Magn. 2005;41(5):2008–11.
13. Michael J, Prideco G, David RH. Intelligent drill pipe creates the drilling network [R]. SPE 80454.
14. Fischer PA. Interactive drilling up-to-data drilling technology[J]. Oil Gas Sci Technol. 2004;59:343–56.
15. Tomazic S. Comments on spectral efficiency of VMSK[J]. IEEE Trans Broadcast. 2002;48 (3):61–2.
16. Warren T. Rotary steerable technology conclusion: implementation issues concern operators. Oil Gas J. 1998;96(12):23–4.
17. Yonezawa T, Cargill EJ, et al. Robotic controlled drilling: a new rotary steerable drilling system for the oil and gas industry, Proceeding IADC/SPE Drilling Conference, Dallas, 2002, pp. 744–758.
18. Poli S, Donaco F, Oppelt J, Ragnitz D. Advanced tools for advanced wells: rotary closed-loop drilling system-results of prototype field testing. SPE Drill Complet. 1998;13(2):67–72.
19. Qilong X, Ruihe W, Feng S, et al. Continuous measurement-while-drilling utilizing strap-down multimodel surveying system. IEEE Trans Instrum Meas. 2014;63(3):650–7.
20. Xue Q, Wang R, Huang L, Sun F. Dynamic solution approach to the inclination and Amizuth of bottom rotating drill string, SPE Western Regional Meeting, Monterey, 2013.
21. Allan DW. Statistics of atomic frequency standards. Proc IEEE. 1966;54(2):221–30.
22. Lawrence CN, Pines DJ. Characterization of ring laser gyro performance using the Allan variance method. J Guid Control Dyn. 1997;20:211–4.
23. Hoffmann OJ, Jain JR, Spencer RW, Makkar N. Drilling dynamics measurements at the drill bit to address today's challenges. IEEE international instrumentation and measurement technology conference: smart measurements for a sustainable environment, Graz, p. 1772–1777, 13–16, 2012.
24. Ali H, Nayfeh, Balakumar B. applied nonlinear dynamics: analytical, computational, and experimental methods. Weinheim: Wiley-VCH; 1995.
25. Ertas D, Bailey JR, et al. Drill string mechanics model for surveillance, root cause analysis, and mitigation of torsional and axial vibrations. Louisiana: Society of Petroleum Engineers; 2013.
26. Jansen JD. Nonlinear rotor dynamics as applied to oil well drillstring vibrations. J Sound Vib. 1991;147(1):115–35.
27. Navarro-Lopez EM, Cortes D. Avoiding harmful oscillations in a drillstring through dynamical analysis. J Sound Vib. 2007;307:152–71.
28. Millheim K, Jordan S, Ritter CJ. Bottom-hole assembly analysis using the finite-element method. SPE J. 1978;30(2):265–74.

29. Costa FS, Rebeiro PR. Finite element modeling of the mechanical behavior of unbalanced drill collars. Rio de Janeiro: Society of Petroleum Engineers; 1997.
30. Ledgerwood LW, Jain JR, Ho®mann OJ, et al. Downhole measurement and monitoring lead to an enhanced understanding of drilling vibrations and polycrystalline diamond compact bit damage. Louisiana: Society of Petroleum Engineers; 2013.
31. Raap C, Craig AD, Graham RB. Drill pipe dynamic measurements provide valuable insight into drill string dysfunctions. Ohio: Society of Petroleum Engineers; 2011.
32. Oueslati H, Jain JR, Reckmann H, et al. New insights into drilling dynamics through high-frequency vibration measurement and modeling. Louisiana: Society of Petroleum Engineers; 2013.
33. Chien ML, Nicholas V, Hamad K, et al. Parametric studies on drill-string motions. Int J Mech Sci. 2012;54(1):260–8.
34. Besaisow AA, Payne ML. A study of excitation mechanisms and resonances inducing BHA vibrations. New Orleans: Society of Petroleum Engineers; 1986.
35. Schen AE, Snell AD, Stanes BH. Optimization of bit drilling performance using a new small vibration logging tool. Amsterdam: Society of Petroleum Engineers; 2005.
36. Field DJ, Swarbrick AJ, Haduch GA. Techniques for successful application of dynamic analysis in the prevention of field-induced vibration damage in MWD tools. Amsterdam: Society of Petroleum Engineers; 1993.
37. Dubinsky V. Reference signal encoding for seismic while drilling measurement: U.S. Patent 6,078,868[P]. 2000-6-20.
38. Petronio L, Poletto F. Seismic-while-drilling by using tunnel boring machine noise [J]. Geophysics. 2002;67(6):1798–809.
39. Poletto F, Malusa M, Miranda F, et al. Seismic-while-drilling by using dual sensors in drill strings [J]. Geophysics. 2004;69(5):1261–71.
40. Masak PC, Malone DL. Inverse vertical seismic profiling using a measurement while drilling tool as a seismic source: U.S. Patent 6,094,401[P]. 2000-7-25.
41. Zhihong B, Ganneng R, Le C. Logging while drilling technology[J]. Fault Block Oil Gas Field. 2001;08(4):22–4.
42. Xinyi Z, Yanjun G, Jingnong W. Yesterday, today and tomorrow of logging while drilling [J]. Logging Technol. 2006;30(6):487–92.
43. Neff JM, Camwell PL. Field test results of an acoustic telemetry MWD system, SPE/IADC drilling conference, 20–22 February, Amsterdam,2007.
44. Reeves ME, Camwell PL, Mcrory J. High speed acoustic telemetry network enables real-time along string measurements, greatly reducing drilling risk[C]. Offshore Europe, 6–8 September, Aberdeen, 2011.
45. Reeves M, Smith DG, Groves D, et al. Unique acoustic telemetry network with distributed pressure measurement nodes enables accurate real-time analysis of sweep effectiveness[C]. SPE Annual Technical Conference and Exhibition, New Orleans, 2013.
46. Qilong X, Henry L, Ruihe W, Baolin L. Continuous real-time measurement of drilling trajectory with new state-space models of kalman filter. IEEE Trans Instrum Meas. 2016;65(1):144–54.
47. Qilong X, Leung H, Ruihe W, Baolin L, Leilei H, Shenglai G. The chaotic dynamics of drilling. Nonlinear Dyn. 2016;83(3):2003–18.
48. Xue Q, Leung H, Huang L, Zhang R, Liu B, Wang J, Li L. Modeling of torsional oscillation of drillstring dynamics. Nonlinear Dyn. 2019;96(1):267–83.
49. Rassenfoss S. Drilling change requires changing drillers. JPT. 2019., https://www.spe.org/en/jpt/jpt-article-detail/?art=5861;71:30–7.
50. Bertocco R, Padmanabhan V. Big data analytics in oil &gas[J]. Bain Brief, 2014.
51. https://www.cannonoperating.com/news/oil-companies-using-data-algorithms-improved-drilling/
52. https://www.isolutions.com/drilling-optimization-project-using-data-analytics/
53. Feblowitz J. The big Deal about big data in upstream oil and gas, paper & presentation, IDC energy insights, 2012.

54. Nicholson R. Big data in the oil & gas Industry, IDC energy insights, September 2012 Seshadri M., 2013, Big data science challenging the oil industry, CTO Global Services, EMC Corporation, March 2013.
55. Drilling optimization through advanced analytics using historical and real-time data. https://www.sas.com/en_id/whitepapers/drilling-optimization-through-advanced-analytics-106635.html.
56. Rommetveit R, Bjorkevoll KS, Fjar E, et al. e-Drilling: a system for real-time drilling simulation, 3D visualization and control[C]//digital energy conference and exhibition. Society of Petroleum Engineers, 2007.

Chapter 2
Signal Detection and Processing of Downhole Information Transmission

Abstract We introduced five types of information transmission method for Measurement While Drilling (MWD): cable method, drilling fluid pulse method, acoustic method and electromagnetic wave method and the latest technology of intelligent drill pipe at present. With the application of advanced drilling technologies such as rotary steering drilling and geosteering drilling, the real-time measurement and transmission of a large number of downhole parameters put forward higher requirements on the information transmission rate. Lower transmission rate has become the bottleneck of parameter expansion for measurement. Intelligent drill pipe (Intelli Pipe) is now the latest downhole signal transmission technology with the highest transmission rate. The communication rate can reach up to 2Mbps, and the stable communication rate can reach 56 kbps, which is too much higher than the current electromagnetic wave or mud pulse transmission applied in MWD, but high manufacturing costs are the main factors that constrain its development.

Keywords Measurement While Drilling (MWD) · Information transmission · Acoustic · Electromagnetic wave · Intelligent drill pipe (Intelli pipe)

2.1 Introduction

In the drilling process, especially in the course of drilling some complex wells such as horizontal wells, extended reach wells and multilateral wells, various downhole parameters that the personnel need to track in real time can be obtained through Measurement While Drilling (MWD) and Logging While Drilling (LWD). The Measurement While Drilling (MWD) and Logging While Drilling (LWD) system is composed of downhole controller, various downhole parameter measuring instruments, MWD information transmission system and a system of receiving, processing and displaying surface information [1]. The development of various subsystem of MWD and LWD system has been extremely uneven for a long time, mainly in the conflict between the low transmission rate in information transmission subsystem and the increasing downhole information requiring transmission.

Q. Xue, *Data Analytics for Drilling Engineering*, Information Fusion and Data Science, https://doi.org/10.1007/978-3-030-34035-3_2

With development and application of various new downhole measuring instruments, more and more parameters for MWD are available, from the initial well inclination, azimuth, tool face and other geometric parameters to drilling pressure, torque, pressure, temperature, natural gamma, formation resistivity and other environmental and geological parameters. At present, there are five types of information transmission method for MWD: cable method, drilling fluid pulse method, acoustic method and electromagnetic wave method and the latest technology of intelligent drill pipe at present, each has its own limitations and scope of application. Among the five methods, drilling fluid pulse method, the most widely applied and reliable one, is the major means for downhole signal transmission. With the application of advanced drilling technologies such as rotary steering drilling [2] and geosteering drilling, the real-time measurement and transmission of a large number of downhole parameters put forward higher requirements on the information transmission rate. Lower transmission rate has become the bottleneck of parameter expansion for measurement.

Drilling fluid pulse signal generator can be divided into positive pulse, negative pulse and continuous wave generator according to the signal type. There have already been some mature products in the signal transmission mode of both positive pulse and negative pulse applied in drilling engineering, their information transmission rate is relatively low, usually less than 3bit/s. However, the transmission rate in continuous wave signal transmission mode is relatively high, which can theoretically reach more than 20bit/s, but with difficult technical implementation. Currently, only a few companies have products for MWD systems [3–5] that transmit information in continuous wave mode. For example, PowerPulser™ of Schlumberger, the continuous wave signal generator with continuous rotary valve, sends signals at a frequency of 24 Hz with the highest data transmission rate of 12bit/s. Baker Hughes' continuous wave signal generator with oscillating shear valve has an experimental transmission rate of 40 bps at the well depth of 3000 ft. and of 15 bps at well depth of 24,000 ft. Halliburton is also working on developing MWD systems that transmit information in a continuous wave mode, aiming at achieving transmission rates of 20–30bit/s. This book mainly introduces the transmission characteristics and attenuation mechanism of continuous wave transmission mode (see Sect. 2.2).

Acoustic data transmission while drilling is a technology that uses acoustic waves to transmit in the channel of drill string. The information transmission does not depend on the traditional mud drilling fluid and formation, so the reliability and applicability of the technology are relatively strong. Besides, it is a technology without any extra drilling cost, because there is no need to embed wires in the drill string which serves as acoustic transmission channel. Also, the technology does not use any special drill string. The acoustic transmission speed in drill string is higher than mud pulse. The carrier frequency of elastic wave which can be used is relatively higher. And its theoretical transmission rate is 1–2 orders of magnitude higher than that of mud pulse and electromagnetic wave data transmission technology. The acoustic data transmission while drilling technology is not yet mature, because of the fast signal decay rate, which seriously limits the commercial application of this technology.

From 2005 to 2007, XACT company successively published the invention patents of related equipment and technology of acoustic remote transmission [6–8]. In 2007, XACT and Extreme Engineering conducted on-site testing of acoustic MWD tools with two diameter specification, one tool with the diameter of 121 mm, well depth of 1205 m and the transmission rate of 10bit/s, another tool with the diameter of 165 mm, well depth of 2489 m and the transmission rate of 20bit/s [9]. Up to 2011, XACT has offered service for more than 400 Wells, among which two relay amplifiers are used to achieve a maximum depth of 4000 meters. However, the use of relay amplifiers makes the reliability of the instrument under great restriction. Its system uses 640-680 Hz single passband and BPSK encoding [10]. In 2013 XACT used six relay amplifiers to offer service for wells at the greatest depth of 4000 m (13,200 ft) [11]. The main difficulty in acoustic transmission lies in the attenuation of acoustic signals in the drill string. The attenuation model will be discussed in 2.3.

The MWD system using the electromagnetic wave transmission has been gradually applied in the oil and gas drilling in recent years, EM-MWD (Electromagnetic Measurement While Drilling) has the advantage of high data transmission rate, free from the influence of drilling medium. And it costs less for data can be transmitted without circulating drilling fluid, which makes up for the deficiency of mud pulse MWD and provides an indispensable technical support for gas drilling and unbalanced drilling. The E-Pulse system developed by Schlumberger can transmit log data with a vertical depth of 4183 m without using extended antenna and repeater technology [12]. In 1987, Russian company Shamara Horizon successfully developed ZTS series of electromagnetic wave logging while drilling instruments [13], which are powered by turbine motor. It has the advantage of high transmitting power and large signal amplitude. The Trend SET system [14], manufactured by Weatherford in 2004, uses signal relay technology and uses cascaded battery packs for power supply. It can last for 500 h with transmission depth of up to 3240 m. It is reliable, heat resistant, and can operate normally in harsh drilling environments.

EM-MWD technology has its own disadvantages: electromagnetic signal propagation in the formation is seriously affected by formation resistivity, especially in the low resistivity formation, signal attenuation is fast, which limits its effective transmission depth. After many years of research and exploration, a number of achievements in research and application has been made in revealing and overcoming the limitation of electromagnetic signal transmission distance while drilling.

Intelligent drill pipe (Intelli Pipe) is now the latest downhole signal transmission technology with the highest transmission rate. The intelligent drill pipe system, successfully developed by the American company Intelli Serv in 2002, uses electromagnetic inductive coupling principle to realize the non-contact signal transmission. It is installed on the inner wall of the drill string the sheathed wires, and "sinks" the wire into the threaded end of the drill pipe at each end of the tool joints, so as to transmit information between different drill pipes through electromagnetic induction. Because of the signal attenuation in the cable when transmiting, relay amplifiers should be attached to the intelligent drill string at intervals of 350 m to 450 m. The technology has been applied in 17 test wells, including vertical and directional gas wells with the depths of more than 14,000 feet (4267.2 meters) in the Arkoma area, southeast of American Oklahoma [15, 16]. The communication rate can reach up to 2Mbps, and the stable communication rate can reach 56 kbps, which is too much higher than the current electromagnetic wave or mud pulse transmission applied in MWD. It tackles the obstacles of data transmission while drilling and realizes the high-rate transmission of downhole information in real time during drilling.

2.2 Characteristics and Processing Methods of Sinusoidal Pressure Wave Signal

At present, drilling fluid positive pulse transmission is commonly used in oil and gas drilling. Its data transmission rate is relatively low, typically from 0.5 to 3bits. And it has a high bit error rate because it can be easily interfered by the external condition. When compared with positive pulse transmission, continuous wave pulse [6], with a relatively higher data transmission rate and stronger anti-interference ability, is a promising drilling fluid pulse transmission technology.

In general, ccontinuous wave pressure generators can be divided into two categories: oscillating shear valves and rotary valves. The shear valve is composed of stator and rotor. The stator is fixed and the rotor oscillates back and forth with respect to the stator at an angle deviation. When the rotor rotates in one direction, the drilling fluid flow area decreases and the pressure increases. When the pressure increases to its maximum, the rotor rotates in reverse, the flow area increases, and the pressure decreases. When the rotor orifice coincides with the stator orifice, the pressure returns to normal. The shear process of the rotor back and forth in relation to the stator produces a continuous pressure change. In addition, the bidirectional rotary characteristics of the shear valve rotor can help effectively reduce the risk of the rotary valve being blocked by solid particles of drilling fluid. The disadvantage of the shear valve is that the speed at the end of the both sides is zero, so the motor needs to be continuously driven through positive and negative rotation. When this occurs, the stepper motor needs to be adopted, but it is difficult for the motor to turn to the set position when encountering great resistance.

Rotary valve and shear valve have the same structure. The difference between them lies in that the rotary valve rotates in one direction and its anti-blocking capability is not as good as the shear valve. However, the motor of the rotary valve can be easily controlled. A lineal motor can be chosen to establish a closed-loop feedback circuit and then generates, through accurate control the rotation of the rotor, continuous pressure wave.

The generation of drilling fluid continuous pressure wave is shown in Fig. 2.1. The drive and control circuit of the continuous pressure wave generator drives the generator rotor to rotate, and the rotor creates an intercepting effect against the stator, which causes the pressure fluctuation of drilling fluid in the drill string and forms a

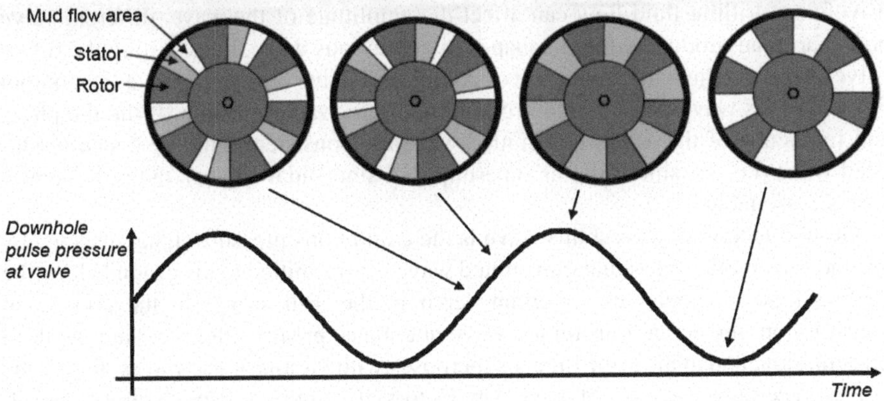

Fig. 2.1 Generation of continuous wave pressure signals

continuous sinusoidal pressure wave. The measured data can be gotten through first, encoding the data measured with downhole sensors, then generating pressure change with modulation system, and finally, processing after detecting pressure signals and decoding. The continuous pressure wave technology has the advantage of the fast rate for data transmission, while the disadvantage lies in the weak signal and high demand for signal processing system.

Adopting the continuous wave pressure pulse signal of drilling fluid, the up-link communication of downhole sensor data is completed. Actual transmitted wave is generated through encoding and modulating the downhole information first, and then controlling the continuous wave generator with downhole controller. The pressure wave is transmitted to the ground through drilling fluid channel, and the pressure signal is collected by the ground receiving sensor (high-precision pressure sensor). After filtering and channel equalization, the ground processed wave is generated, and the real-time data processing is conducted after demodulation and decoding. The whole process includes four waveforms: coded modulated wave, actual transmitted wave, ground receiving wave and ground processed wave.

(1) Coded modulated wave: this waveform converts downhole measurement information into continuous waves for transmission (such as sinusoidal or superimposed sinusoidal waves) according to certain encoding and modulation methods. This wave describes the data frame to be sent, which is an ideal waveform and can be strictly expressed through mathematics. The wave has a great impact on the amount of information sent per unit time (related to pre-coding, such as differential coding), signal transmission rate, signal anti-interference, signal detection synchronization and error detection.

Actual transmitted wave: the wave is the drilling fluid pressure wave generated by the continuous wave generator. It is a continuous wave nearly close to the code modulation wave, generated through regularly changing the flow area of the rotary valve port after controlling the speed and position of the continuous wave generator according to the code modulation waveform. There are many factors affecting the waveform: drilling fluid flow can affect the amplitude of the wave, different valve port shape can produce different shapes of continuous wave, by adjusting the rotary valve speed to adjust the frequency of continuous wave output, and to adjust output of continuous wave phase by controlling the rotary valve position. If the the phase and frequency of the actual transmitted wave are consistent with the coded modulated wave (i.e. for sinusoidal or superimposed sinusoidal waves), the pulse generator is reasonable.

Ground receiving waves: this wave is the continuous pressure signal detected by ground sensor after the actual transmitted wave is transmitted to the ground along the drilling fluid channel with a certain depth in the drill string. In the process of transmission, the actual transmitted wave attenuates greatly after a certain depth of transmission, and at the same time it superposes with various noise waves, finally the ground receiving wave is obtained. The factors affecting the wave include controllable factors and uncontrollable factors. The controllable factors are mainly the gas content of drilling fluid, carrier frequency and the noise of mud pump. Uncontrollable factors include signal attenuation due to well depth, reflection wave caused by drill string structure, noise with similar frequency and carrier frequency, etc. The requirements for this wave are high signal-to-noise ratio, non-distortion, and high power spectral density under the carrier frequency of the actual transmitted wave (which is related to the motor speed).

Ground processed wave. This wave is a pressure wave obtained through filtering and channel equalization of ground receiving wave. It can be an analog signal or a digital signal. The main factors affecting the wave are filterring method and channel equalization method, which can be described by mathematical model. The requirement of the wave is to eliminate or reduce the noise and inter-symbol interference as much as possible, and to realize spectrum analysis, filtering in frequency domain and other functions. The wave can be demodulated and decoded after synchronous processing of clock signal.

2.2.1 Rotary Valve Mud Pulse Generator

Rotary valve mud pulse generator, as the core component of generating continuous wave pulse signal [5], can control the rotation of rotary valve to generate drilling fluid pressure pulse with certain rules. Hydraulic drive and motor drive are two main forms.

Driven by the upper impeller, the rotor of the hydraulic-drive rotary signal generator with pilot control rotates in a predetermined coding mode through the regulation of pilot control mechanism, thus generating pressure wave signal in the drill pipe. After confirming the runner parameters of the signal generator, if the drilling fluid flow deviates from the design flow, or the rotation speed deviates from the design rotation speed, the efficiency will decline and erosion will be intensified. This signal generator has low power dissipation.

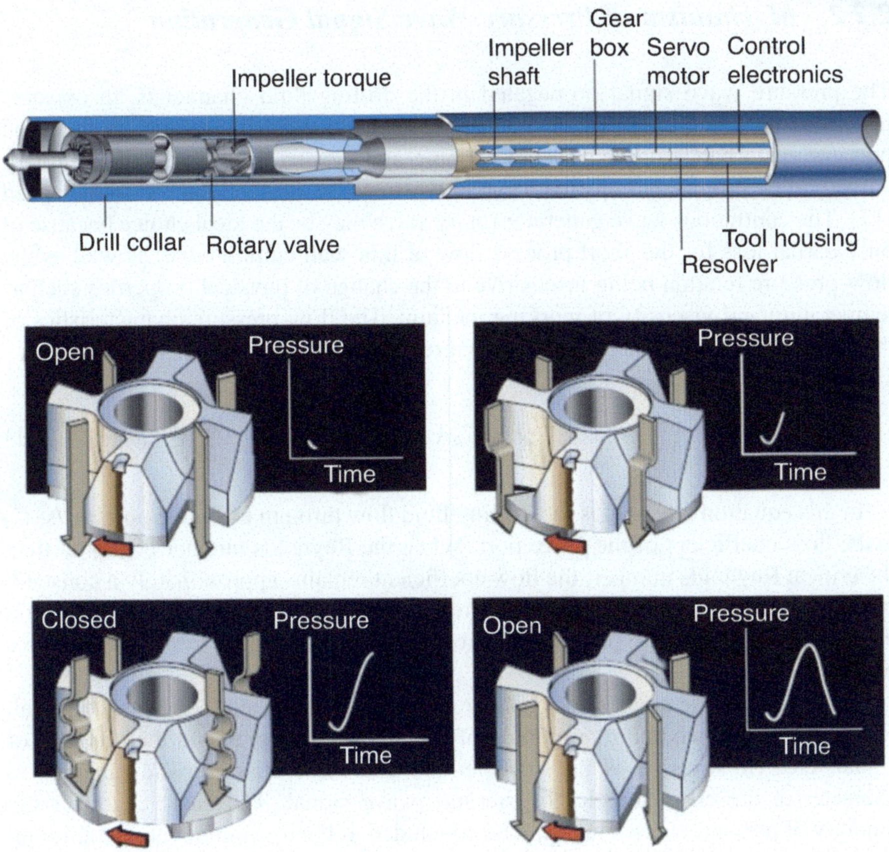

Fig. 2.2 Frame diagram of slimpulse-MWD

The rotary signal generator directly driven by motor is the drilling fluid pulse generator, which is widely applied in field at present. The drilling fluid pressure signal is generated by controlling the rotary valve speed or rotation frequency according to the digital baseband signal.

The PowerPulser developed by Schlumberger uses the rotary valve in continuous rotation, thus generating pressure waves. Blades are set under the rotary valve and on the drive shaft, as shown in Fig. 2.2. The permissible operating temperature is 177 °C, and it can work in drilling fluid with the flow rate of 2.2 L/s to 75.7 L/s. The servo control system adopted improves the data transmission reliability, transmission rate and signal intensity, and it has stronger anti-interference ability.

The key of pulse generator research lies in the structure design of the stator and rotor and the motor control scheme. The parameters of rotary valve models with different structures, such as flow rate, valve port shape, axial clearance of stator and rotor, and speed, all have a great impact on the pulse signal.

2.2.2 Mechanism of Pressure Wave Signal Generation

The pressure wave signal propagated in the drilling fluid channel is, in essence, generated by compressibility of the drilling fluid. In the drilling fluid channel, the pressure disturbance caused by the opening and closing the throttle of the rotary valve will be propagated upstream and downstream respectively at the sound speed [17]. The continuous wave generator rotary valve may be the ideal choice because of only partial loss for the short process flow of thin-wall cutting edge, as well as the flow-pressure relation being insensitive to the change of physical properties such as temperature and viscosity of working medium. The flow-pressure characteristics of the continuous wave generator can be expressed as:

$$Q = C_d A \sqrt{\frac{2\Delta p}{\rho}} \tag{2.1}$$

In this equation where Q is the drilling fluid flow through the valve port, m^3/s; C_d is the flow coefficient of the valve port. When the Reynolds number is greater than the critical Reynolds number, the flow coefficient remains approximately a constant, usually from 0.6 to 0.8. A is the flow area of the valve port, m^2; ρ is drilling fluid density, kg/m^3; Δp is the pressure difference between the valve front and the valve back, Pa.

The characteristic quantity of pressure wave signal is related to the size of signal, for example, the maximum amplitude of pressure wave signal and the duration of signal, etc. However, it is not sufficient to use these characteristic quantity to characterize the characteristics of pressure wave signal. The basic characteristic quantity of pressure wave signal can be concluded as the maximum value, minimum value, average value, power and frequency spectrum of signal.

Where the average value of the pressure wave signal is M_p,

$$M_p = \frac{1}{T}\int_{-T/2}^{T/2} \Delta p(t)dt \tag{2.2}$$

The average value reflects the magnitude of direct current component in pressure wave signal.

The power of the pressure wave signal, P_p.

$$P_p = \frac{1}{T}\int_{-T/2}^{T/2} \Delta p(t)^2 dt \tag{2.3}$$

Through the following analysis, it can be found that the power of pressure wave signal is approximately proportional to the torque of the rotary valve and the water power.

Pressure wave signal, when travelling in the drilling fluid channel, will be reflected in areas such as downhole bit and the surface drilling pump. Different propagation paths will lead to different phase shifts of different frequency components, making at the receiver point the direct wave of some frequency components is counteracting the reflected wave, while part of the frequency components enhance each other, thus forming the so-called selective attenuation phenomenon. Attenuation during transmission is a key problem of downhole signal transmission, which will be discussed in detail in 2.2.4

There is a large deviation between waveform of the pressure wave signal generated from the quadrant valve port and the sinusoidal pressure wave signal, containing large harmonic components, which attenuates rapidly in the drilling fluid channel. So only a few pressure wave signals generated from the quadrant valve port can be useful. In the continuous carrier modulation of digital signal, sinusoidal wave is used as the carrier signal. Therefore, the expected sinusoidal wave pressure signal in the design of rotary valve is:

$$\Delta p_0(t) = \frac{p_{\max} - p_{\min}}{2} \sin\left(n\omega t - \frac{\pi}{2}\right) + \frac{p_{\max} + p_{\min}}{2} \tag{2.4}$$

The correlation coefficient of pressure wave signal $\Delta p(t)$ and $\Delta p_0(t)$within a period:

$$\rho_{xy} = \frac{\int_0^T \Delta p(t)\Delta p_0(t)dt}{\sqrt{\int_0^T \Delta p^2(t)dt \int_0^T \Delta p_0^2(t)dt}} \tag{2.5}$$

The generation of pressure wave signals can be expressed by the simple model shown in Fig. 2.3. The area of Sect. 1.1 is A_1, the area of orifice section is A_0, and the area of jet contraction Sect. 2.2 is A_2. Different from the free jet, the downstream of the orifice is not connected with the atmosphere. So the liquid flow will diffuse after passing through the orifice, and it is at its highest velocity at the jet section with the minimum pressure. With the diffusion of the jet flow, the velocity decreases and the pressure has shown a little recovery. However, due to the resistance loss at the orifice, the pressure can not be completely restored.

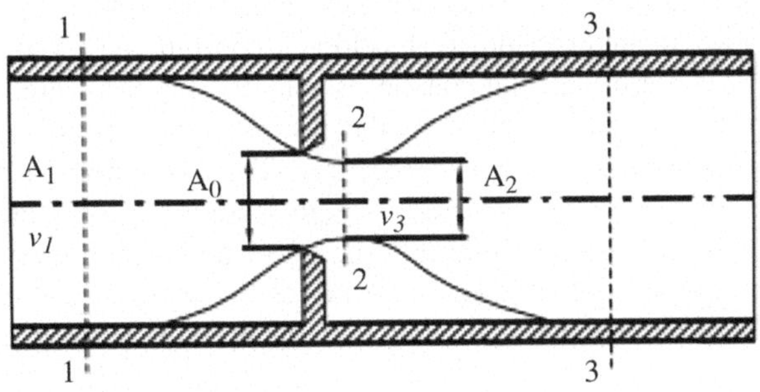

Fig. 2.3 Unsteady flow of orifice

For the incompressible liquid between Sect. 1.1 and jet Sect. 2.2, Bernoulli's equation is listed as follows:

$$\rho\frac{\partial\varphi_1}{\partial t}+\frac{\rho\alpha_1 v_1^2}{2}+\rho g z_1+p_1=\rho\frac{\partial\varphi_2}{\partial t}+\frac{\rho\alpha_2 v_2^2}{2}+\rho g z_2+p_2+\zeta\frac{\rho v_2^2}{2} \tag{2.6}$$

In above equation, $z_1 = z_2$, φ is the velocity potential function, $\partial\varphi/\partial x = v$. $\varphi_1 = v_1 x_1$, $\varphi_2 = v_2 x_2$ at the Sect. 1.1 and Sect. 2.2. Also, it can be known from the continuity equation that $v_1 A_1 = v_2 A_2 = v_2 C_c A_0$, by substituting the above equation into (2.6), it can be drawn:

$$P_1 - P_2 = \rho\frac{\partial(v_2 x_2 - v_1 x_1)}{\partial t}+\frac{\rho v_1^2}{2}\left((\alpha_2+\xi)\left(\frac{A_1}{C_c A_0}\right)-\alpha_1\right) \tag{2.7}$$

Assume that the upstream velocity v_1 is constant, the orifice area changes with time. Because of the turbulent flow at the jet section, so $\alpha_2 = 1$. Usually, A_0 is much smaller than A_1, so α_1 can be ignored, while the maximum flow area of the rotary valve of continuous wave generator, cannot be ignored when compared with the flow area of drill pipe, so α_1 should be retained in the above equation.

2.2.3 *Transmission Characteristics of Continuous Pulse Signal*

The transmission characteristics of continuous wave signal mainly include signal reflection, signal transmission and signal attenuation. At present, there are two main theories on the attenuation of pressure waves in pipelines: one is the acoustic attenuation model based on Lamb law [18], and the other is the attenuation model based on wall friction which is commonly used in the pipeline water hammer theory. The acoustic attenuation model based on Lamb's law is simple and easy to use. Therefore, almost all oil companies use this formula for attenuation estimation. The attenuation model based on wall friction, in the course of its development, has undergone the stages of quasi-static wall shear model, frequency-related friction of laminar flow [19], and frequency-related friction of turbulent flow [20]. Although the water hammer theory is developing continuously, it is limited to Newtonian fluid.

The drilling fluid channel is actually a multilayered cylindrical waveguide composed of drilling fluid in the drill string, drill string, annulus drilling fluid and formation, as shown in Fig. 2.4. The formation in the figure can be open-hole formation, casing, marine risers in offshore drilling, etc. The multilayered coupling model is very complex and needs to be appropriately simplified according to the problems studied [21]: ①not taking viscosity of drilling fluid into account; ②drilling fluid velocity compared with the pressure wave propagation speed is very small,

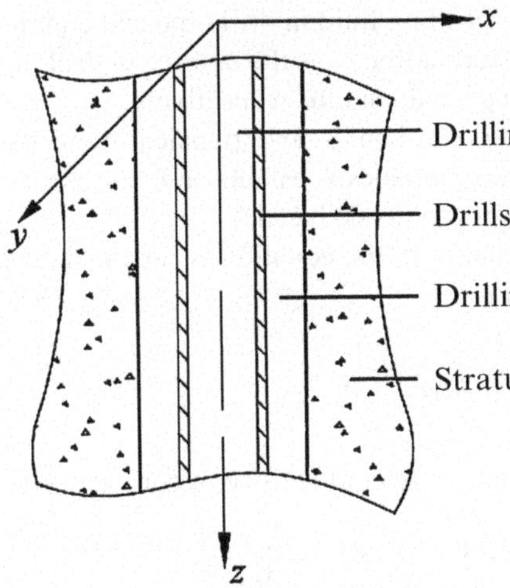

Fig. 2.4 Multi-layered waveguide model of drilling fluid channel

which can be ignored; ③the wave length of pressure wave signal is greater than the radius of the drill string. ④drill string is an axisymmetric uniform elastomer.

In the cylindrical coordinate system, the motion equation of the multilayered coupling model is [22]:

1. Longitudinal vibration equation of drilling fluid in drill string:

$$B_i \frac{\partial^2 u_i}{\partial z^2} = 2\zeta B_i \frac{\rho_p}{E} \frac{\partial^2 u_p}{\partial t^2} + \left(1 + \frac{2B_i}{E}\left(\frac{2A_i}{A_p} + 1 + \nu\right)\right) \rho_i \frac{\partial^2 u_i}{\partial t^2} - \frac{4B_0}{E} \frac{A_i + A_p}{A_p} \rho_0 \frac{\partial^2 u_0}{\partial t^2} \tag{2.8}$$

2. Vibration equation of annular drilling fluid:

$$B_0 \frac{\partial^2 u_0}{\partial z^2} = -2\zeta B_0 \frac{\rho_p}{E} \frac{A_t + A_p}{A_0} \frac{\partial^2 u_p}{\partial t^2} - \frac{4B_i \rho_i}{E} \frac{A_i}{A_p} \frac{A_i + A_p}{A_0} \frac{\partial^2 u_i}{\partial t^2} + \left(1 + \frac{2B_0}{E} \frac{A_i + A_p}{A_0}\left(\frac{2A_i}{A_p} + 1 - \nu\right) + \frac{B_0}{G}\left(\frac{A_i + A_p}{A_0} + 1\right)\right) \rho_0 \frac{\partial^2 u_0}{\partial t^2} \tag{2.9}$$

In the equation, u_p, u_i and u_0 are respectively the axial displacement of the drill string, drilling fluid in the drill string and annular drilling fluid. A_p, A_i and A_0 are respectively the cross-sectional areas of drill string, in drill pipe and annulus. ζ is Poisson's ratio of drill string. E is the elastic modulus of drill string. B_i and B_0 is the equivalent volume elastic modulus of drilling fluid in drill pipe and annulus respectively. G is the shear modulus of the formation. ρ_p is the density of drill pipe. ρ_i and ρ_0 are drilling fluid density in drill pipe and annulus respectively.

Based on hydromechanics, elastic mechanics and transient flow theory, the calculation model of transmission characteristics is established, and the basic equation of continuous pulse signal transmission is deduced.

After analyzing the fluid differential unit, the equation is obtained according to the law of mass conservation, as follows:

$$\frac{d}{dt}(\rho A \Delta x) = 0 \tag{2.10}$$

Where A is sectional area of the pipe, ρ is fluid density, it can be expanded as:

$$\frac{1}{\rho}\frac{d\rho}{dt} + \frac{1}{A}\frac{dA}{dt} + \frac{1}{\Delta x}\frac{d\Delta x}{dt} = 0 \tag{2.11}$$

According to the definition of fluid compressibility, it can be:

$$\frac{1}{\rho}\frac{d\rho}{dt} = \frac{1}{K_l}\frac{dP}{dt} \tag{2.12}$$

Relative change rate of the fluid differential unit can be expressed:

$$\frac{1}{\Delta x}\frac{d\Delta x}{dt} = \frac{1}{\Delta x}\lim_{\Delta t \to 0}\frac{(V + \frac{\partial V}{\partial x}\Delta x)\Delta t - V\Delta t}{\Delta t} = \frac{\partial V}{\partial x} \tag{2.13}$$

The relative change rate of the fluid differential unit is related to the elasticity of the pipe wall and the support of the pipe.

Set ε_2 as the circumferential stress of the pipeline. When the pressure increases, the increment of pipeline circumference is $\pi D \varepsilon_2$, and the diameter is $D(1 + \varepsilon_2)$. Therefore, the corresponding increment of circulation area and relative change rate are $\Delta A \approx \frac{\pi}{4} D^2 2\varepsilon_2 = \frac{\pi}{2} D^2 \varepsilon_2$, so:

$$\frac{1}{A}\frac{dA}{dt} \approx 2\frac{d\varepsilon_2}{dt} \tag{2.14}$$

According to hooke's law, the stress analysis of the pipe wall is shown in Fig. 2.5.

$$\varepsilon_2 = \frac{1}{E}(\sigma_2 - \mu_0 \sigma_1)$$

$$\frac{d\varepsilon_2}{dt} = \frac{1}{E}\left(\frac{d\sigma_2}{dt} - \mu_0 \frac{d\sigma_1}{dt}\right) \tag{2.15}$$

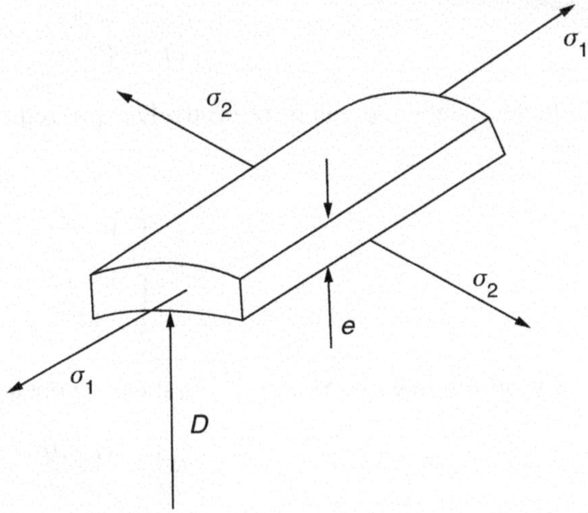

Fig. 2.5 Stress distribution of pipe wall in transmission

For thin tube walls, circumferential stress $\sigma_2 = PD/2e$, it can be obtained by doing time derivation:

$$\frac{d\sigma_2}{dt} = \frac{D}{2e}\frac{dP}{dt} \tag{2.16}$$

Axial stress of pipeline can be divided into the following three cases:
①Only the upstream end of the pipe is fixed;

$$\frac{d\sigma_1}{dt} \approx \frac{A}{\pi De}\frac{dP}{dt} = \frac{D}{4e}\frac{dP}{dt} = \frac{1}{2}\frac{d\sigma_2}{dt} \tag{2.17}$$

②The whole pipe is fixed, the axial stress is 0, then $\varepsilon_1 = 0$;

$$\varepsilon_1 = \frac{1}{E}(\sigma_1 - \mu_0\sigma_2) = 0 \tag{2.18}$$

$$\frac{d\sigma_1}{dt} = \mu_0\frac{d\sigma_2}{dt} \tag{2.19}$$

③The whole pipe being connected by expansion joint, the pipe is free from axial force;

$$\frac{d\sigma_1}{dt} = 0 \tag{2.20}$$

By substituting the three cases into the Eq. (2.15), it can be obtained as follows:

$$\frac{d\varepsilon_2}{dt} = \frac{\psi}{E}\frac{d\sigma_2}{dt} \tag{2.21}$$

In the equation, ψ can be respectively expressed as follows according to the three cases.

$$\psi = \begin{cases} 1 - \dfrac{\mu_0}{2} \\ 1 - {\mu_0}^2 \\ 1 \end{cases} \tag{2.22}$$

By substituting (2.15) into (2.13), it can be obtained as follows:

$$\frac{1}{A}\frac{dA}{dt} = \frac{D\psi}{Ee}\frac{dP}{dt} \tag{2.23}$$

Substitute (2.12), (2.13), (2.23) into (2.11), the following equation can be obtained, which is one-dimensional continuous equation of unsteady flow in the pipeline

$$\frac{1}{K_l}\frac{dP}{dt} + \frac{D\psi}{Ee}\frac{dP}{dt} + \frac{\partial V}{\partial x} = 0 \tag{2.24}$$

For thick wall pipes, the influence of circumferential stress uniformity should be taken into account, the influence factor can be expressed as follows:

$$\psi = \begin{cases} \dfrac{1}{1+\dfrac{e}{D}}\left[\left(1 - \dfrac{\mu_0}{2}\right) + 2\dfrac{e}{D}1 + \mu_0\left(1 + \dfrac{e}{D}\right)\right] \\ \dfrac{1}{1+\dfrac{e}{D}}\left[\left(1 - {\mu_0}^2\right) + 2\dfrac{e}{D}1 + \mu_0\left(1 + \dfrac{e}{D}\right)\right] \\ \dfrac{1}{1+\dfrac{e}{D}}\left[1 + 2\dfrac{e}{D}1 + \mu_0\left(1 + \dfrac{e}{D}\right)\right] \end{cases} \tag{2.25}$$

Apparently, when e/D is 0, the calculation formula of the thick wall pipe is almost the same as that of the corresponding thin wall pipe. It is generally assumed that D/e = 25 is the dividing line between thin wall pipe and thick wall pipe. In general, when D/e > 40, the transmission speed calculated by the two formulas differs slightly. In drilling engineering, both drill pipe and drill collar should belong to thick wall pipe, so it is recommended to use the thick wall pipe formula.

Because drilling fluid belongs to gas-liquid two-phase flow, it can be treated as single-phase flow when the gas content is low. As shown in Fig. 2.6, select a pipe with length of $l = c_0 t$ and the volume of $V = Sl$.

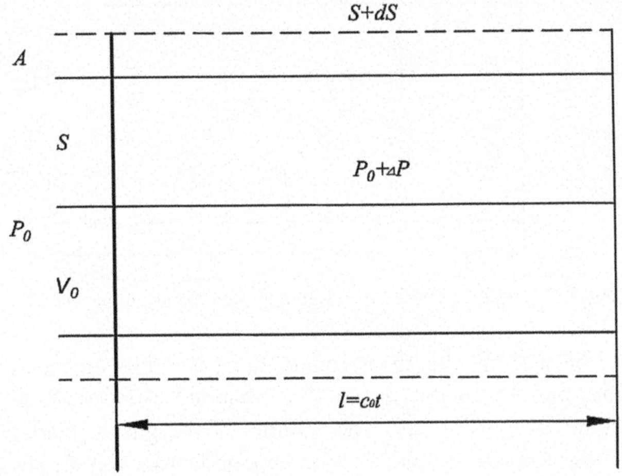

Fig. 2.6 Pipeline fluid transmission model

In Fig. 2.6, the initial velocity of continuous pulse signal is V_0, the fluid density in the pipeline is ρ, and the pressure increase in the pipeline is ΔP. Using momentum conservation law, we can get:

$$-\Delta PSt = -\rho V_0 Sl \tag{2.26}$$

$$\Delta P = \frac{\rho V_0 l}{t} = c_0 \rho V_0 \tag{2.27}$$

The pipe is influenced by ΔP, so the volume changes $dV = \Delta PV/K_e$. Due to liquid compressibility and pipeline expansion, some space will be left. At this time, the same volume of liquid will be added, which is expressed as V_0St.

$$V_0 St = \frac{\Delta PV}{K_e} = \frac{\Delta PSl}{K_e} = \frac{\Delta PSc_0 t}{K_e} \tag{2.28}$$

$$V_0 = \frac{c_0 \Delta P}{K_e} \tag{2.29}$$

It can be obtained through simultaneous Eqs. (2.28) and (2.29):

$$c_0 = \sqrt{\frac{K_e}{\rho}} \tag{2.30}$$

According to the Eq.2.30, the main parameter affecting the transmission speed of continuous pulse signal is the apparent elastic modulus of the system K_e. The following Fig. 2.7 is used for studing K_e. Assuming that there is a piston on the right side of the container, the volume of the liquid in the container is V_l, the volume of the gas is V_g, and the system pressure is P.

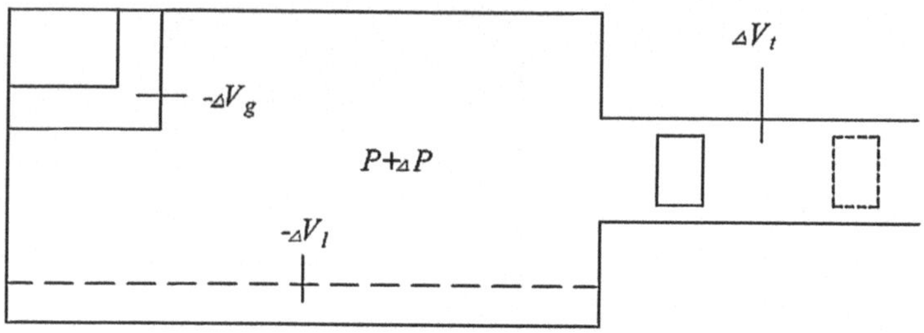

Fig. 2.7 Analysis model of liquid apparent elastic modulus

Now move the piston to the left by a certain amount of volume ΔV_t. At this point, the pressure of the system P is changed to $P + \Delta P$, the volume of the liquid V_l is changed to $V_l - \Delta V_l$, the volume of the gas is changed V_g to $V_g - \Delta V_g$, and the volume of the container V_c is changed to $V_c + \Delta V_c$, so it can be obtained:

$$\Delta V_t = -\Delta V_l - \Delta V_g + \Delta V_c \tag{2.31}$$

It can be calculated by substituting the Eq. (2.31) into $dV = \Delta PV/K_e$, shown as follows:

$$\frac{1}{K_e} = \frac{\Delta V_t}{\Delta PV_t} = \frac{V_l}{V_t}\left(-\frac{\Delta V_l}{\Delta PV_l}\right) + \frac{V_g}{V_t}\left(-\frac{\Delta V_g}{\Delta PV_g}\right) + \frac{\Delta V_c}{\Delta PV_c} \tag{2.32}$$

Because the volume elastic modulus of liquid, gas and pipe can be $K_l = -\frac{\Delta PV_l}{\Delta V_l}, K_g = -\frac{\Delta PV_g}{\Delta V_g}$ and $K_c = -\frac{\Delta PV_c}{\Delta V_c}$ respectively and substitutes them into Eq. (2.32):

$$\frac{1}{K_e} = \left(\frac{1}{K_c} + \frac{1}{K_l}\right) + \beta_g\left(\frac{1}{K_g} - \frac{1}{K_l}\right) \tag{2.33}$$

The volume elastic modulus of the pipeline K_c K_c is mainly generated by deformation. Let $K_c = \Delta PV_c/\Delta V_c$, then:

$$\frac{\Delta A}{A} = \frac{\Delta PD}{Ee}\psi \tag{2.34}$$

$$K_e = \frac{\Delta PV_t}{\Delta V_t} = \frac{\Delta PA}{\Delta A} = \frac{Ee}{D\psi} \tag{2.35}$$

It can be obtained through simultaneous Eq. (2.34) and (2.35):

$$\frac{1}{K_e} = \frac{1}{K_l}\left[1 + \psi\frac{K_l D}{Ee} + \beta_g\left(\frac{K_l}{K_g} - 1\right)\right] \tag{2.36}$$

In addition, the density calculation formula of the transmission medium is shown as follows:

$$\rho = (1 - \beta_g)\rho_l + \beta_g\rho_g \tag{2.37}$$

Therefore, it can be obtained from the Eq. (2.36) and (2.37):

$$c_0 = \sqrt{\frac{K_e}{\rho}} = \sqrt{\frac{\frac{K_l}{1-\beta_g\rho_l+\beta_g\rho_g}}{1 + \psi\frac{K_l D}{Ee} + \beta_g\left(\frac{K_l}{K_g} - 1\right)}} \tag{2.38}$$

In the equation, the density of the gas is determined by the gas state equation $\rho_g = P/ZRT$, where: P is the absolute pressure, Pa; Z is the compressibility coefficient of real gas; R is the gas constant, let $R = 287.4\ N{\cdot}m/kg{\cdot}K$; T is Kelvin, K; In the boundary condition of top fixed, $\psi = \frac{1}{1+\frac{e}{D}}\left[\left(1 - \frac{\mu_0}{2}\right) + 2\frac{e}{D}(1 + \mu_0)\left(1 + \frac{e}{D}\right)\right]$.

2.2.4 *Drilling Fluid Pulse Signal Transmission Characteristics and Noise Analysis*

In the upload process of downhole information, susceptible to the external interference, useful signals are severely attenuated and mixed with a lot of noise, which results in the drowning of useful signals collected at the wellhead in the noise. Therefore, it is an important part of extracting the weak drilling fluid cw signal from strong noise.

The premise of flow field analysis of rotary valve is to make clear the intensity and frequency of pressure wave generated by pulse signal generator so that the signal can be detected when transmitted onto the ground. Therefore, it is necessary to analyze the transmission characteristics of pulse signal to provide some boundary conditions like pressure and speed for flow field of rotary valve. The study of drilling fluid pressure pulse signal transmission characteristics can be used to evaluate various factors influencing signal transmission speed and signal baud property. The influencing factors such as signal transmission speed, signal energy loss and waveform characteristics during signal transmission can all be analyzed.

The downhole pressure pulse signal travels along the drilling fluid in the drill string to the signal receiver, and its amplitude decreases as the transmission distance increases. The attenuation of drilling fluid pulse amplitude is related to the characteristics of drilling fluid and its transmission distance. Field tests show that the signal decreases by about half at each well depth of 450 m to 900 m. The attenuation of drilling fluid pulse signal increases with the decrease of drill string diameter, and increases with the increase of drilling fluid compressibility, well depth, drilling fluid viscosity and signal frequency. The gas is not completely removed from the fault pump or drilling fluid, resulting in the presence of gas in the drilling fluid, which will improve the compressibility of the drilling fluid, greatly reducing the amplitude of the pulse signal sent onto the ground.

The attenuation of continuous pulse signals is related to the transmission distance in the pipeline and the characteristics of the transmission medium. The loss of continuous pulse signal propagating along the pipeline mostly comes from the friction of the pipe wall. Similar to other physical transmission phenomena, continuous pulse signal also conforms to the law of exponential decay. In a pipeline filled with gas-liquid two-phase flow, the quantitative relationship between the intensity of continuous pulse signal and the transmission distance is expressed as:

$$p = p_0 \cdot e^{-\frac{x}{L}} \tag{2.39}$$

In the equation, p is the intensity of pulse signal, Pa; p_0 is the initial intensity of pulse signal, Pa; x is the transmission distance, m; L is the attenuation factor, that is, the transmission distance when the pulse signal attenuates to the its initial intensity 1/e, m.

The expression of defining attenuation is shown as follows:

$$\alpha = \frac{p}{p_0} = e^{-\frac{x}{L}} \tag{2.40}$$

Where the attenuation factor can be expressed as:

$$L = \frac{cD}{2}\sqrt{\frac{\rho}{\pi f \mu}} \tag{2.41}$$

For the relation between the angular frequency and the frequency of the pulse signal is, it can be obtained by solving the two simultaneous Eqs. (2.40) and (2.41).

$$L = \frac{D}{[1 + \left(\frac{R_f}{\omega}\right)^2]^{\frac{1}{4}} \cos[\frac{1}{2}\left(\tan^{-1}\frac{R_f}{\omega}\right)]}\sqrt{\frac{K_l}{2\mu\omega\left[1 + \psi\frac{K_l D}{Ee} + \beta_g\left(\frac{K_l}{K_g} - 1\right)\right]}} \tag{2.42}$$

By substituting the Eq. (2.42) into the (2.39), the formula for attenuation can be obtained:

$$\alpha = exp\left\{-\frac{\left[1+\left(\frac{R_f}{\omega}\right)^2\right]^{\frac{1}{4}}\cos\left[\frac{1}{2}\left(\tan^{-1}\frac{R_f}{\omega}\right)\right]}{D}\sqrt{\frac{2\mu\omega\left[1+\psi\frac{K_l D}{Ee}+\beta_g\left(\frac{K_l}{K_g}-1\right)\right]}{K_l}}x\right\} \quad (2.43)$$

For deep wells or with viscous drilling fluids, it is not easy to increase the signal transmission rate, because higher signal frequency will greatly reduce the signal amplitude. Therefore, the signal transmitted to the ground can be detected only by reducing the signal frequency, reducing the signal attenuation and increasing the pulse signal amplitude.

There are many factors affecting pulse attenuation. Some are uncontrollable in the drilling process except the carrier frequency and drilling fluid gas content, but those uncontrollable factors can still make a great impact. The selection of the drilling fluid in continuous wave communication system is as critical as the selection of carrier frequency. Reasonable selection of carrier frequency and carrier amplitude is one of the key points of studying drilling fluid continuous pulse transmission, which could be achieved by analyzing the influence of well depth, drilling fluid viscosity and signal frequency on signal attenuation, and combining signal coding modulation method.

Noise may flood the drilling fluid pulse signal and distort the signal so that even after filtering, the original information of the signal will be missed. Therefore, noise has a great impact on the structure and working conditions of the signal generator. By analyzing the influence of noise factors and the drilling fluid transmission characteristics, the selection of the appropriate signal strength and frequency range can be thus made, which can improve the reliability of signal transmission. From the perspective of rotary valve flow field, only the rotary valve flow field analysis based on noise analysis has its practical significance.

Part of the noise is produced by downhole instruments, and its transmission direction is the same as that of the transmitted pulse signal. And some noise comes from ground instruments such as a mud pump, which transmits in the opposite direction to the transmitted pulse signal. There are in the drilling fluid channel many noise sources with variable frequencies and distribution ranging over wide spectrum. Even for the same sources of noise, their frequency may change over time during drilling. The frequency of the drilling fluid pulse communication system usually operates below 100 Hz. The main noise sources include: mud pump, the interaction of the drill bit with the bottom hole, the interaction of the drill string with the wellbore, downhole turbine generator, etc. Among all the noises, the noise generated by the mud pump has the greatest impact on the drilling fluid pulse signal. The noise source and the frequency range are shown in Fig. 2.8. Study the signal propagation, signal attenuation and the influence of noise, and select the optimal frequency and amplitude of the continuous wave carrier. The purpose of channel transmission characteristics analysis is to analyze the attenuation law and noise influence of the actual transmitted wave to the ground receiving wave, and to

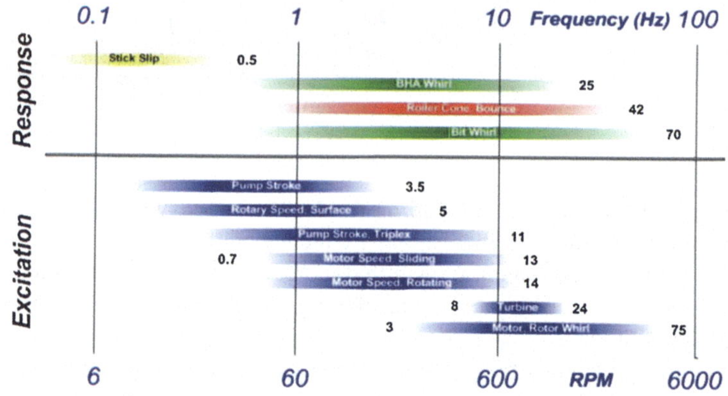

Fig. 2.8 Drilling noise source and the frequency range

solve such problems like: under what intensity and frequency, pressure waves are less disturbed by noise.

2.2.5 *The Encoding and Modulation of the Pulse Signal Carrier Data*

Data coding refers to the binary number composed of 0 and 1 by conversion of downhole sensor signals. Signal modulation loads the coded signals onto carrier signals, thus forming continuous signals. Its drive pulse generator generates pressure waves that transmit coded information along the drilling fluid in the drill string. Coded modulation signal serves as the driving source of pulse signal generation, and as the data basis ensuring the normal operation of pulse signal generator. Therefore, studying data coded modulation is a prerequisite for rotary valve flow field analysis of drilling fluid pulse signal generator, which can provide feasible and correct driving information for flow field analysis.

As shown in Fig. 2.9, downhole information data frame of continuous wave communication system adopts PSK modulation with the data transmission rate of 6bit/s, and each has a fixed time width (also known as bit period), which is in PSK modulation the time period of sending a continuous pressure wave (carrier frequency) to represent binary "0" or "1". To change the binary "0" to the binary "1", the speed of turning valve can be reduced and the phase of continuous wave signal can be converted. In the figure, each downhole variable parameter has a different update cycle. For example, resistivity transmission takes 8bits and updates every 8 s. In addition, control bits such as data frame synchronization and error detection are included in the coded modulation data frame to improve the reliability of information transmission. Some coding modulation methods in modern communication technology are applied to drilling fluid pulse communication system, which

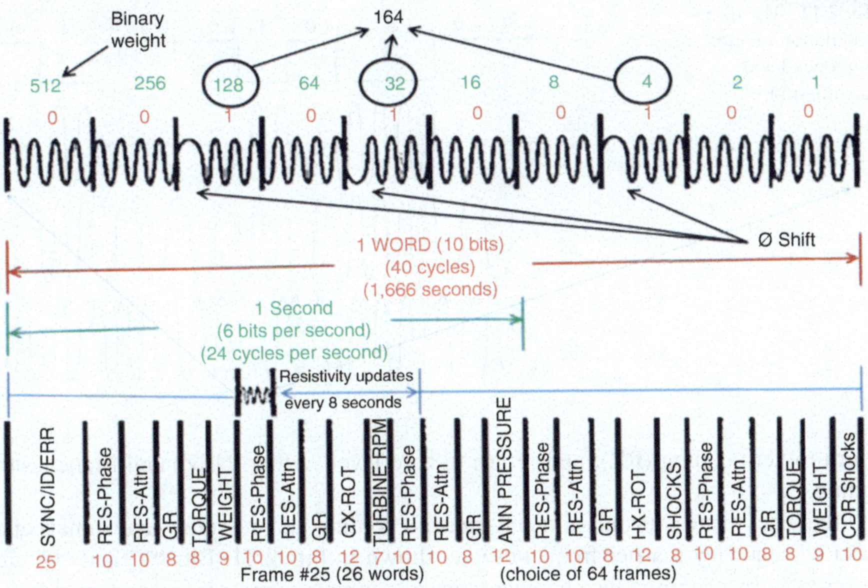

Fig. 2.9 The continuous wave pulse signal adopting PSK modulation data frame

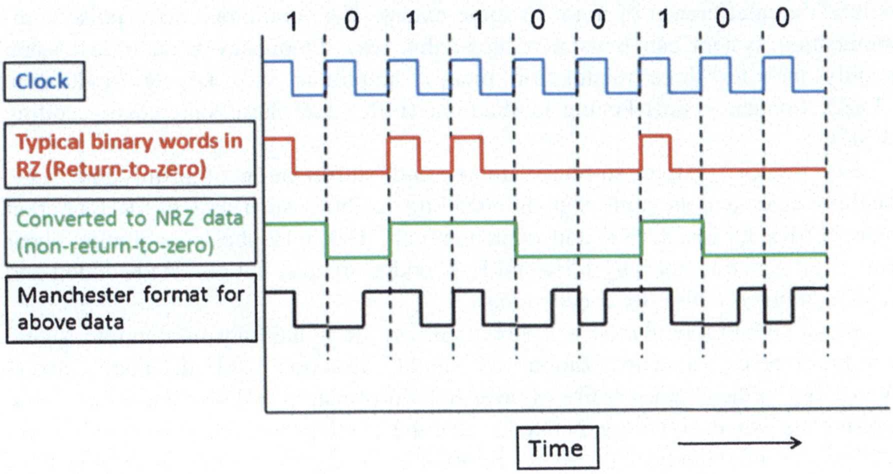

Fig. 2.10 Different coding modulation methods for baseband transmission

can transmit downhole information onto the ground. Data communication methods mainly include baseband transmission and frequency band transmission.

Baseband transmission can transmit discrete pressure pulse signals, as shown in Fig. 2.10, but it cannot transmit continuous pressure wave signals. And baseband transmission can be applied in modes of positive pulse and negative pulse. There are many encoding modulation methods applied in baseband transmission, such as

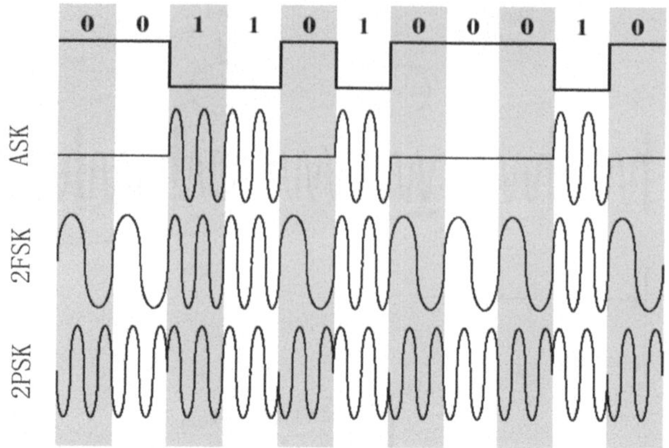

Fig. 2.11 The three-modulation method for frequency band transmission

return-to-zero coding (RZ), non-return to zero level coding (NRZ) and Manchester coding.

Frequency band transmission can transmit continuous pressure wave signals at a specific frequency (carrier frequency), as shown in Fig. 2.11. Bits "1" and "0" are represented by changing the characteristics of continuous pressure wave (amplitude, phase, frequency). Compared with baseband transmission, frequency band transmission can transmit data by adopting carrier frequency different from noise, which can reduce the interference of noise to some extent. The continuous wave pulse communication system can transmit data in this way. Frequency band transmission mainly includes three modulation modes: amplitude shift keying modulation (ASK), frequency shift keying modulation (FSK) and phase shift keying coding (PSK).

To improve signal-to-noise ratio and information transmission rate, Schlumberger adopts some signals encoding methods such as CPFSK (coherent phase FSK), QPSK, OPSK and other methods. Baker Hughes's oscillating shear valve can generate not only FSK and PSK coded signals, but also ASK coded and ASK combined with FSK coded signals.

Signal coding modulation will directly affect the signal transmission rate, signal anti-interference, synchronization of signal detection and detection errors. According to the characteristics of downhole information and characteristics of the continuous wave signals generated, selecting appropriate encoding modulation method can effectively improve the information transmission rate of drilling fluid continuous wave communication system.

2.2.6 Ground Signal Receiving and Data Processing

The extraction and identification of mud pulse signal involves different disciplines, such as electronics, fluid mechanics, vibration, signal identification theory and technology, etc. It is a typical interdisciplinary problem. The research work of ground decoding system mainly includes the following four aspects:

1. The mechanism analysis of mud pulse signal transmission. The factors affecting the viscosity velocity of mud pulse signal and the transmission velocity of drilling fluid under ideal conditions are systematically analyzed. The attenuation degree of mud pulse signal amplitude along the drill string and its influencing factors are systematically analyzed.
2. Research on mud pulse signal noise. Different de-noising algorithms are systematically studied. Under different noises, the optimal de-noising algorithm is selected to de-noise in the mud pulse signal, and the mud pulse signal is effectively identified.
3. Research on mud pulse signal recognition algorithm. The mud pulse signal recognition model is established systematically, and the theoretical basis of mud pulse signal recognition algorithm is proposed.
4. Design and development of ground signal acquisition box. Develop ground signal acquisition box. The ground signal acquisition box adopts high-precision 16-bit A/D conversion chip to collect mud pulse signal, and the acquisition circuit is transmitted to the computer through USB output interface for signal processing.

The wireless MWD ground system should collect the mud pressure pulse signals in real time and accurately, and convert them into various parameters that can be directly used by the staff on the ground. Preliminary research lays the basis for the successful interpretation of information by the ground decoding system. Sections 2.2.1–2.2.4 of this chapter comprehensively analyzes the generation mechanism, transmission characteristics and noise model of mud pulse signal, providing a theoretical basis for the recognition and noise reduction algorithm of the ground acquisition system.

The traditional wavelet decomposition can decompose the signal into different frequency bands, and eliminate the noise of high frequency band through threshold processing. But the low frequency noise will remain in the reconstructed signal. For weak continuous wave signals, part of the noise is mixed with useful signals. Although some effect has been achieved by using wavelet threshold, this part of the noise remains in the reconstructed signal, affecting the further optimization of denoising results. Therefore, to improve the precision of denoising, the low-frequency noise must be eliminated from the reconstructed signal as much as possible.

Compared with traditional wavelet transform, empirical mode decomposition has the advantage of self-adaptability. The number of decomposition layers depends on the characteristics of the signal. The decomposed modal components can reveal the local characteristics of the signal and reconstruct the signal by analyzing the instantaneous frequency of the modal components. However, the number of decomposition layers of traditional wavelet transform needs to be set. If there are too many decomposition layers, the coefficients on large scale are easy to be distorted. If there are too few layers, useful signals and noises cannot be separated well. Moreover, the

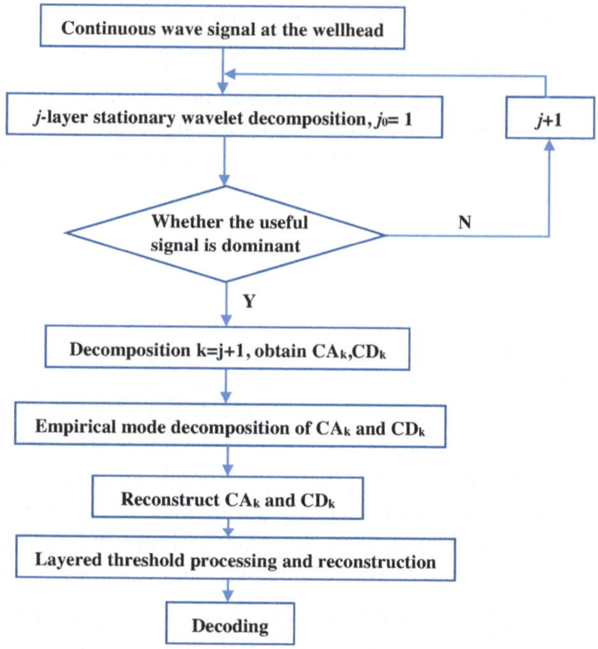

Fig. 2.12 Combined de-noising scheme decomposition flow chart

selection of threshold value is based on the empirical formula, and the signal with different characteristics will also produce a different effect.

For the weak continuous wave under strong noise, the wavelet transforms and empirical mode decomposition when used alone cannot achieve the ideal denoising effect, but the two methods in the process show their own advantages of denoising. Wavelet denoising can effectively remove high frequency noise while the empirical mode decomposition under high signal-to-noise ratio has a good decomposition performance. The two methods need to be combined for signal processing. Firstly, the signal is decomposed to different scales by means of wavelet decomposition, and then empirical mode decomposition is carried out after selecting decomposition layer coefficients dominated by useful signals, which is equivalent to empirical mode decomposition processing signals under high signal-to-noise ratio. This well satisfies the conditions of empirical mode decomposition. The specific operation process is shown in Fig. 2.12.

1. Stationary wavelet decomposition. j layer wavelet decomposition for sufficiently long signals is carried out (the length is defined to delete boundary signals when suppressing end effect, so as not to affect normal information), with j starting from 1. Do the calculation of frequency of the low frequency coefficient decomposed each time, and the assessment of the frequency of useful signal absolutely dominant in the low frequency coefficients, if not, further

decomposition is needed until it is decomposed to the appropriate decomposition layers. Then denote this layer as j, and do signal decomposition of K $= j + 1$ times.

2. Do empirical mode decomposition of the low frequency coefficient CA_k and high frequency coefficient CD_k at the K layer. The signal to be decomposed is set as $x(t)$, and its specific decomposition process is:
   a. Determine all local extremum points of the signal $x(t)$, do the fitting of the local maximum value and minimum value respectively, and generate corresponding upper envelope E_1 and lower envelope E_2. The upper and lower envelope should cover all data points.
   b. Calculate the average value of the upper envolope and lower envelope $m(t) = (E_1(t) + E_2(t))/2$
   c. Calculate processing sequence $h_1(t) = x(t) - m(t)$. For nonlinear non-stationary process, the initial calculated $h_1(t)$ generally do not satisfy the conditions of internal modelfunction, so repeat the steps (a) and (b) until the conditions are met, then $c_1(t) = h_1(t)$.
   d. Separate the internal modelfunction obtained in step (3) to obtain the remaining time sequence $r_1(t) = x(t) - c_1(t)$
   e. Following the below equation, repeat steps a-d to obtain the internal modelfunction in turn until the remaining sequence cannot be extracted, then the cycle ends.

$$\begin{cases} r_2(t) = r_1(t) - c_2(t) \\ r_3(t) = r_2(t) - c_3(t) \\ \cdots \\ r_n(t) = r_{n-1}(t) - c_n(t) \end{cases} \tag{2.44}$$

Until now, time series x(t) is decomposed into the form of multiple internal modelfunctions $c_i(t)$ and a superposition of residual $r_n(t)$. And $c_i(t)$ represents the different frequency bands of the data sequence from high to low, that is:

$$x(t) = \sum_{i=1}^{n} c_i(t) + r_n(t) \tag{2.45}$$

3. The instantaneous frequency of each internal modelfunction is calculated, and select the internal model components distributed in the frequency band of useful signals to reconstruct the coefficient at the K layer according to the frequency of the original signal.
4. The high frequency coefficients at other layers are denoised by multi-layered threshold [23], and then the stationary wavelet is used to reconstruct the signal, and delete the boundary part at both ends of the reconstructed signal.

2.3 Acoustic Signal Transmission Technology

The acoustic transmission system [24] uses sound waves to transmit signals through the drill pipe. Since the signal decays rapidly in the drill pipe string, relay stations should be installed every 400–500 m in the drill pipe. Because of the diameter change at drill string joint and the influence of screw assembly, the signal will be attenuated by reflection and refraction. The advantages of acoustic transmission are simple structure and low cost. The disadvantages are as follows: less information transmitted, interference of the noise on sound wave when drilling, weak signal, fast signal attenuation with depth, low reliability of relay device and high cost.

2.3.1 Acoustic Transmission System Model of Data while Drilling

Acoustic wireless transmission of data while drilling uses the acoustic wave as carrier, and drill string formed by periodic cascade of drill pipe and coupling as channel, so that downhole logging data can reach the ground control system quickly and effectively. Figure 2.13 shows the acoustic wireless transmission system model of logging data while drilling.

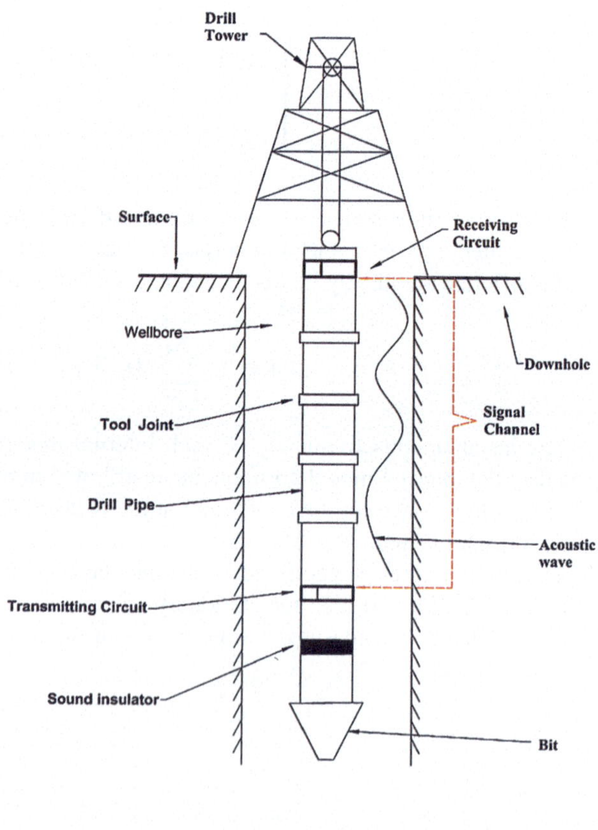

Fig. 2.13 The acoustic wireless transmission system model of logging data while drilling

In the course of the drilling, the drill string composed of a cascade of drill pipe and coupling goes through the wellbore and directly to the ground, forming a channel for acoustic transmission. Downhole transmission circuit modulates the logging data and by vibrator converts it into acoustic waves, which is loaded into the channel of the drill string. After attenuation and noise interference, the acoustic waves are transmitted to the receiver end acceleration sensor. Finally the acceleration sensor converts the received acoustic signal into electrical signals, which are sent to the receiving circuit and are restored to the original sending data after processing. By cascaded sound insulation between the transmitting end and the drill bit, the interference of the strong noise from the drill bit to the downhole circuit can be effectively reduced.

LWD data from downhole transmitter reaches the ground receiving end through the channel, and the control command from the ground reaches the downhole receiving end through the channel. The acoustic wireless transmission of data while drilling string realizes the two-way communication between surface and downhole by using the channel of the drill string. In the uplink transmission of logging data, X_d represents the logging data signal at the sending end, Y_d represents the signal at the receiving end, H_c represents the frequency domain response of the channel, N_c, N_b and N_s represents downhole environmental noise, bit coupling noise and wellbore environmental noise respectively. The uplink transmission system of logging data can be expressed as:

$$Y_d = H_c(X_d + N_c + N_b) + N_s \tag{2.46}$$

Where X_c is the downlink logging command signal, and Y_c is the downlink receiving command signal.

2.3.2 *Types and Characteristics of Acoustic Wave through Pipeline Transmission*

The transmission of acoustic wave in elastic solid medium can be equivalent to the transmission of elastic wave in solid medium. Therefore, the transmission characteristics of acoustic wave in drill pipe can be analyzed according to the wave theory of elastic wave in solid medium. Due to different boundary conditions, the transmission form of acoustic wave in drill pipe is different from that in infinite solid medium. Only approximate solutions of acoustic wave equation can be obtained under special conditions. For the drill pipe unit, it can be considered that the infinite long uniform thin rod in vacuum or in air is isotropic and its cross-sectional area is much smaller than the acoustic wavelength.

Acoustic wave is transmitted through tubing strings [25], and three types of acoustic waves can be observed at frequencies ranging from 1 Hz to 2 k Hz: extensional wave, torsional wave, and bending wave. Bending waves are the slowest of the three and can disperse even in pipes of the same diameter, which leads to great attenuation of the signal. Therefore, generally bending waves will not be applied for transmission, while extensional waves and torsional waves can do this.

For the above frequencies, the wavelengths of the extensional and torsional waves are longer than the diameters of ordinary pipes, so they will not scatter in pipes of the same diameter. The problem of acoustic transmission would be easily solved if there are uniform connectors in the actual pipeline. But actually, tubing strings are combined and assembled through toll joints. Therefore, the structure of tubing string is periodically divided and joint with a certain cross-sectional area, and the diameter of the tube is quite different from the size of the cross-sectional area. Extensional and torsional waves will be partially reflected at each toll joint, with the reflection coefficient of extensional waves depending on the proportion of the cross-sectional area in the whole tube, and the torsional waves depending on the polar moment of inertia of the cross section. The torsional waves can be strongly reflected at the joint. That's why we choose extensional wave. The elastic waveform phenomenon of the extensional wave and torsional wave in tubing string [26] can be expressed as follows:

$$\cos k(d_1 + d_2) = \cos \frac{\omega d_1}{c} \cos \frac{\omega d_2}{c} + M \sin \frac{\omega d_1}{c} \sin \frac{\omega d_2}{c} \tag{2.47}$$

Where is c is the velocity of acoustic wave, k is the wave number, and M is the reflection coefficient.

$$M = \frac{1}{2}\begin{cases} a_1/a_2 + a_2/a_1 & \textit{extensional wave} \\ I_1/I_2 + I_2/I_1 & \textit{torsional wave} \end{cases}$$

Where, d_1 and d_2 represents the length, a_1 and a_2 represents the cross-sectional area, $I_1$and I_2 is the polar moment of inertia between the joint and the pipe.

Before analyzing the transmission characteristics, set the following:

Drill string density as $\rho(x)$, joint area as $a(x)$, velocity of longitudinal wave as c, impedance as $z = \rho ac$, and u_n^j as the transmission location $u(x, t)$, where $x = x_n$, $t = j\Delta t$, n and j are indexes of location and time respectively (Fig. 2.14).

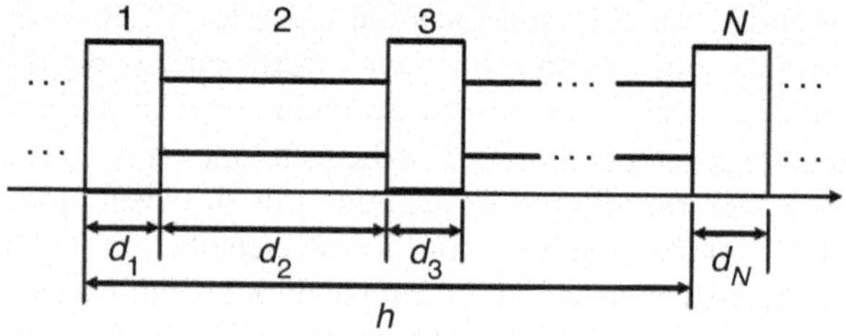

Fig. 2.14 The physical model diagram of general tubing string

The transmission pipeline model is shown in the Fig. 2.14:

For the N pipe joint, the potential function can be expressed as:

$$\Phi = \Phi_i e^{jkx} + \Phi_r e^{-jkx} \tag{2.48}$$

Where Φ represents the potential function, Φ_i represents the amplitude of potential function that directly reaches the pipe joint, Φ_r represents the amplitude of potential function that reaches the pipe joint through reflection, and k represents the wave number of the longitudinal wave.

The relation can be expressed as follows:

$$u = \frac{\partial \Phi}{\partial x} = jk[\cos(kx)(\Phi_i - \Phi_r) + j\sin(kx)(\Phi_i + \Phi_r)] \tag{2.49}$$

$$F = -\rho ac^2 \frac{\partial u}{\partial x} = j\rho ac^2 k^2 [\sin(kx)(\Phi_i - \Phi_r) - j\cos(kx)(\Phi_i + \Phi_r)] \tag{2.50}$$

Angle of matrix can be expressed as:

$$\begin{bmatrix} u \\ F \end{bmatrix} = A(x) \begin{bmatrix} \Phi_i + \Phi_r \\ \Phi_i - \Phi_r \end{bmatrix} \tag{2.51}$$

Where $A(x)$ is the matrix of x, the elements of the matrix can be expressed as: $a_{11} = -k\sin(kx)$, $a_{12} = jk\cos(kx)$, $a_{21} = \rho ac^2k^2\cos(kx)$ and $a_{22} = j\rho ac^2k^2\sin(kx)$.

Studing the transmission effect of the tubing string, denote Φ_i and Φ_r as Φ_I and Φ_R at the receiving end, it can be obtained from the above equation:

$$\begin{bmatrix} u \\ F \end{bmatrix}_{x=d_n^+} = A(0) \begin{bmatrix} \Phi_I + \Phi_R \\ \Phi_I - \Phi_R \end{bmatrix} \tag{2.52}$$

Besides, there is no reflected acoustic wave at the transmitting end, so $\Phi_r = 0$, Φ_i is the amplitude of potential function of transmitting sound wave.

Reflection coefficient R and transmission coefficient T can be defined as $R = \frac{\Phi_R}{\Phi_I}$, $I = \frac{\Phi_i}{\Phi_I}$. Let $M_N = A(d_N)A^{-1}(0)$, then:

$$A(0) = \begin{pmatrix} 1+R \\ 1-R \end{pmatrix} = MA(0)\begin{pmatrix} T \\ T \end{pmatrix} \tag{2.53}$$

Since:

$$A(0) = \begin{pmatrix} 0 & jk \\ \rho ac^2 & 0 \end{pmatrix} \tag{2.54}$$

Let $M = [M_{ij}]_{2\times 2}$, then it can be obtained:

$$\begin{pmatrix} jk & jm_{11}k + m_{12}\rho ac^2k^2 \\ -\rho ac^2k^2 & jm_{21}k + m_{22}\rho ac^2k^2 \end{pmatrix} \cdot \begin{pmatrix} R \\ T \end{pmatrix} = \begin{pmatrix} jk \\ \rho ac^2k^2 \end{pmatrix} \tag{2.55}$$

From the above Eq. 2.55, transmission coefficient and reflection coefficient of tubing string can thus be obtained.

Drill string has roughly the same basic structure, but for different types of drill string, its joint thread and joint cross-sectional area are not the same. So even with the same drill string channel structure, the acoustic transmission characteristics also matters in it. As the proportion of drill string pipe body and joint cross-sectional area gradually increases, the cross-sectional area of the two gradually approaches, and the reflection of acoustic transmission in the channel decreases. At this time, the higher the transmission coefficient is, the smaller the curve curvature can be and the smaller the attenuation of acoustic transmission can be. Therefore, the variation of drill string pipe body and cross-sectional area affects the attenuation range of acoustic transmission. The greater the difference, the more serious the attenuation will be.

2.3.3 Attenuation Characteristics of Acoustic Propagation along Tubing String

When the acoustic wave propagates in the tubing string [27], its signal strength will gradually decline. The coupling, acoustic frequency and the damping of surrounding media are the main causes of the attenuation. Generally speaking, the propagation attenuation of acoustic signal can be divided into the following three forms:

1. Absorption attenuation inside the drill pipe: the acoustic waves are transmitted in the drill pipe. Due to the viscous effect of the medium, the internal friction between particles is caused, which converts acoustic energy into heat energy. Heat energy is transmitted through the medium, resulting in energy loss.
2. Diffusion-type attenuation at the coupling: when there exists granular structure, defective structure and the structure containing dopants in the drill pipe, it will cause attenuation of acoustic wave intensity. Generally, when the granule size is much smaller than the wave length, the scattering attenuation is proportional to the fourth power of the acoustic wave frequency. When granule size is close to wavelength, scattering attenuation is proportional to the square of acoustic wave frequency.
3. Scattering attenuation: as the acoustic wave travels around, the energy of the acoustic wave gradually diffuses, which reduces the energy stored per unit area. The acoustic energy per unit area decreases with the increase of the square value of the distance from the sound source.

Among them, absorption attenuation is actually the energy conversion when acoustic waves travel along the tubing string. Part of it becomes heat energy, the other part is converted to the energy needed for intramolecular motion. The absorption attenuation law can be expressed as follows:

$$A = A_0 e^{-\delta L} \tag{2.56}$$

Where A represents the acoustic wave intensity at the distance from the sound source L, with the unit dB; L represents the propagation distance, with unit m; e is a constant of 2.3; δ is attenuation coefficient of acoustic wave in tubing string.

The attenuation coefficient of acoustic wave is usually related to the viscosity coefficient of surrounding medium, the transmission distance of acoustic wave signal and the transmission frequency. The law of it can be expressed as: the greater the viscosity coefficient of the surrounding medium is, the greater δ is. In addition, the attenuation coefficient also increases with the increase of transmission distance and frequency. It can be seen from the Eq. (2.56) that the amplitude of acoustic wave travelling along the tubing string can be approximately regarded as attenuation by the exponential law.

Similarly, the discontinuity of the transmission medium at the joint between the tubing and the coupling also lead to reflection and scattering attenuation of acoustic waves as they travel along the tubing string. Acoustic waves attenuate slightly in the same oil tube but attenuate greatly at the coupling. This is because the signal strength of acoustic waves becomes relatively weak after some energy is reflected by screw threaded during the transmission of acoustic waves along the tubing string. Then the refracted acoustic waves are weaker than than incident acoustic waves. In the whole transmission process, the refracted wave will always be smaller than the incident wave, which makes acoustic energy is being attenuating in the whole process of transmission.

The wave equation corresponding to the acoustic transmission can be expressed as follows:

$$A = A_0 e^{-\alpha x} e^{j(\omega t - kx)} \tag{2.57}$$

In the equation, is envelope curve function in the amplitude wave Eq. (2.57). In the process of acoustic wave propagation along tubing string, its signal amplitude attenuates exponentially with the increase of transmission distance.

The attenuation coefficient α can be expressed as follows:

$$\alpha = \frac{1}{x} \ln \frac{A_0}{A} = \frac{2.3026}{x} \lg \frac{A_0}{A} \tag{2.58}$$

In which A_0 is the initial wave amplitude, and A is the wave amplitude at the wave distance x.

Assuming that there are two different well depths x_1, x_2 ($x_1 < x_2$), the amplitude of acoustic signals in the two different well depths is expressed as A_1 and A_2 respectively, and the attenuation coefficient can be approximatively expressed as:

$$\alpha = \frac{1}{x_2 - x_1} \ln \frac{A_1}{A_2} \quad (2.59)$$

2.3.4 Noise Interference Characteristics and Model

Based on different research objects, noise and useful signals are on opposite sides. Noise is considered as a kind of "pollution" to useful signals, and its interference to useful signals can be divided into additive interference and multiplicative interference, which correspond to additive noise and multiplicative noise respectively. In most cases, noise interference can be considered as additive noise. The characteristics of noise determines its interference characteristics. According to the inherent characteristics of noise, noise can be divided into white noise, colored noise and impulse noise, Among which white noise is one of the simplest and most common noise. It mainly features in the noise component containing all frequency, and the probability of each noise component is the same. The noise spectrum is a straight line. Compared with the concept of white noise, the concept of colored noise does not include the noise components of all frequency, and the corresponding noise spectrum is not a straight line. Impulse noise is a kind of impulse interference which appears in a very short time with a large peak value in time domain and noise components at all frequency points of frequency domain.

Downhole environmental noise is generated by a series of downhole drilling logging, including the contact between drill pipe and borehole wall, circulation flow of drilling fluid, and working noise of LWD instruments, etc. [28]. The characteristics of such noise are closely related to drilling pipe parameters, formation characteristics, the type of drilling fluid, flow rate, the type of logging instrument and working state. The noise generated by the collision of drill pipe and borehole wall is related to the characteristics of borehole wall. The harder the wellbore rock layer is, the higher the noise frequency is. The circulating flow of drilling fluid can produce relative motion relatively to the drill pipe, and its viscous coupling effect reduces the acoustic energy. Meanwhile, the drilling fluid carries a large number of crushed rock in the fractured stratum, which will excite noise interference when the gravel collides with the drill pipe. When the bit drilling stops, the noise frequency of drilling fluid circulation is mainly in the low frequency band with the frequency ranging from 1 to 2300Hz.

When noise intensity X_n is Gaussian distribution, the probability density function of X_n is:

$$f_{X_n} = \frac{1}{\sqrt{2\pi}\sigma_n} \exp\left[-\frac{(x_n - u_n)^2}{2\sigma_n^2}\right] \quad (2.60)$$

Where u_n and σ_n^2 are respectively the average and variance of X_n. When the noise intensity obeys Gaussian distribution and the power spectral density is a constant, the white noise is called Gaussian white noise. Gaussian white noise is very simple and easy to deal with in time domain and frequency domain, so it is often used as a mathematical model for theoretical analysis of noise.

Compared with Gaussian white noise, impulse noise is featuring its fast burst, short duration, high energy and fast attenuation, which causes serious interference to signal transmission. Bernoulli-Gaussian noise model is widely used in Power Line Communication (PLC) channels [29, 30]. This kind of noise w_k consists of Gaussian noise z_k and impulse noise $q_k i_k$, which can easily and clearly describe the influence of Gaussian noise and impulse noise. Similarly, in the acoustic transmission system while drilling, to simplify the interference of impulse noise, the mathematical model of impulse noise should be established. Most researchers apply Bernoulli-Gaussian noise model for the modeling of impulse noise.

Bernoulli-Gaussian noise model is expressed as follows:

$$w_k = z_k + q_k i_k, \quad k = 0, 1, \cdots \tag{2.61}$$

Where z_k and i_k are mutually independent Gaussian white noise sequences, and q_k is Bernoulli sequences, and the expression is:

$$q_k = \begin{cases} 0, & 1 - P_I \\ 1, & P_I \end{cases} \tag{2.62}$$

P_I is the probability of the impulse occurring.

The variance of Gaussian sequence z_k is denoised as σ_z^2, the variance of Gaussian sequence i_k is denoised as σ_I^2, and the two SNR are defined as:

$$\begin{cases} SNR^g = 10 \cdot \log_{10}\left(\dfrac{\sigma_s^2}{\sigma_z^2}\right) \\ SNR^I = 10 \cdot \log_{10}\left(\dfrac{\sigma_s^2}{\sigma_I^2}\right) \end{cases} \tag{2.63}$$

Where SNR^g is the power ratio of Gaussian white noise z_k to signal, and SNR^I is the power ratio of Gaussian white noise I_k to signal. The probability density w_k of is the joint probability of two Gaussian white noises, the expression of it is:

$$P_{w_k}(w_k) = \frac{1}{\sqrt{2\pi}}\left[P_I \frac{1}{\sqrt{\left(\sigma_z^2 + \sigma_q^2\right)}} \exp\left(-\frac{w_k^2}{2\left(\sigma_z^2 + \sigma_q^2\right)}\right. + (1 - P_I)\frac{1}{\sqrt{\left(\sigma_z^2\right)}} \exp\left(\frac{w_k^2}{2\sigma_z^2}\right)\right] \tag{2.64}$$

Downhole environmental noise, bit coupling noise and wellbore environmental noise have caused serious interference to acoustic transmission while drilling. Effectively suppressing noise interference plays a key role in improving transmission quality. The methods of noise suppression mainly include active noise suppression and passive noise suppression, which can be expressed as noise source control and propagation control. At present, passive noise suppression schemes are mostly adopted in noise suppression by installing mufflers, sound absorbers, sound insulation and other multi-technology combination schemes.

Downhole environmental noise and bit coupling noise can reach the downhole transmitter through drill collar, and the noise can be effectively reduced by cascading sound insulation between the drill bit and the acoustic transmitter. For wellbore noise, the interference caused by such noise to the receiving end can be effectively reduced by installing sound insulation wall and muffler at the side of diesel generator set. Meanwhile, for the remaining noise interference, a series of signal processing means can be used to suppress noise, thus improving the quality of signal transmission.

2.4 Electromagnetic Transmission Mode

Electromagnetic wave MWD system realizes electromagnetic signal transmission using the drill pipe as an antenna, excites on both ends of bare conductors pipe in the drilling hole extremely low frequency electric current that carries the transmitted data, which goes through the surrounding strata to form electric field or magnetic field, and then is transmitted to the ground by long-distance transmission. Through the receiving electrode at the wellhead, the attenuated electromagnetic signal from the ground can thus be received. During the drilling process, the exposed drill pipe and borehole wall, the space between them and the surrounding strata all constitute the electromagnetic transmission channel. The propagation distance of electromagnetic wave in formation is seriously affected by formation resistivity. The formation resistivity of different region and depth varies greatly, which leads to different transmission depths of EM-MWD instrument signals. In addition, contact resistance exists between each section of drill pipe and between drill pipe and stratum, and the size of contact resistance changes dynamically with the vibration of drill pipe, thus affecting the stability of signal transmission.

2.4.1 EM-MWD Electromagnetic Communication Channel

According to the attenuation characteristics of electromagnetic signals during channel transmission [31–33], the following conclusions can be drawn:(1) electromagnetic signals are seriously attenuated by formation, and the formation resistivity suitable for EM-MWD operation is usually 2–200$\Omega \cdot m$; (2) if the transmission distance of EM-MWD signals reaches thousands of meters, the working frequency of electromagnetic signals should not be more than 100 Hz; (3) the working frequency, transmission distance, formation resistivity and drill pipe resistivity of electromagnetic signals can all affect the signal voltage and signal current between ground receiving electrodes.

Assuming that the resistivity of homogeneous formation around the borehole ρ is 2–2000$\Omega \cdot m$, the relative dielectric constant is not more than 200, and the highest working frequency of electromagnetic signal is 100 Hz, then:

$$\frac{\sigma}{\omega\varepsilon} = \frac{1}{2\pi f \rho \varepsilon_r \varepsilon_0} \tag{2.65}$$

In the equation, σ is formation resistivity, f is working frequency of electromagnetic signal, and free space dielectric constant $\varepsilon_0 = 8.85 \times 10^{-12}$F/m. Then the conduction current of electromagnetic signal propagating in the homogeneous formation is much larger than that of displacement current. Therefore, EM-MWD usually uses low frequency current field for data communication.

EM-MWD electromagnetic communication channel has the following characteristics:

1. The low frequency current signal is seriously attenuated. When inner diameter and outer diameter of the drill pipe, drill pipe resistivity, insulation end length and lower drill pipe length are fixed, the low-frequency current signal is seriously attenuated due to the influence of formation resistivity, signal working frequency and transmission distance when propagating in homogeneous formation, and the conduction current is much larger than the displacement current. So the ground EM-MWDS or downhole EM-MWDS communication system modeling need not consider shadow fading and multipath fading, and EM-MWD electromagnetic communication channel is approximated as Gaussian white noise AWGN channel.
2. Low frequency channel. To make the transmission distance reach thousands of meters, the working frequency of electromagnetic communication signals is usually less than 100 Hz [34]. In addition to weak electromagnetic communication signals, voltage signals between ground receiving electrodes are also mixed with self-potential and wellsite power frequency interference signals. If the electromagnetic signal communication frequency band is higher than the self-potential frequency band and lower than the power frequency band, filter method can be used to extract weak useful communication signals from received signals.
3. Parametric variation channel. In the drilling process, the dynamic contact of lower drill pipe with the formation results in dynamic change of contact resistance; The formation resistivity around the lower drill pipe varies with different drilling depths, and the signal attenuation degree is also different. The composition of drilling fluid around the insulation nodes affects the resistivity and thus affects the emission current. For the above factors, the transmitting signal current changes dynamically with drilling fluid, formation and contact resistance around the

insulation nodes, resulting in the dynamic change of the receiving signal voltage amplitude with time. This requires the EM-MWD receiver possessing the automatic gain control function.

2.4.2 The Transmission Characteristics of Electromagnetic Wave in Formation

In the through-the-earth communication, the main factor affecting the propagation of electromagnetic wave is the electromagnetic parameters of the earth medium, including: conductivity, permeability and dielectric constant. The three parameters jointly affect the propagation of electromagnetic wave in the stratum. The propagation coefficient is expressed as follows:

$$k = \sqrt{j\omega\mu(\sigma + j\omega\varepsilon)} = j\omega\sqrt{\mu(\varepsilon - j\sigma/\omega)} = \alpha + j\beta \tag{2.66}$$

Where,

$$\alpha = \omega\left\{\frac{\mu\varepsilon}{2}\left[-1 + \sqrt{1 + (\sigma/\omega\varepsilon)^2}\right]\right\}^{1/2}$$

$$\beta = \omega\left\{\frac{\mu\varepsilon}{2}\left[1 + \sqrt{1 + (\sigma/\omega\varepsilon)^2}\right]\right\}^{1/2}$$

When $|E/E_0| = 1/e$, the skin depth can be expressed as:

$$\delta = 1/\alpha = \sqrt{2}/\left\{\omega^2\mu\varepsilon\left[-1 + \sqrt{1 + (\sigma/\omega\varepsilon)^2}\right]\right\}^{1/2} \tag{2.67}$$

Considering that the permeability of the formation is the same as that of free space, the propagation coefficient in the formation can be written as:

$$k = j\frac{2\pi}{\lambda}\sqrt{\varepsilon_r - 60j\sigma\lambda} \tag{2.68}$$

In the electric conduction of the medium, when $\sigma/\omega\varepsilon \gg 1$, the skin depth in Eq. (2.67) can be simplified as $\delta = 1/\sqrt{\pi f\mu\sigma}$. Skin depth reflects the penetration ability of electromagnetic wave, indicating that the penetration ability of electromagnetic wave is related to the frequency (f) of electromagnetic wave and the conductivity of medium. The higher the frequency is, the greater the conductivity of medium is and the smaller the penetration ability of electromagnetic wave is. So the more serious the signal attenuation is, and the shorter the propagation distance will be. Due to the serious attenuation of electromagnetic wave caused by the formation, the signal of receiving point is weak, and the available frequency band for wireless communication through the formation only includes very low frequency band or even below it.

2.4.3 The Transmission Characteristics of Electromagnetic Wave in Drill Pipe

Due to the narrow drilling space, the rotating drill pipe is used as the transmission channel of electromagnetic waves. Only the vertical electric antenna (axial current along the drill pipe) and the vertical magnetic antenna (horizontal current ring around the drill pipe provokes the magnetic field alongside) are feasible to stimulate the electric field and magnetic fields respectively. Solve the modular equation of the propagation of these two fields along the conductive cylinder [35]. The result shows that the attenuation rate of the magnetic field corresponding TE wave is 3–4 orders of magnitude higher than that of the TM wave. Therefore, the most suitable excitation mode is the vertical electric antenna.

There are two kinds of analysis methods for conducting electromagnetic wave: field-based analysis method and path-based analysis method. According to the electromagnetic wave transmission mode in the drill pipe, it is more convenient to adopt the "path-based" analysis method, that is, the drill pipe is equivalent to uniform transmission line. For uniform transmission line, dz. can be taken from any point z of transmission line for research. Considering the physical significance of distributed parameter resistance, inductance, leakage conductance and capacitance, there exist parallel distributed capacitance $C_1 dz$, series distributed inductance $L_1 dz$, series distributed resistance $R_1 dz$ and parallel distributed leakage conductance $G_1 dz$ on the transmission line of dz length. Thus, the equivalent circuit of the transmission line can be drawn according to the "path-based" analysis method, as shown in Fig. 2.15.

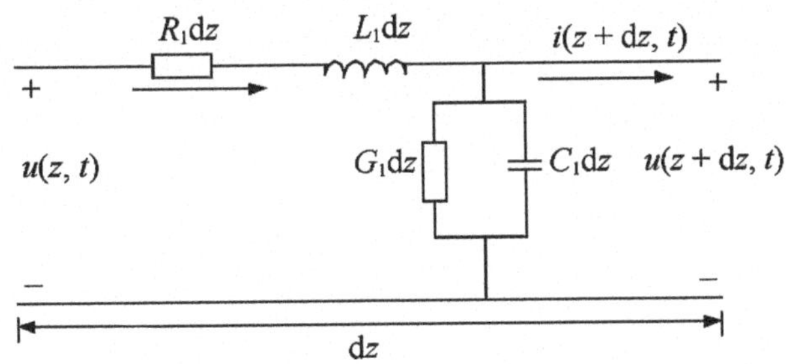

Fig. 2.15 The equivalent circuit model of the transmission line

It can be drawn from Kirchhoff's law:

$$u(z,t) - R_1 i(z,t)dz - L_1 \frac{\partial i(z,t)}{\partial t} dz - u(z+dz,t) = 0 \tag{2.69}$$

$$i(z,t) - G_1 i(z,t)dz - C_1 \frac{\partial u(z,t)}{\partial t} dz - i(z+dz,t) = 0 \tag{2.70}$$

Since:

$$\frac{\partial u(z,t)}{\partial z} = \frac{u(z+dz,t) - u(z,t)}{dz}, \frac{\partial i(z,t)}{\partial z} = \frac{i(z+dz,t) - i(z,t)}{dz}$$

By substituting the Eqs. (2.67) and (2.70), it can be obtained:

$$-\frac{\partial u(z,t)}{\partial z} = R_1 i(z,t) + L_1 \frac{\partial i(z,t)}{\partial t} \tag{2.71}$$

$$-\frac{\partial i(z,t)}{\partial z} = G_1 u(z,t) + C_1 \frac{\partial u(z,t)}{\partial t} \tag{2.72}$$

Equations (2.70) and (2.72) are the differential equations of voltage and current when electromagnetic wave propagates in drill pipe, in which equation $R_1 = \rho/2b\tau$ $(1 - \tau/2b)$, $L_1 = \mu_0/2\pi \ln(D/b)$, $G_1 = 2\pi\sigma/\ln(D/b)$, $C_1 = 2\pi\varepsilon/\ln(D/b)$.

To determine the size of D, the total parallel admittance of the equivalent transmission line can be set as equal to the admittance from the drill pipe of the infinite distance when the drill pipe exists independently, namely:

$$g = (hR_\infty)^{-1} = 2\pi\sigma/\ln(2h/b) \tag{2.73}$$

Thus, the equivalent radius of the transmission line is $D = 2h$. If $U(z)$ and $I(z)$are respectively complex voltage and complex current at z, the plural form of Eqs. (2.71) and (2.72) is:

$$-\frac{dU(z)}{dz} = (R_1 + i\omega L_1)I(z) \tag{2.74}$$

$$-\frac{dI(z)}{dz} = (G_1 + i\omega C_1)U(z) \tag{2.75}$$

By derivation of the Eqs. (2.74) and (2.75), it can be obtained:

$$\frac{d^2U(z)}{dz^2} = \Gamma^2 U(z) \tag{2.76}$$

$$\frac{d^2I(z)}{dz^2} = \Gamma^2 I(z) \tag{2.77}$$

Where Γ is defined as propagation coefficient, $\Gamma = \sqrt{(R_1 + i\omega L_1)(G_1 + i\omega C_1)} = \psi + j\varphi$.

Equations (2.76) and (2.77) are called wave equations of the transmission line. Their general solution form is as follows:

$$U(z) = A_1 e^{-\Gamma z} + A_2 e^{\Gamma z} \tag{2.78}$$

$$I(z) = \frac{1}{Z_0}\left(A_1 e^{-\Gamma z} - A_2 e^{\Gamma z}\right) \tag{2.79}$$

The characteristic impedance of the transmission line is $Z_0 = \sqrt{(R_1 + i\omega L_1)/(G_1 + i\omega C_1)}$, integration constant A_1 and A_2 can be determined by the boundary condition of the transmission line.

2.4.4 The Attenuation Model of Downhole Electromagnetic Signal

Electromagnetic waves can penetrate and propagate through most nonmetallic medium, and the depth of penetration is inversely proportional to the frequency of the wave and the conductivity of the medium. Electromagnetic MWD system realizes the information transmission from this characteristic of electromagnetic wave. AJ Mansure and others [36] proposed the signal model of "long electrode embedded in the formation ideal conductor". According to the signal model, the distribution of signal current on the drill string in the formation can be calculated as:

$$I_r(z) = I_t \exp(-zs) \tag{2.80}$$

Where z is the distance from a point on the drill string to the signal source point, m; $I_r(z)$ is the intensity of signal current at point z, A; I_t is the maximum amplitude of source point signal current, A; s is the skin inversion coefficient of electric current in the formation, $s = \sqrt{\pi f \mu / \rho}$, where μ represents the magnetic conductivity of the formation, generally $4\pi \times 10^{-7}$ H/m. ρ represents the resistivity of electric current through the formation, $\Omega \cdot m$ and f is the emission signal frequency, Hz.

It can be seen from the Eq. (2.80) that the current intensity at a point on the drill string is related to formation resistivity ρ, signal frequency f and the distance from this point to the signal source z, on the premise that the amplitude of transmitting signal current I_t remains unchanged. If the signal frequency and formation resistivity remain unchanged, the signal current on the drill string $I_r(z)$ decreases exponentially with the increase of distance z, as shown in Fig. 2.16 (the vertical axis is the logarithm).

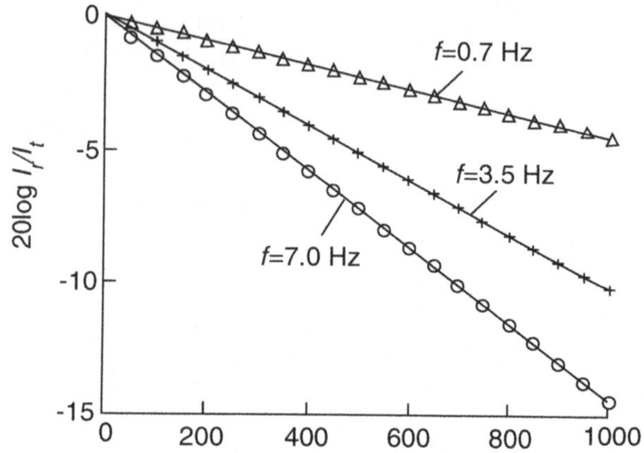

Fig. 2.16 Current attenuation of electromagnetic waves of different frequencies in single formation

The actual formation tends to be stratified. If the drill string passes through the formation with different resistivity $\rho_1, \rho_1, \cdots, \rho_n$ in n layers, and the corresponding thickness of each layer is $z(1), z(2), \cdots, z(n)$ respectively, then the current distribution on the drill string can be expressed as:

$$I_r(z) = I_t \exp\left[-\sum_{i=1}^{n} z(i)s(i)\right] \tag{2.81}$$

Where, $s(i) = \sqrt{\pi f \mu / \rho(i)}$

It can be seen from Eq. (2.81) that, in the multi-stratum environment, the law of signal attenuation is basically consistent with that of a single formation. In single formation, the higher the transmitted signal frequency is, the stronger the attenuation is and the smaller the signal amplitude received on the ground is. In order to improve the MWD depth, signal frequency must be reduced, which means the reduction of data transmission rate, affecting the real-time performance of MWD measurement data. The lower the resistivity of the drill string through the formation is, the faster the signal current decayed. Electromagnetic MWD in drilling engineering are committed through different formation with large difference in resistivity. Therefore, transmission frequency must be dynamically adjusted according to formation resistivity and working depth without increasing downhole transmission power, so as to improve MWD depth and data transmission rate.

2.5 Remote Transmission System of the IntelliPipe Data

In 2002, the IntelliPipe [37] system, successfully developed by American Intelli Serv, realizes the non-contact transmission of signals by using the electromagnetic induction coupling principle, and solved a series of problems such as line wear and poor contact in the wired transmission mode [15]. Since the intelligent drill pipe constitutes a network system, downhole instruments can also communicate, receive and execute instructions with each other. Information transmission can thus be achieved with various downhole tools and measurement sensors in the drilling process, so as to effectively evaluate oil and gas reservoirs in real time and guide drilling tracks accurately. The induction coil unit is installed in the joint, which facilitates the electrical conduction of the drill string for the full length. Downhole instruments such as sensors is installed in intelligent drill collars that are connected to the lower end of the drill pipe via armored cables. At the top of the drill pipe is a rotating ring connected to the ground instruments. The advantages of this transmission mode are fast data transmission and simple two-way communication. The disadvantages lie in that the transmission drill pipe needs special processing and costs a lot. Moreover, the transmission drill pipe can only transmit downhole measurement information and cannot transmit power (electricity) to downhole instruments.

2.5.1 Principle of Signal Inductive Coupling Transmission

The signal can be transmitted along the cable in the drill pipe and transmitted through an inductive coupler at the joint of the drill pipe. The signal transmission principle of inductive coupler can be analyzed according to electromagnetic induction theorem.

The number of turns of the coupler coil is set as N, the magnitude of the current passing through the coil is set as i, and the magnitude of the magnetic flux generated by each turn of the coil is set as Φ, which is proportional to the magnitude of the current i. If the magnetic induction line can pass through all N turns of the coil, namely, there is no leakage flux between adjacent turns of the coil, then the full flux $\psi = N\Phi$, which is also called the flux linkage. The relation of it and current is:

$$\psi = Li \tag{2.82}$$

Where L is the self-inductance of the coupling coil.

Start studying from two coils that are close to each other, as shown in Fig. 2.17, coil 1 and coil 2 are close to each other. Coil 1 is called primary coil, and let its number of turns be N_1, the self-inductance, L_1. Coil 2 is called secondary coil, and let its number of turns be N_2, the self-inductance coil, L_2.

The current i_1 generates the flux Φ_{11} through the primary coil and generates the flux Φ_{21} through the secondary coil. The flux Φ_{1n} that does not pass through the secondary coil is called the flux leakage. The transfer of magnetism from one coil to another is called magnetic coupling. The flux linkage generated by i_1 through the secondary coil is:

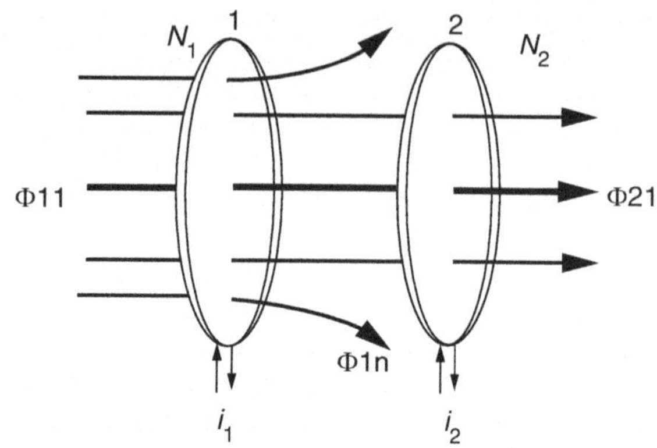

Fig. 2.17 Mutual inductance coil magnetic coupling

$$\psi_{21} = N_2\Phi_{21} = M_{i_1} \tag{2.83}$$

Where M is the coefficient describing the coupling effect between two coils, which is called mutual inductance. The size of it is related to the shape, position, distance and other factors of the two coils. Due to the existence of flux leakage Φ_{1n}, the two coils in Fig. 2.17 cannot be fully coupled. The coupling coefficient k ($0 \leq k \leq 1$) is defined as the ratio of coupling flux to total flux, i.e.

$$k = \frac{M}{\sqrt{L_1 L_2}} \tag{2.84}$$

Then $k = 0$ represents no coupling; $k = 1$ represents full coupling. At this time, $M_{\max} = \sqrt{L_1 L_2}$.

According to the law of electromagnetic induction, if the area of the closed coil remains unchanged, when the magnetic flux across its area changes, the coil will generate induced electromotive force, magnitude of which is the rate of flux change. Thus, the mutual inductance electromotive force e of the secondary coil is:

$$e = -\frac{d\psi_{21}}{dt} = -M\frac{di_1}{dt} \tag{2.85}$$

From the above equation, we can know the magnitude of mutual induced electromotive force. Firstly, it is related to the speed of current change rate of coil 1, namely *di/dt*. The second is related to the mutual inductance M between the two coils.

2.5.2 *The Circuit Model of an Inductive Coupler*

The magnetic induction drill pipe simulation prototype is shown in Fig. 2.18. The inductive coupler structure is similar to that of a transformer with high working

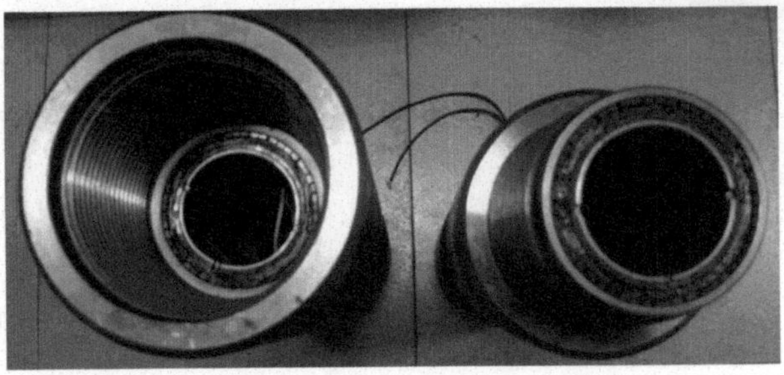

Fig. 2.18 Inductively coupled with simulation drill pipe

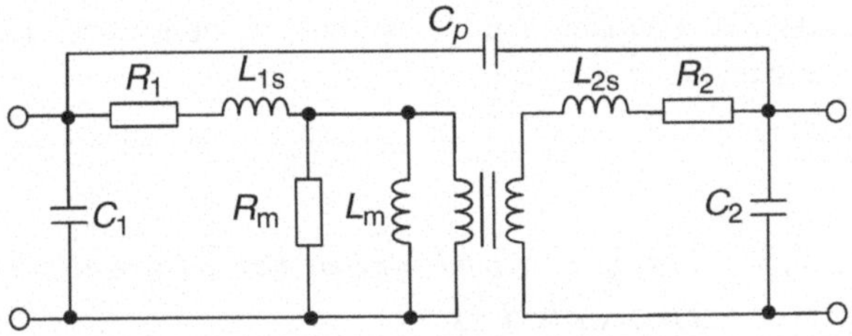

Fig. 2.19 Model of high-frequency transformer circuit

frequency. The model of the inductive coupler can be obtained from the general model of high-frequency transformer. The high frequency transformer model takes into account the parasitic effect of components at high frequency and can be represented by a network of lumped parameter elements (inductor, capacitor, resistor and double-port ideal transformer), as shown in Fig. 2.19.

In Fig. 2.19, L_{1s} and L_m represents leakage inductance and magnetizing inductance respectively. The resistance R_1 and R_2 represent the direct current loss of transformer coil at low frequency. R_m represents hysteresis loss and eddy current loss of magnetic core at high frequency. C_1 and C_2 are distributed capacitors generated by primary and secondary coils respectively. C_p is the stray capacitance between the two coils of the transformer. These parameters are all related to working frequency [38].

The current transformer model is not too different from the high-frequency transformer model, and it is only equivalent to the circuit model after a switch between the primary side and secondary side in Fig. 2.19. L_{1s}, L_{2s} and L_m are not real in the circuit, and it is difficult to measure. To make it more intuitive, the three parameters are represented by coil self-inductance and mutual inductance. At the same time, for the model in the figure above, it does not involve the secondary loss of the transformer, but the secondary loss of the inductive coupler should be considered due to its symmetry. Therefore, on the basis of high-frequency transformer, the circuit model of inductive coupler is given, as shown in Fig. 2.20.

The resistance R represents the dc loss of the coil, the capacitance C represents the distributed capacitance of the coil; R_m, the hysteresis loss and eddy current loss of the magnetic core at high frequency; C_p, an the stray capacitance between the two coils. The ideal transformer model is not adopted in the transformer. The coupling

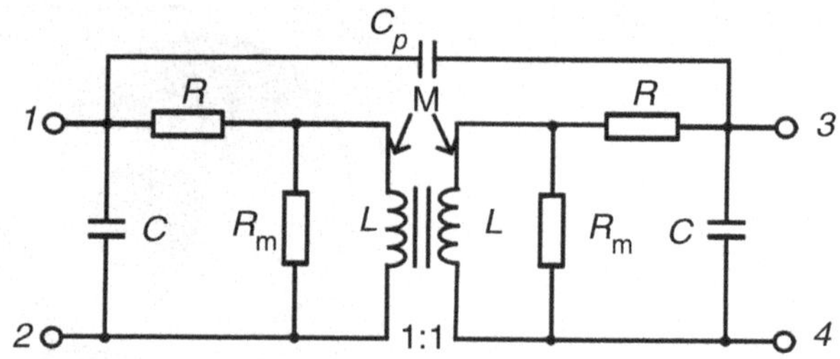

Fig. 2.20 Model of inductive coupling circuit

coefficient $k < 1$, L is the self-inductance of the primary coil and secondary coil of the inductive coupler, and M is mutual inductance. End 1 and end 3 are of the same name.

2.5.3 *The Circuit Model of the Magnetic Inductive Transmission*

The cable used in the channel is radio-frequency coaxial cable, which can be equivalent to a uniform and unbalanced transmission line filled with medium in the middle of two concentric cylindrical conductors. The characteristic impedance is denoted as $Z_0 = 75\Omega$, and every 100 m attenuation is denoted as $\alpha_{100} = 2.2dB$ when the signal frequency is less than 5 MHz. The attenuation of signal propagation can manifest as the decrease of signal output amplitude [39]. The signal output amplitude changes with the transmission distance, which can be expressed as follows:

$$u(l) = u_{in} \cdot 10^{-kx/20} \tag{2.86}$$

In the equation, l is the position of a point on the transmission line, and $u(l)$ is the signal output amplitude at l on the transmission line, u_{in} is the amplitude of the input signal, and α is the attenuation per unit length.

By substituting $\alpha_{100} = 2.2dB$ into the equation, the voltage attenuation ratio of 100 m long coaxial cable is 97%. It can be seen that the attenuation of 10 m long coaxial cable is much more smaller. The cable can then be equivalent to the simplest zero-order circuit, in which case the cable serves as a capacitor. From its main electrical characteristic, coaxial cable serves as the capacitor C_t. Considering the capacitance effect of coaxial cable and compensating the capacitance of inductive coupler, the transmission model of magnetic induction channel is shown in Fig. 2.21:

In Fig. 2.21, S is the equivalent circuit model of inductive coupler; T is the zero-order circuit model of coaxial cable, which manifests as the capacitance connected at both ends of the coil.

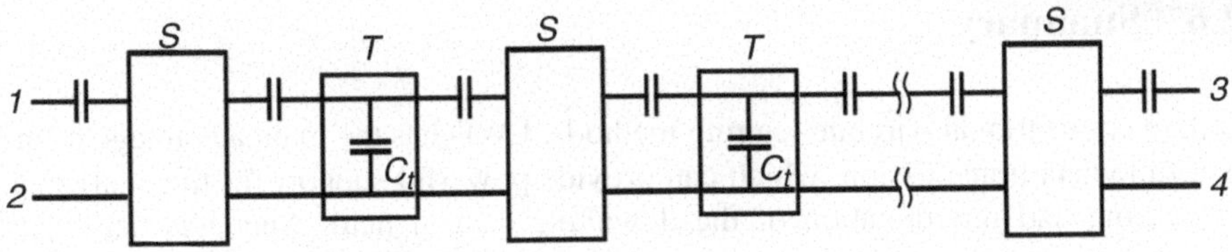

Fig. 2.21 Transmission model of magnetic induction channel

2.5.4 *Advantages and Disadvantages of Intelligent Drillpipe Transmission Technology*

Intelligent drill pipe transmission technology has the advantages of high data transmission speed, large capacity, real-time and two-way transmission, which is suitable for conventional drilling, underbalanced drilling and gas drilling. It is a breakthrough in downhole signal transmission technology.

Compared with mud pulse, electromagnetic wave and acoustic wave transmission mode, intelligent drill pipe transmission mode has obvious advantages, which are mainly shown in the following aspects:

1. high data transmission speed, large capacity and real-time. The data transmission rate is up to 57,600 bits /s, which enables real-time monitoring of drilling process without any interference.
2. two-way communication. Through downward transmission functions and specialized downhole interface joints, the rotary steering drilling system can be remotely operated while drilling continuously, thereby improving the control accuracy and efficiency of well track, improving wellbore smoothness and wellbore quality, reducing non-production time and reducing overall drilling costs.
3. wide range of application. It can be used for data transmission under any downhole conditions, including underbalanced drilling and gas drilling.
4. real-time monitoring of the whole wellbore, which is conducive to timely prevention of downhole complex conditions.
5. mud pulse can be used as an alternative transmission mode when the intelligent drill pipe is applied. In case the transmission of drill string is interrupted, mud pulse transmission mode can be enabled without lifting the drill string.
6. with the capability of high-speed transmission and two-way communication. It will greatly promote the further development of monitoring, evaluation, diagnosis and prediction technologies while drilling, such as MWD, LWD, geological guidance, formation testing while drilling and seismic while drilling.

In addition to the above advantages, intelligent drill pipe telemetry system also has some shortcomings. For example, the cost of intelligent drill pipe is much higher than that of ordinary drill pipe. It is not possible to supply downhole electricity through such soft-connected cables.

2.6 Summary

Unlike conventional wireline logging methods, LWD has the main advantage of the real-time data transmission, which can provide powerful support for the analyzing, processing and interpretation of the downhole data in field. And it is helpful to evaluate formation timely and effectively, update the formation model, optimize well track and completion well design, thus improving the efficiency and effectiveness of oil and gas exploration and development.

Data transmission technology is the core as well as bottleneck of LWD. At present, the widely used LWD data transmission technologies mainly include mud pulse transmission, electromagnetic transmission, acoustic transmission and intelligent drill pipe transmission. In general, the wireless transmission mode has a simple structure and relatively low technical cost, but with a low data transmission rate, limited data transmission volume and poor anti-interference ability. The wired transmission mode has obvious advantages, including high transmission rate (up to 1-2Mbit /s) and two-way transmission, but with complex instrument design and high cost. The main trend of development for later research lies in the following three aspects:

1. in the future, the main goal of mud pulse transmission is to achieve higher data transmission rate, improve working reliability and environmental adaptability, so as to achieve the best transmission effect under the factors such as mechanical drilling speed, mud type, well depth, drill pipe diameter and others.
2. electromagnetic and acoustic transmission technologies need to solve the problems of signal attenuation and noise interference, improve the reliability of signal transmission, increase the transmission depth of measured data and effectively reduce the costs.
3. the transmission of intelligent drill pipe continues to achieve the two-way, high-rate and real-time communication. Continuous improvement need to be make in reduced costs, further prevention of the operational risks, improved drilling efficiency and reduced non-operational time.

References

1. Wallace RG, High Data Rate MWD. Mud pulse telemetry [Z]. U.S. Department of Energy's natural gas conference. Houston: Texas; 1997.
2. Warren T. Rotary steerable technology conclusion: implementation issues concern operators [J]. Oil and gas Journal. 1998;96(12):23–4.
3. Perry CA, Burgess DE, Turner WE. Rotary Pulser for transmitting information to the surface from a drill string down hole in a well [P]. In: United States: US7327634B2; 2008.
4. Malone D, Johnson M. Logging while drilling tools, systems, and methods capable of transmitting Data at a plurality of different frequencies [P]. United States. 1994;5375098
5. Hahn D, Peters V, Rouatbi C, Eggers H. Oscillating shear valve for mud pulse telemetry and associated methods of use [P]. In: United States: US6975244B2; 2005.

6. Shah Vimal GW, Johnson DH, Sinanovic S. Design Considerations for a New High Data LWD Acoustic Telemetry System[R]. SPE. 88636
7. Dopf AR, Camwell PL, Siemens WL, et al. Apparatus for receiving downhole acoustic signals: US; 2005.
8. Camwell PL, Neff JM. Telemetry transmitter optimization using time domain reflectometry: US; 2011.
9. James Michael Neff, Paul Leonard Camwell. Field Test Results of An Acoustic Telemetry MWD System, SPE/IADC Drilling Conference, 20–22 February, Amsterdam, The Netherlands,2007.
10. Reeves ME, Camwell PL, Mcrory J. High speed acoustic telemetry network enables real-time along string measurements, greatly reducing drilling risk[C]. In: Offshore Europe, vol. 6-8. Aberdeen: September; 2011.
11. Reeves M, Smith D G, Groves D, et al. Unique Acoustic Telemetry Network With Distributed Pressure Measurement Nodes Enables Accurate Real-Time Analysis of Sweep Effectiveness [C]. SPE Annual Technical Conference and Exhibition, New Orleans,,2013.
12. Weisbeck D, Blackwell G, Park D, et al. Case history of first use of extended-range EM-MWD in offshore. Underbalanced Drilling[R]. SPE/IADC. 2002;74461
13. Anonymous. New tool extends MWD to underbalanced wells[J]. Drilling Contractor, March/ April 2005, issue:24–25.
14. Bhagwan J. Trofimenkoff. Electric drill stem telemetry. IEEE trans geoscience and remote sensing. IEEE. 1982;GE-20:193–7.
15. Perry A. Fischer. Interactive drilling up-to-data drilling technology[J]. Oil&Gas Science and Technology. 2004;59:343–56.
16. Montaron BA, Hache JMD. Improvements in MWD telemetry: “Right Data at the right time” [A]. PE Asia Pacific oil and gas conference[C]. Society of Petroleum Engineers. 1993:337–46.
17. Chin WC. MWD Siren Pulser Fluid Mechanics[J]. Petrophysics. 2004;45(4):363–79.
18. Hutin R, Tennet RW, Kashikar SV. New mud pulse telemetry techniques for Deepwater applications and improved real-time data capabilities [R]. SPE. 2001:67762.
19. Zielke W. Frequency-dependent friction in transient pipe flow [J]. ASME J Basic Eng. 90 (1):109–15.
20. Vardy AE, Brown JMB. Transient turbulent friction in smooth pipe flows[J]. J Sound Vib. 2003;259(5):1011–36.
21. Lee HY. Drillstring axial vibration and wave propagation in boreholes [D]. Gambridge. Massachusetts institute of Technology. 1991;
22. Lea SH. A propagation of coupled pressure waves in borehole with Drillstring[C]. SPE. 1996;37156:963–72.
23. Acharyya M, Kundu MK. Document image segmentation using wavelet scale-space features [J]. IEEE Tran On Circuits and Systems for Video Technology. 2002;12(12):1117–27.
24. Han JH, Kim YJ, Karkoub M. Modeling of wave propagation in drill strings using vibration transfer matrix methods[J]. J Acoust Soc Am. 2013;134(3):1920–31.
25. Douglas S. Drumheller, Acoustical Properties of Drill Strings, Jul. 1988, 5–25.
26. Douglas S. Drumheller, attenuation of sound waves in drill strings, 2387 J. Acoust Soc Am N (4), October 1993.
27. Jia D, Cui C, Wang S. Propagation properties of acoustic waves inside periodic pipelines [J]. Inf Sci. 2014;275(12):360–9.
28. Pighi R, Franceschini M, Ferrari G, et al. Fundamental performance limits of communications systems impaired by impulse noise[J]. IEEE Trans Commun. 2009;57(1):171–82.
29. Ghosh M. Analysis of the effect of impulse noise on multicarrier and single carrier QAM systems [J]. IEEE Trans Commun. 1996;44(2):145–7.
30. Trofimenkoff FN, Segal M, Klassen A, et al. Characterization of EM downhole-to-surface communication links [J]. Geoscience and remote sensing. TEES Transactions on. 2000;38 (6):2539–48.

31. Wei L, Zaiping N, Xiangyang S, et al. Numerical modeling for excitation and coupling transmission of near field around the metal drilling pipe in Lossy formation [J]. Geoscience and remote sensing. IEEE Transactions on. 2014;52(7):3862–71.
32. Poh Kheong V, Rodger D, Marshall A. Modeling an electromagnetic telemetry system for signal transmission in oil fields [J]. Magnetics, IEEE Transactions on. 2005;41(5):2008–11.
33. Wait JR, Hill DA. Theory of transmission of electromagnetic waves along a drill rod in conducting rock [J]. Geoscience Electronics, IEEE Transactions on. 1979;17(2):21–4.
34. Jing C, Yong L. Study on underground transmission of electromagnetic measurement while drilling (EM-MWD) [J]. Electronic measurement Technology. 2009;32(10):4–7.
35. Mansure AJ, Keefe RG, Caffey TWH, Bartel LC, Ballard S. The surface area modulation downhole telemetry system for measurement while drilling. United States: N. p. Web. 1998;
36. Michael J, Prideco G, David RH. Intelligent Drill Pipe Creates the Drilling Network [R]. SPE. 80454
37. Fischer PA. Interactive Drilling Up-to-data Drilling Technology[J]. Oil&Gas. Science and Technology. 2004;59:343–56.
38. Janzisz B, Darizisz C. High frequency transformer modeling[A]. 2001. IEEE International Symposium on Circuits and Systems[C] Sydney. 2001:676–9.
39. Tomazic S. Comments on spectral efficiency of VMSK[J]. IEEE Trans Broadcast. 2002;48 (3):61–2.

Chapter 3
Dynamic Measurement of Spatial Attitude at the Bottom Rotating Drillstring

Abstract In oil and gas directional drilling technology and application, how to accurately measurement the spatial attitude of the bottom drillstring in real time while the drillstring rotating is a challenging problem. We developed a set of "strap-down" measurement system using the triaxial accelerometer and triaxial magnetometers installed near the bit, and real-time well deviation and azimuth can been measured even when the drillstring rotates. Although magnetic based system is the classical, we will use this system to achieve continuous measurement-while-drilling relying on software algorithms. We developed the novel state space models to establish the Kalman filter, improving the accuracy of dynamic measurements. Simulation and experiments results show that the continuous survey system with Kalman filter approach could effectively enhance the measurement precision, and deduce the error that produced by the drillstring vibration. The algorithm greatly improved the accuracy of well-trajectory measurements and is expected to be applied to ordinary magnetic surveying systems, which are more widely used in drilling engineering.

Keywords Directional drilling · Continuous measurement-while-drilling · Kalman · State space model

3.1 Introduction

Directional drilling technology involves directing a wellbore along a predefined trajectory, which dramatically reduces costs and saves time during drilling operations [1–3]. During the last several years, more and more attention is been paid to the development of directional well-drilling technologies. Technology for directional-drilling navigation is currently based on an integrated magnetometer and accelerometer triad [4]. To compute the bottom-hole assembly (BHA) position, measuring data of the earth's magnetic field and the force of gravity are employed. The surveying system is executed along the well trajectory at stationary survey stations. In drilling engineering, the bottom drillstring attitude (inclination and azimuth) measurement is usually carried out when the drillstring is not rotating.

Q. Xue, *Data Analytics for Drilling Engineering*, Information Fusion and Data Science, https://doi.org/10.1007/978-3-030-34035-3_3

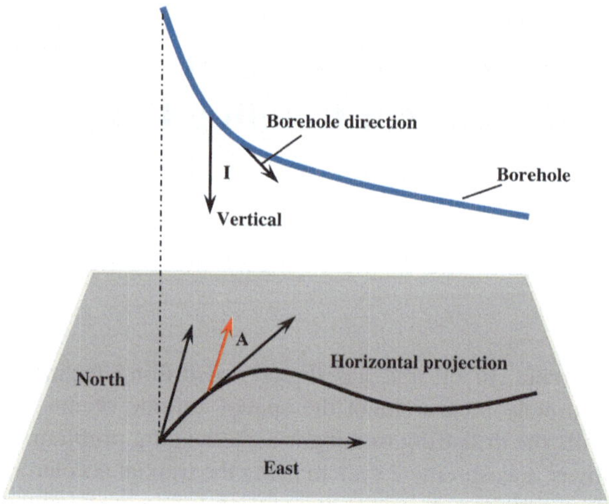

Fig. 3.1 Diagram of Inclination and Azimuth

Besides the conventional drilling assembly, directional drilling operations require position sensors to provide estimations of the inclination expressed as I (deviation from the vertical direction as show Fig. 3.1) and azimuth expressed as A (deviation from the north direction in the horizontal plane as shown in Fig. 3.1). These sensors are part of the MWD (measurement while drilling) tool, which is installed several feet behind the drill bit to monitor all physical parameters that affect the drilling operation.

There are two conventional systems for measurements while drilling based on magnetometers and gyroscope. The system is called Gyroscope based system which uses gyro and accelerometer, and the one is called Magnetometers based system using flux sensors and accelerometers, because in the drilling engineering the accelerometer is always used in various measuring systems.

Current MWD surveying is performed along the well path at stationary survey stations. That is to say, usually in drilling engineering, the bottom drill string attitude (inclination and azimuth) measurement is carried out in the case that the drill string does not rotate.

However, as drilling technology improves, continuous measurement of the well trajectory becomes increasingly important. It is also essential in a rotary-steerable system (RSS). Rotary steering system is a mechatronics tool generated by the development of directional drilling, which can drill more economical and smoother borehole. However, to measure down hole tools posture with the drill string rotating is one of the technical difficulties, since this is completely different from the current MWD surveying method. Actually, in the control process, assume that measured values under stationary state remain unchanged during drilling, could still achieve the closed-loop control of rotary steerable. However, the disadvantage is obvious, as shown in Fig. 3.2 actual drilling trajectory fluctuate with the design trajectory. This phenomenon cannot be avoided because of the hysteresis measurement.

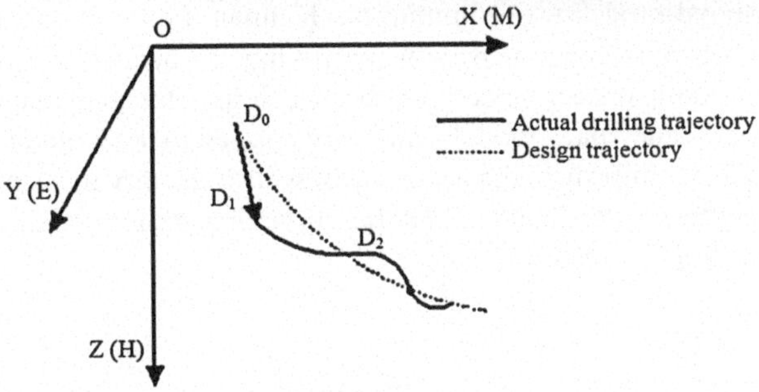

Fig. 3.2 Schematic diagram of traditional trajectory control

An RSS [5–7] is a mechatronics tool developed for directional drilling. It can drill a more economical and smoother borehole. Since the introduction of RSS, rotary-steerable technology has achieved notable progress in reliability and has become a standard drilling tool in many worldwide applications. The application of RSS is restricted to high-cost offshore sites, and is becoming more common in cost-sensitive land work, especially in shale-gas and shale-oil drilling [8]. Despite popular use of the RSS, field-trial results of continuous measurement of inclination and azimuth have not been well documented in the literature. However, measuring the posture of downhole tools as the drillstring rotates is essential because of its closed-loop control structure [9, 10]. A bottom-drilling tool shows complex dynamics while rotating owing to the combined effects of nonlinear vibration, such as vertical vibration, horizontal vibration, eddy, and sticky slip [11, 12]. The effects of such vibrations cause to measurement sensors to generate large errors. This is a huge challenge for signal processing [9], which is completely different from the measurement-while-drilling (MWD) survey systems in current oil and gas industry.

Continuous MWD are studied under laboratory conditions using a gyroscope-based system [13–15]. They proposed an advanced inclination and direction sensor package based on an inertial navigation system (INS). They verified the reliability of the algorithm through simulation, which used INS to achieve continuous MWD with high accuracy. The influences of vibration and temperature on MWD were also analyzed [16–18]. Literature [19, 20] conducted a similar study by developing a MWD instrument based on a predigested inertial measurement unit. However, they did not consider the downhole complex situations, severe vibration, and high temperature. These are great challenges for measurement accuracy and sensors lifetime. Drillstring vibration can greatly affect the life of the gyroscope, and an increasing temperature can cause an error drift in the gyroscope.

The approaches mentioned above can increase the accuracy of measurements to a certain extent. However, some effort still needs to be made in order to improve the performance of the MWD instrument. We developed a strap-down multimode surveying system in the Ref. [9]. We gave more consideration to the actual situation of the drilling process and used field test data to study the measurement algorithm. This system is more conducive to field applications. Moreover, improvement in accuracy is crucial, especially for the continuous MWD of an RSS, for precise and efficient measurement of wellbores drilled for oil and gas exploration. Within the signal-processing community, Kalman Filter remains a very active topic. This chapter is concerned with improving the accuracy and stability of inclination and azimuth measurements of the accelerometer and magnetometer. We developed a new state-space models that were applied to the Kalman Filter model. The algorithm greatly improved the accuracy of well-trajectory measurements and is expected to be applied to ordinary magnetic surveying systems, which are more widely used in drilling engineering.

3.2 Construction of Measurement System

We develop a "strap-down" MWD system here that incorporates three-axis magnetometers and three-axis accelerometers arranged in three mutually orthogonal directions [9, 10], as show in the Fig. 3.3(a). The magnetic measuring tools are installed in the interiors of nonmagnetic drill collars. These special drill collars are usually designed from monel metal to avoid external interference with measurements taken by magnetic MWD surveying tools [3, 9]. Figure 3.3(b) shows the structure of the downhole measurement system, and Table 3.1 shows the characteristics of the sensors. a_x, a_y and a_z are defined as survey signals of the triaxial accelerometers on the x, y, and z axes, respectively. Moreover, m_x, m_y and m_z are defined as survey signals of the triaxial magnetometers on the x, y, and z axes, respectively. Assume that the earth's magnetic field strength is M. Obviously, $M = \sqrt{m_x^2 + m_y^2 + m_z^2}$. Under a certain sample frequency (100 Hz), the measuring signal can be consider as a time series.

Theoretically, that is enough in measuring subsystem only have three accelerometers and three fluxgate sensors which were installed in the instrument coordinate system, additional requires a temperature sensor and an angular rate gyroscope.

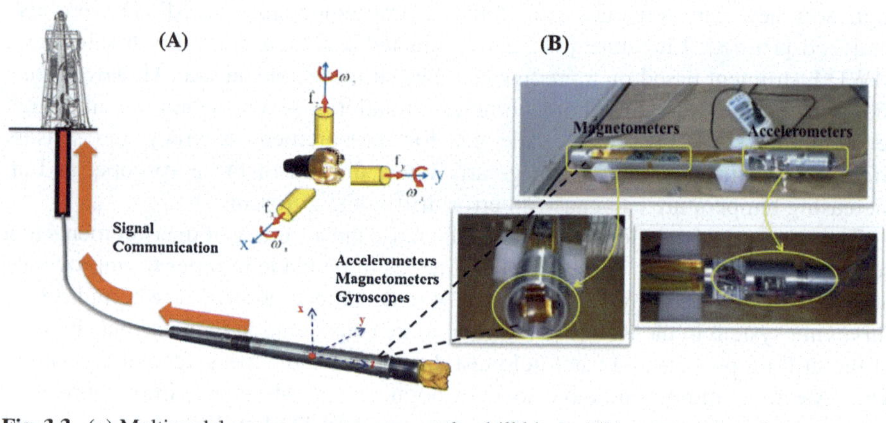

Fig. 3.3 (**a**) Multimodel measurement system at the drill bit, and (**b**) construction of measurement system

Table 3.1 Characteristics of sensors

Parameter	Accelerometers (MS9010)	Fluxgate magnetometers (CTM-DT06)
Range	±10 g	±100,000 nT
Scale factor	200 mV/g	5 V/G ± 5%
Nonlinearity	0.8% of SF	–
Calibration	<50 mg	±0.005 G
Noise	0.140 mg/√Hz	≤0.1 nT
Bandwidth	1000 Hz	350 Hz

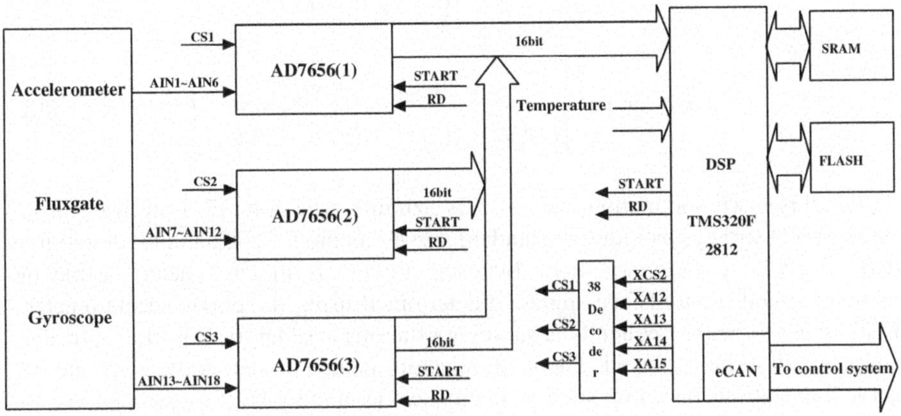

Fig. 3.4 Data acquisition board hardware block diagram

Actually, we designed with a redundancy scheme in order to effectively improve the system reliability.

1. 7 accelerometers (Colibrys MS9010): AX, AX11, AX12, AY, AY11, AY12, AZ. Of which AX, AX11 and AX12 were installed in the axes of the X axis of the instrument; AY, AY11 and AY12 were installed in the axes of the Y axis of the instrument; AZ was installed in the axes of the Z axis of the instrument.
2. 8 fluxgate sensors(CTM-DT06): FX1, FX2, FX3, FY1, FY2, FY3, FZ1, FZ2. Of which FX1, FX2 and FX3 were installed in the axes of the X axis of the instrument; FY1, FY2, FY3 were installed in the axes of the X axis of the instrument; FZ1, FZ2 were installed in the axes of the X axis of the instrument.
3. 3 Temperature sensors: Tem1, Tem2, Tem3.
4. 3 Gyroscopes (ADXRS150): G1, G2, G3.(take the gyro measurement signals as a reference)

In the module design, the AD mode using two DSP chips (TMS320F2812) as the processor, combine the three AD7656 chips to complete the data collection. Data acquisition board hardware block diagram is shown in Fig. 3.4.

Assume the acceleration of gravity is G, and that g_x, g_y and g_z are defined as survey signals of gravity acceleration on the x, y, and z axes, respectively. Then, $G = \sqrt{g_x^2 + g_y^2 + g_z^2}$. The survey signals ($a_x$, a_y, a_z) of the triaxial accelerometers include not only the acceleration of gravity but also the acceleration of the drillstring vibration. Electronics instruments consisting of three-axis accelerometers and three-axis magnetometers are shown to relate to measure inclination and azimuth. The drillstring posture is determined via the following known equations:

$$Inc = \arctan\left(\frac{\sqrt{g_x^2 + g_y^2}}{g_z}\right) \tag{3.1}$$

$$Azi = \arctan\left[\frac{G \cdot (g_x m_y - g_y m_x)}{m_z\left(g_x^2 + g_y^2\right) - g_z(g_x m_x - g_y m_y)}\right] \tag{3.2}$$

The three-axis inclination and six-axis azimuth equations [22] are used in the static MWD surveys as industry-standard survey methods. The azimuth direction is determined in a stationary mode by using three-axis magnetometers, while the inclination and the tool face angle are determined using three-axis accelerometers. Drilling has to pause frequently at surveying stations in order to allow the inclination and azimuth to be surveyed. The well trajectory is then computed between the two surveying stations based on some mathematical assumptions; It is assumed that the drilled distance is a smooth arc. In Eqs. (3.1) and (3.2), only the acceleration of gravity is concerned, and the drillstring vibration signals is considered as noise. The determination of azimuth while drilling (while the sensors are rotating) was not demonstrated owing to limitations in the capabilities of downhole processing. The survey signals of triaxial accelerometers do not include only the acceleration of gravity but also the acceleration of drillstring vibration. Unlike the conventional MWD, the inclination and azimuth should be solved dynamically when the drillstring rotates. Accelerometers are hypersensitive to vibrations, so the violent vibration of a bottom-hole assembly leads the measured signals to be submerged in noise signals. The continuous MWD in This chapter is based on a strap-down multi-model surveying system [9].

According to the field test data, we completed a frequency analysis of the vibration signals to evaluate if the white noise hypothesis. Frequency analysis is indispensable when discussing the most advanced approach that involves dynamic spectral analysis, and to analyze how the spectrum of a signal evolves in time. This analysis is performed with Gabor transforms [25], which are close relatives of wavelet transforms. Gabor transforms are particularly useful in analyzing drillstring vibration signals. The transform is employed here to determine the sinusoidal frequency and phase content of local sections of a signal as it varies with time. The function is first multiplied by a Gaussian window function. The resulting function is then transformed by the Fourier transform.

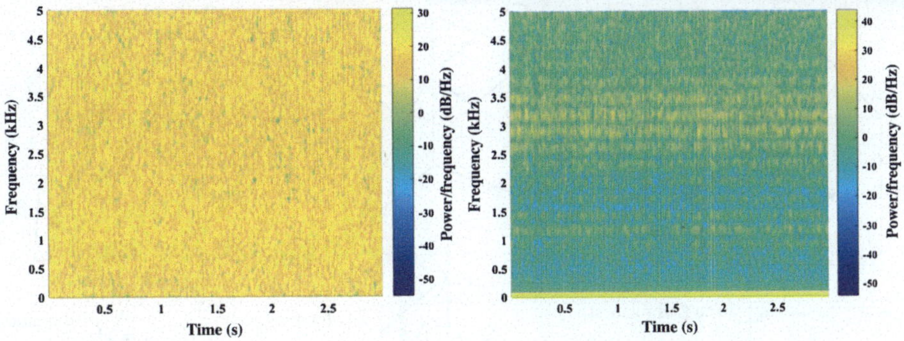

Fig. 3.5 Energy spectrum of lateral vibration signals (left) and longitudinal vibration signals (right)

We use a time-frequency box to express the energy distribution of the Gabor function. The time-frequency spectrum of the vibration signal is plotted in Fig. 3.5. The Gabor transform can help us discriminate the rotational movement pattern. Define $G_{u,\xi}(t)$ as the time window function, where u is center point, σ is the width, the Fourier transform of $G_{u,\xi}(t)$ can be expressed by

$$\widehat{G}_{u,\xi}(\omega) = \widehat{G}(\omega - \xi)e^{iu(\omega-\xi)} \quad (3.3)$$

As shown in Fig. 3.5, from the energy spectrum of lateral vibration signals (left) and longitudinal vibration signals (right), we can find the spectrum not particularly has energy concentration point, the signals can be considered as random.

3.3 Stationary Surveying

The present MWD surveying systems incorporate three-axis magnetometers and three-axis accelerometers arranged in three mutually orthogonal directions as show above. In the three-dimensional Cartesian coordinate system, the geographic coordinate system (North West) XYZ and the probe carrier coordinate system xyz are selected respectively, wherein the XYZ axes of the geographic coordinate system point to the geographic North (N), West (W), and Upper (U) respectively. The z-axis of the probe coordinate system is the axial direction of the drill, pointing upwards, the x and y axes are perpendicular to each other on the cross-section of the drill, and the three axes satisfy the right-hand rule. According to the Euler rotation transformation theorem, the three-dimensional attitude coordinates of the carrier in space can be obtained by a finite number of rotations relative to the geographic coordinate system.

As shown in Fig. 3.6. The first two coordinate systems coincide with each other, first rotating the azimuth around the OZ axis to obtain the $x_1y_1z_1$; Next, the $x_1y_1z_1$ system rotates the tilt angle θ around y_1 to obtain the $x_2y_2z_2$; system. Finally, $x_2y_2z_2$

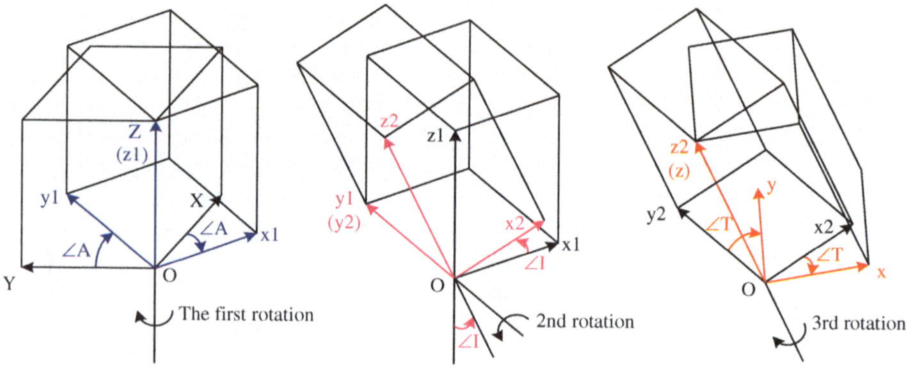

Fig. 3.6 Euler transformation proncipple

rotates the tool face angle φ around the z_2 axis to obtain the $x_3y_3z_3$ system, so that $x_3y_3z_3$ is the xyz system. According to the relative rotation angle relationship between the above coordinate systems, the three rotation transformation matrices can be obtained as $R_Z(\psi)$, $R_Y(\theta)$ and $R_Z(\varphi)$, respectively.

Let the azimuth cosine array between the geographic coordinate system and the exploration system be $C_t^d = \mathrm{R}_Z(\varphi)\mathrm{R}_Y(\theta)\mathrm{R}_Z(\psi)$, after a certain spatial transformation, the following formula is obtained.

$$\begin{aligned} C_t^d &= \mathrm{R}_Z(\varphi)\mathrm{R}_Y(\theta)\mathrm{R}_Z(\psi) \\ &= \begin{bmatrix} \cos\varphi & -\sin\varphi & 0 \\ \sin\varphi & \cos\varphi & 0 \\ 0 & 0 & 1 \end{bmatrix} \begin{bmatrix} \cos\theta & 0 & \sin\theta \\ 0 & 1 & 0 \\ -\sin\theta & 0 & \cos\theta \end{bmatrix} \\ &\times \begin{bmatrix} \cos\psi & -\sin\psi & 0 \\ \sin\psi & \cos\psi & 0 \\ 0 & 0 & 1 \end{bmatrix} \end{aligned} \tag{3.4}$$

Then,

$$\mathrm{C}_\mathrm{t}^\mathrm{d} = \begin{bmatrix} \cos\varphi\cos\theta\cos\psi - \sin\varphi\sin\psi & -\cos\varphi\cos\theta\sin\psi - \sin\varphi\cos\psi & \cos\varphi\sin\theta \\ \sin\varphi\cos\theta\cos\psi + \cos\varphi\sin\psi & -\sin\varphi\cos\theta\sin\psi + \cos\varphi\cos\psi & \sin\varphi\sin\theta \\ -\sin\theta\cos\psi & \sin\theta\sin\psi & \cos\theta \end{bmatrix} \tag{3.5}$$

Then the coordinate conversion relationship between the geographic coordinate system and the probe coordinate system is as shown in Eq. (2.6):

$$[x\ y\ z]^T = C_t^d [X\ Y\ Z]^T \tag{3.6}$$

Wherein $[x\ \ y\ \ z]^T$ indicates the probe coordinate system, $[X\ \ Y\ \ Z]^T$ indicates the geographic coordinate system. If the accelerometer and the fluxgate are used, the attitude of the target carrier in the probe coordinate system is $[x\ \ y\ \ z]^T$, Then multiply both ends of Eq. (3.6) by $\left[C_t^d\right]^{-1}$, obtaining the attitude of the carrier in the earth coordinate system is $[X\ \ Y\ \ Z]^T$, then,

$$[X\ Y\ Z]^T = R_Z^{-1}(A) R_Y^{-1}(I) R_Z^{-1}(T) [x\ y\ z]^T \tag{3.7}$$

Equations (3.5) and (3.7) are the basic transformation relations between the two coordinate systems, and are also the theoretical basis for the attitude calculation of the measurement system while drilling [26].

The magnitude of gravity G and the strength of the Earth's magnetic field B is known in the geographic coordinate system and is expressed as follows:

$$[X\ Y\ Z]_G^T = [0\ 0\ \ -G]^T \tag{3.8}$$

$$[X\ Y\ Z]_B^T = [B_H\ 0\ \ B_V]^T \tag{3.9}$$

wherein, $B_H = B\cos\Phi$, a northward horizontal component representing the strength of the Earth's magnetic field; $B_V = B\sin\Phi$ is a vertical component representing the strength of the Earth's magnetic field (up); Φ is the local geomagnetic dip, generally 45.5°.

Then the attitude angles ψ, θ, φ are determined separately, the relationship between the theoretical value of the attitude angle of the bottom hole coordinate system and the target body in the earth coordinate system is:

$$\left[G_x G_y G_Z\right]^T = C_t^d [0\,0 - G]^T \tag{3.9}$$

$$\left[B_x B_y B_Z\right]^T = C_t^d [B_H\, 0\, B_V]^T \tag{3.10}$$

Define tool face as φ_G, inclination as θ, azimuth as ψ. About the static algorithm, there are several different formulas, basically they are the same on the principle.

$$\begin{bmatrix} G_x \\ G_y \\ G_z \end{bmatrix} = \begin{bmatrix} G\cos\varphi\sin\theta \\ -G\sin\varphi\sin\theta \\ -G\cos\theta \end{bmatrix} \tag{3.11}$$

$$\begin{bmatrix} B_x \\ B_y \\ B_z \end{bmatrix} = \begin{bmatrix} B_H(\cos\psi\cos\varphi\cos\theta - \sin\psi\sin\varphi) - B_V\cos\varphi\sin\theta \\ -B_H(\cos\psi\sin\varphi\cos\theta + \sin\psi\cos\varphi) + B_V\sin\varphi\sin\theta \\ B_H\cos\psi\sin\theta + B_V\cos\theta \end{bmatrix} \tag{3.12}$$

Where $B_H = B \cos \phi$ is the horizontal component of the earth's magnetic field; $B_V = B \sin \phi$ is the vertical component of the earth's magnetic field (ϕ is the local magnetic inclination angle). Then the toolface, inclination and azimuth will be obtained:

$$\varphi_G = tg^{-1}\left(\frac{-G_y}{G_x}\right) \tag{3.13}$$

$$\theta = \mathrm{tg}^{-1}\left(\frac{\sqrt{G_x^2 + G_y^2}}{-G_z}\right) \tag{3.14}$$

$$\psi = \mathrm{tg}^{-1}\left(\frac{G(B_y G_x - B_x G_y)}{B_z\left(G_x^2 + G_y^2\right) - G_z(B_x G_x + B_y G_y)}\right) \tag{3.15}$$

Richard D. DiPersio et al. [27] given the following calculation formulas from another perspective. They provide useful inclination and azimuth equations, which can be computed with the axial accelerometer and magnetometer readings. The azimuth can be obtained as follows.

$$\lambda = \arcsin\left[\frac{G_x B_x + G_y B_y + G_z B_z}{G_{total} M}\right] \tag{3.16}$$

$$\psi = \frac{\cos(\phi) - \cos(\theta)\sin(\lambda)}{\sin(\theta)\cos(\lambda)} \tag{3.17}$$

When the angle of inclination θ is small, it is difficult to calculate the face angle of the gravity tool, If $\theta = 0$, the measurement G_x and G_y should be zero, then Eq. (3.13) cannot be used. Even if θ is not zero, when θ is small, the component G_x, G_y of the gravitational acceleration applied to the measuring sensor will be small, the actual measurement signal becomes very weak and the noise effect is significantly enhanced. At this time, if the gravity acceleration component is still used to calculate the tool face φ_G, it will be very inaccurate. The magnetic component B_x and B_y measured by the magnetic sensor is mainly determined by the horizontal component of the geomagnetic field, and the geomagnetic component is not affected by the attitude of the carrier. Therefore, in the case of a small Inclination well, the magnetic tool face angle φ_M is used instead of the required gravity tool face angle.

$$\varphi_M = tg^{-1}\left(\frac{B_y}{B_x}\right) \tag{3.18}$$

In practical applications, in order to improve the calculation accuracy of the bottom hole's spatial attitude, it is often solved by the mean value of the gravity component and the geomagnetic component signal in a window for a period of time.

3.4 Dynamic Solver of Inclination and Azimuth

3.4.1 Magnetic Basis Measurement System

The magnetic surveying tools are installed inside nonmagnetic drill collars (Fig. 3.7). These special drill collars are usually designed from monel metal to avoid external interferences with the measurements taken by the magnetic MWD surveying tools.

Measurement system is conducted by triaxial accelerometer and triaxial fluxgate, and redundancy design is adopted, due to the severe downhole vibration, the reliability of gyroscopes will be greatly reduced. In order to ensure the system reliability in the entire drilling process, we just take gyroscopes measurement signals as a reference. Nonmagnetic drill collar is used (Fig. 3.7) to separate the electronic survey instrumentation from the magnetic fields of drillstring both above and below and prevent the distortion of the earth's magnetic field at the sensors.

The stationary surveying method as show above, but when the drillstring rotated, the above formulas were no longer applicable. When the sensors are installed in the center of the rotating drillstring, the measurement x and y axis signal will exhibit a sine wave, with zero mean. So, we need to develop a new set of formulas to meet the needs of the dynamic rotation.

Firstly, give the azimuth measurement method when the inclination angle is small. When the inclination angle I is close to zero, $\cos I \approx 1$, $\sin I \approx 0$, according to Eq. (3.12), then:

$$\mathrm{Bx} \approx \mathrm{B}\cos\Phi\cos(A+T) \tag{3.19}$$

$$\mathrm{By} \approx -\mathrm{B}\cos\Phi\sin(A+T) \tag{3.20}$$

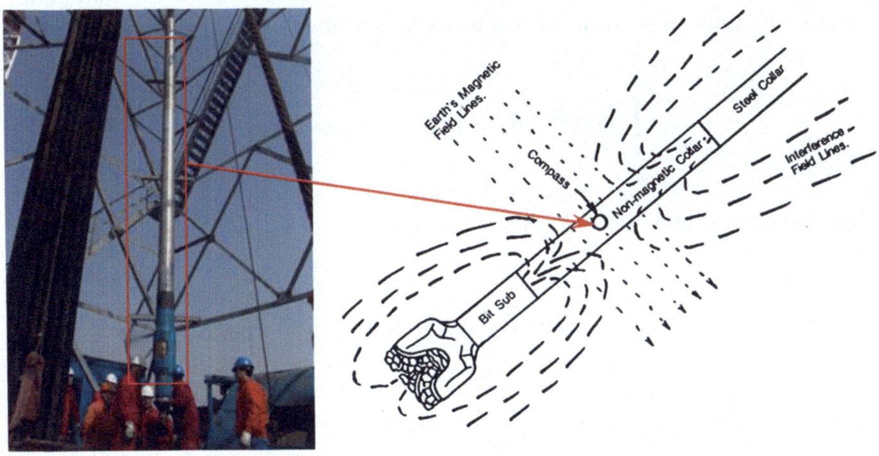

Fig. 3.7 Reduce the magnetic interference using nonmagnetic drill collar

The value of Gx and Gy are very small when the well inclination is small, at this time sin$I$ cannot be approximated as 0. Multiplying Gx, Gy and Bx, By:

$$\begin{bmatrix} \mathrm{Gx} \\ \mathrm{Gy} \end{bmatrix} [\mathrm{Bx\ By}] \approx \mathrm{k} \begin{bmatrix} \cos(A+T) \\ \sin(A+T) \end{bmatrix} [\cos T \ \ \sin T] \tag{3.21}$$

where $k = G \cdot \sin I \cdot B \cdot \cos \Phi$, in this Eq. sin$I$ also cannot be approximated as 0 using trigonometric formulas:

$$\begin{aligned} \mathrm{Gy} \cdot \mathrm{Bx} - \mathrm{Gx} \cdot \mathrm{By} &\approx k \cdot (\cos(A+T) \cdot \sin T - \sin(A+T) \cdot \cos T) \\ &= k \sin A \end{aligned} \tag{3.22}$$

$$\begin{aligned} \mathrm{Gx} \cdot \mathrm{Bx} + \mathrm{Gy} \cdot \mathrm{By} &\approx k \cdot (\cos(A+T) \cdot \cos T + \sin(A+T) \cdot \sin T) \\ &= k \cos A \end{aligned} \tag{3.23}$$

Gain the azimuth A_{small} when the inclination angle is small:

$$A_{small} \approx \mathrm{tg}^{-1}\left(\frac{\mathrm{Gy} \cdot \mathrm{Bx} - \mathrm{Gx} \cdot \mathrm{By}}{\mathrm{Gx} \cdot \mathrm{Bx} + \mathrm{Gy} \cdot \mathrm{By}}\right) \tag{3.24}$$

When the drillstring is rotating, the measurement x and y axis signal will exhibit a sine wave, the formulas above need to be further deduced, by the Eq. (3.20) we can obtained:

$$\begin{aligned} \mathrm{Gx} \cdot \mathrm{By} &\approx k \cdot \cos(A+T) \cdot \sin T \\ &= k \cdot (\cos A \cos T - \sin A \sin T) \cdot \sin T \\ &= \frac{k}{2}(-\sin A + \cos A \sin 2T + \sin A \cos 2T) \end{aligned} \tag{3.25}$$

Integrating both sides of the above formula with the tool face angle T, obtained:

$$\int_{T=0}^{2N\pi} \mathrm{Gx}(T) \cdot \mathrm{By}(T) \mathrm{d}T \approx -N\pi k \cdot \sin A \tag{3.26}$$

Similarly, integrating Gx · Bxcan be obtained:

$$\int_{T=0}^{2N\pi} \mathrm{Gx}(T) \cdot \mathrm{Bx}(T) \mathrm{d}T \approx N\pi k \cdot \cos A \tag{3.27}$$

Then the azimuth A'_{small} can be obtained when the drilling string is rotating.

$$A'_{small} = tg^{-1}\left(\frac{-\int_{T=0}^{2N\pi} \mathrm{Gx}(T)\cdot \mathrm{By}(T)\mathrm{d}T}{\int_{T=0}^{2N\pi} \mathrm{Gx}(T)\cdot \mathrm{Bx}(T)\mathrm{d}T}\right) \tag{3.28}$$

And though the Eq. (3.14) get the inclination angle I_{small}:

$$I_{small} = \sin^{-1}\left(\frac{\sqrt{\frac{1}{2N\pi}\left(\int_{T=0}^{2N\pi} \mathrm{Gx}(T)\cdot \cos T\cdot \mathrm{d}T\right)^2 + \left(\int_{T=0}^{2N\pi} \mathrm{Gx}(T)\cdot \sin T\cdot \mathrm{d}T\right)^2}}{G}\right) \tag{3.29}$$

wherein T is the magnetic tool face angle, g is the earth gravitational acceleration.

When the inclination angle changed from 0° to 90°, using Eqs. (3.14) and (3.15), integration throughout the cycle for magnetic toolface angle can be obtained the other continuous formula under rotation condition. This formula applies to the entire well inclination conditions. Computational formulas of deflection "I_r" and azimuth "A_{mr}" are given as following:

$$I_r = \mathrm{tg}^{-1}\frac{\sqrt{\left(\int_{T_M=0}^{2N\pi} \mathrm{Gx}dT_M\right)^2 + \left(\int_{T_M=0}^{2N\pi} \mathrm{Gy}dT_M\right)^2}}{-\int_{T_M=0}^{2N\pi} \mathrm{Gz}dT_M} \tag{3.30}$$

$$A_{\mathrm{mr}} = \mathrm{tg}^{-1}\left(\frac{\int_{T_M=0}^{2N\pi} \mathrm{G(ByGx - BxGy)}dT_M}{\int_{T_M=0}^{2N\pi} \mathrm{Bz(Gx^2 + Gy^2)}dT_M - \int_{T_M=0}^{2N\pi} \mathrm{Gz(BxGx + ByGy)}dT_M}\right) \tag{3.31}$$

Where T_M is magnetic tool-face angle: $T_M = \mathrm{tg}^{-1}\left(\frac{-\mathrm{By}}{\mathrm{Bx}}\right)$.

3.4.2 Gyro Based Measurement System

The three-axis accelerometer and the three-axis gyro are mounted along the axis of the drill as shown in Fig. 3.8. The measuring sensor rotates along with the drill string to form a strap-down inertial navigation while drilling measurement system.

Assume that the three-axis accelerometer output signal is: $f^b = [f_x\, f_y\, f_z]^T$, The output signal of the three-axis gyroscope is: $\omega^b = [\omega_x\, \omega_y\, \omega_z]^T$, Using these six output

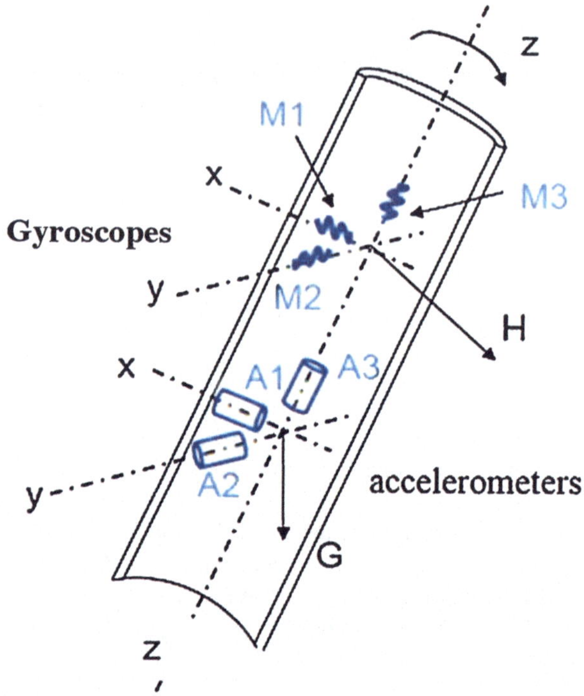

Fig. 3.8 Install schematic diagram of triaxial accelerometer and gyroscope

signals to effectively solve the attitude of the drilling tool belongs to the scope of the solution of the Strapdown Inertial Navigation System. Next, we apply the solution method of the inertial navigation system to solve the drilling posture and realize continuous measurement while drilling.

Accelerometers and gyroscopes are mounted in a non-magnetic drill collar of a rotary steering system with their sensitive axes along the tool's forward axis (y), cross-sectional direction (x) and z-axis perpendicular to the xy plane, forming a triaxial axis of the carrier coordinate system. As shown, the accelerometer and gyroscope can provide linear acceleration and angular velocity in the coordinate system. However, if the tool pose is to be solved, this coordinate system needs to be converted to another coordinate system.

As shown in the Fig. 3.8, The three axes X^e, Y^e, Z^e represent the Earth coordinate system. The navigation coordinate system is used to calculate the carrier position, velocity and attitude, Because the navigation coordinate system is along the local north, east, and vertical directions, as shown in the Fig. 3.9, "*N*, *E*, *UP*", wherein λ is longitude angle, φ is latitude angle. In the carrier coordinate system, the well inclination, orientation and tool face angle can be obtained by the INS mechanical equation.

The acceleration and angular velocity values measured in the b-system are converted to the n-system by the transformation matrix R_b^n, We can solve the spatial position and attitude [28] of the bottom hole tool. Define the longitude of the bottom

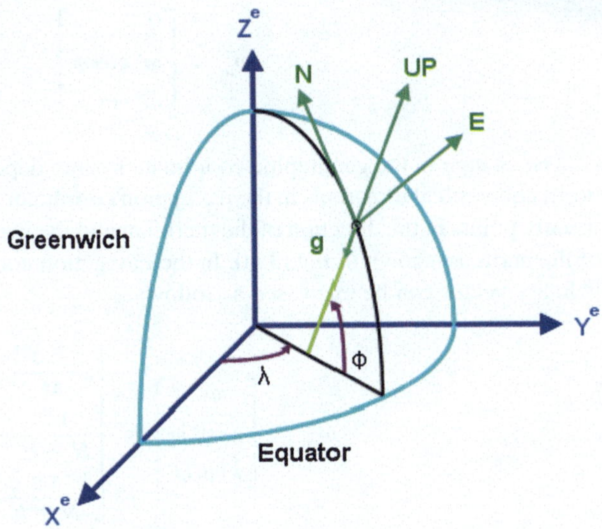

Fig. 3.9 Navigation frame of a given point relative to the earth-fixed frame

hole as λ, latitude is φ, Altitude is h, then its position can be expressed as: $r^n = [\varphi\, \lambda\, h]^T$. Defining the velocity component of the n-system, northbound speed is V^n, eastward speed is V^e, vertical speed is V^u. Then define the speed of the n system as: $V^n = [V^e\ V^n\ V^u]^T$.

The velocity component can be expressed as a derivative of the position component versus time and can be expressed as follows:

$$\dot{r}^n = \begin{bmatrix} \dot{\varphi} \\ \dot{\lambda} \\ \dot{h} \end{bmatrix} = \begin{bmatrix} 0 & 1/_{M+h} & 0 \\ 1/_{(N+h)\cos\varphi} & 0 & 0 \\ 0 & 0 & 1 \end{bmatrix} \begin{bmatrix} V^e \\ V^n \\ V^u \end{bmatrix} = D^{-1} V^n \tag{3.32}$$

Wherein, M is the radius of curvature of the meridian and N is the radius of curvature of the ellipse of the earth.

In the carrier coordinate system, the measured value of the accelerometer $f^b = [f_x f_y f_z]^T$ converting to a geographic coordinate system through a transformation matrix R_b^n is:

$$f^n = \begin{bmatrix} f^e \\ f^n \\ f^u \end{bmatrix} = R_b^n f^b = R_b^n \begin{bmatrix} f_x \\ f_y \\ f_z \end{bmatrix} \tag{3.33}$$

Acceleration component f^n in the n-series can integrate the velocity component v^n. However, due to the existence of the Earth itself, it will affect the solution process. The rotation speed of the earth is $\omega^e = 15$ deg/hr, the angular velocity vector expressed in the n-system is as follows:

$$\omega_{ie}^{n} = \begin{bmatrix} 0 \\ \omega^{e}\cos\varphi \\ \omega^{e}\sin\varphi \end{bmatrix} \tag{3.34}$$

The change in the geographic coordinate system depends on the definition of the north and vertical directions in the navigation coordinate system. The north direction usually points in the direction of the meridian and the vertical direction to the surface of the earth, as shown in Fig. 3.10. In the navigation coordinate system, the angular velocity vector can be expressed as follows:

$$\omega_{en}^{n} = \begin{bmatrix} -\dot{\varphi} \\ \dot{\lambda}\cos\varphi \\ \dot{\lambda}\sin\varphi \end{bmatrix} = \begin{bmatrix} -\dfrac{V^{n}}{M+h} \\ \dfrac{V^{e}}{N+h} \\ \dfrac{V^{e}\tan\varphi}{N+h} \end{bmatrix} \tag{3.35}$$

Earth's gravity also affects the IMU's acceleration measurements. We can use the gravity model to make corrections. The Earth's gravity field [29] can be expressed as:

$$g = a_1\left(1 + a_2\sin^2\varphi + a_3\sin^4\varphi\right) + \left(a_4 + a_5\sin^2\varphi\right)h + a_6h^2 \tag{3.36}$$

Wherein $a_1 = 9.7803267715$ m/sec^2; $a_2 = 0.0052790414$; $a_3 = 0.0000232718$; $a_4 = -0.000003087691089$ 1/sec^2; $a_5 = 0.000000004397731$ 1/sec^2; $a_6 = 0.000000000000721$ 1/(m$*$sec^2).

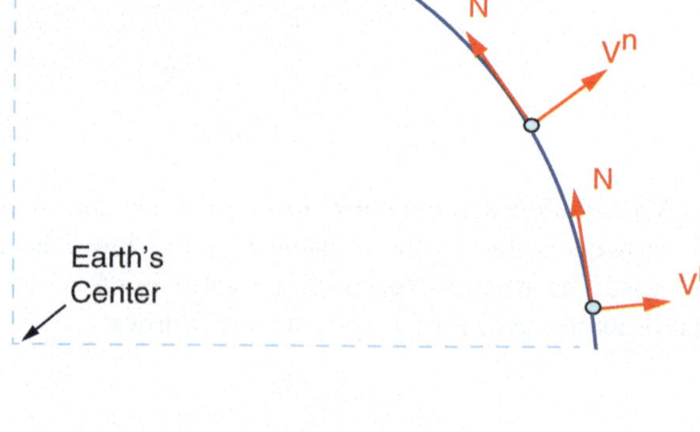

Fig. 3.10 Change of orientation of the navigation frame

The earth's gravitational field in the *n*-series can be expressed as:

$$g^n = [0\,0 - g] \tag{3.37}$$

Taking into account the influence of the Earth's factors, the rate of change of the velocity component V^n can be expressed as follows:

$$\dot{V}^n = R_b^n f^b - \left(2\Omega_{ie}^n + \Omega_{en}^n\right) V^n + g^n \tag{3.38}$$

Wherein:

$$\Omega_{ie}^n = \begin{bmatrix} 0 & -\omega^e \sin\varphi & \omega^e \cos\varphi \\ \omega^e \sin\varphi & 0 & 0 \\ -\omega^e \cos\varphi & 0 & 0 \end{bmatrix} \tag{3.39}$$

$$\Omega_{en}^n = \begin{bmatrix} 0 & -\frac{V^e \tan\varphi}{N+h} & \frac{V^e}{N+h} \\ \frac{V^e \tan\varphi}{N+h} & 0 & \frac{V^n}{M+h} \\ -\frac{V^e}{N+h} & -\frac{V^n}{M+h} & 0 \end{bmatrix} \tag{3.40}$$

Transformation matrix R_b^n can be obtained by the following differential equation:

$$\dot{R}_b^n = R_b^n \Omega_{nb}^b = R_b^n \left(\Omega_{ib}^b - \Omega_{in}^b\right) \tag{3.41}$$

wherein Ω_{ib}^b is the antisymmetric matrix of angular velocity measured by the gyro, the angular velocity vector ω_{ib}^b can be used as follows:

$$\Omega_{ib}^b = \begin{bmatrix} 0 & -\omega_z & \omega_y \\ \omega_z & 0 & \omega_x \\ -\omega_y & \omega_x & 0 \end{bmatrix} \tag{3.42}$$

The gyro measures the angular velocity of the bottom drilling tool, and also measures the angular velocity of the Earth's rotation and the direction of the navigation coordinate system. Therefore, the angular velocity Ω_{in}^b needs to be subtracted from Ω_{ib}^b to eliminate the effects of these two factors. Angular velocity vector Ω_{in}^b contains two parts, the first is the earth's rotation speed Ω_{ie}^b and the speed of change of the direction of the navigation coordinate system Ω_{en}^b, as shown below:

$$\Omega_{in}^b = \Omega_{ie}^b + \Omega_{en}^b \tag{3.43}$$

The antisymmetric matrix of the velocity matrix can be expressed as follows:

$$\omega_{in}^{b} = \omega_{ie}^{b} + _{en}^{b} = R_n^b \omega_{ie}^n + R_n^b \omega_{en}^n = R_n^b \left(\omega_{ie}^n + \omega_{en}^n\right)$$
$$= R_n^b \left[\begin{bmatrix} 0 \\ \omega^e \cos\varphi \\ \omega^e \sin\varphi \end{bmatrix} + \begin{bmatrix} -\dfrac{V^n}{M+h} \\ \dfrac{V^e}{N+h} \\ \dfrac{V^e \tan\varphi}{N+h} \end{bmatrix} \right] = R_n^b \begin{bmatrix} -\dfrac{V^n}{M+h} \\ \dfrac{V^e}{N+h} + \omega^e \cos\varphi \\ \dfrac{V^e \tan\varphi}{N+h} + \omega^e \sin\varphi \end{bmatrix} \tag{3.44}$$

Finally, the transformation matrix can be obtained as follows:

$$\dot{R}_b^n = R_b^n \left[\begin{bmatrix} \omega^x \\ \omega^y \\ \omega^z \end{bmatrix} - R_n^b \begin{bmatrix} -\dfrac{V^n}{M+h} \\ \dfrac{V^e}{N+h} + \omega^e \cos\varphi \\ \dfrac{V^e \tan\varphi}{N+h} + \omega^e \sin\varphi \end{bmatrix} \right] \tag{3.45}$$

Define the well angle as θ, Azimuth angle is ψ, tool face is ϕ, transformation matrix R_b^n can be expressed as follows:

$$R_b^n = \begin{bmatrix} \cos\psi\cos\phi + \sin\psi\sin\theta\sin\phi & \sin\psi\cos\theta & \cos\psi\sin\phi - \sin\psi\sin\theta\cos\phi \\ -\sin\psi\cos\phi + \cos\psi\sin\theta\sin\phi & \cos\psi\cos\theta & -\sin\psi\sin\phi - \cos\psi\sin\theta\cos\phi \\ -\cos\theta\sin\phi & \sin\theta & \cos\theta\cos\phi \end{bmatrix} \tag{3.46}$$

According to the previous assumption, the output signal of the three-axis accelerometer is: $f^b = [f_x\, f_y\, f_z]^T$; The output signals of the three-axis gyroscope is: $\omega_{ib}^b = \left[\omega_x\, \omega_y\, \omega_z\right]^T$. The amount of angular change can be calculated from the measured angular velocity:

$$\theta_{ib}^b = \begin{bmatrix} \Delta\theta_x \\ \Delta\theta_y \\ \Delta\theta_z \end{bmatrix}_{ib}^b = \begin{bmatrix} \omega_x \\ \omega_y \\ \omega_z \end{bmatrix} \Delta t \tag{3.47}$$

where Δt is the data sampling time.

Similarly, the line speed can be calculated from the measured value of the acceleration:

$$\Delta v^b = \begin{bmatrix} \Delta v_x \\ \Delta v_y \\ \Delta v_z \end{bmatrix} = \begin{bmatrix} f_x \\ f_y \\ f_z \end{bmatrix} \Delta t \tag{3.48}$$

Considering the effects of the Earth's rotation and the change in the direction of the navigational coordinate system, when at the time t_k, the increase in angle can be expressed as:

$$\theta_{in}^b(t_k) = \omega_{in}^b(t_k)\Delta t = R_n^b(t_k)\begin{Bmatrix} \dfrac{V^n(t_k)}{M+h} \\ \dfrac{V^e(t_k)}{N+h} + \omega^e \cos\varphi \\ \dfrac{V^e(t_k)\tan\varphi}{N+h} + \omega^e \sin\varphi \end{Bmatrix}\Delta t \tag{3.49}$$

Then,

$$\theta_{nb}^b(t_k) = \theta_{ib}^b(t_k) - \theta_{in}^b(t_k) = \left[\Delta\theta_x\ \Delta\theta_y\ \Delta\theta_z\right]^T \tag{3.50}$$

In summary, the mechanical equations of the strapdown inertial navigation system based on the geographic coordinate system are as follows:

$$\dot{x}^n = \begin{bmatrix} \dot{r}^n \\ \dot{V}^n \\ \dot{R}_b^n \end{bmatrix} = \begin{bmatrix} D^{-1}V^n \\ R_b^n f^b - \left(2\Omega_{ie}^n + \Omega_{en}^n\right)V^n + g^n \\ R_b^n\left(\Omega_{ib}^b - \Omega_{in}^b\right) \end{bmatrix} \tag{3.51}$$

The solution of Eq. (3.51) can be solved by Euler angle method, direction cosine method or quaternion method. In this way, by mounting a three-axis accelerometer and a three-axis gyroscope on the carrier axis, information such as the attitude, velocity and position of the carrier can be obtained by the measured values.

3.4.3 *Quaternion Solution Method*

To solve the mechanical equation of the inertial navigation system, the attitude matrix R_b^n needs to be obtained by the differential equation of attitude rate. Quaternion is one of the commonly used methods [30]. As shown in Fig. 3.11, assuming that a rigid body rotates at the origin, according to Euler's theorem, the orientation of the moving coordinate system (b-series) relative to the navigation coordinate system (n-series), Equivalent to b system rotating an angle Θ around an equivalent axis. Therefore, the quaternion $Q = [q_1\ q_2\ q_3\ q_4]^T$ used to describe the rotation of the b system relative to the n system, defined as follows:

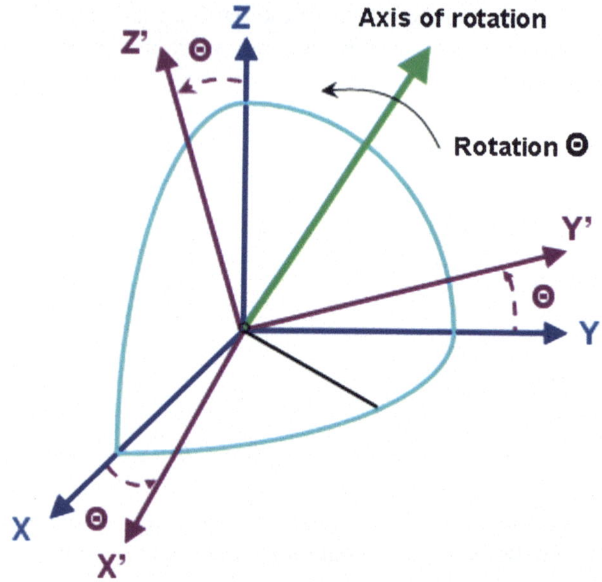

Fig. 3.11 Quaternion rotation

$$Q = \begin{bmatrix} q_1 \\ q_2 \\ q_3 \\ q_4 \end{bmatrix} = \begin{bmatrix} (\Theta_x/\Theta)\sin(\Theta/2) \\ (\Theta_y/\Theta)\sin(\Theta/2) \\ (\Theta_z/\Theta)\sin(\Theta/2) \\ \cos(\Theta/2) \end{bmatrix} \tag{3.52}$$

Wherein the rotation angle is $\Theta = \sqrt{\Theta_x^2 + \Theta_y^2 + \Theta_z^2}$, Θ_x/Θ,Θ_y/Θ and Θ_z/Θ are the cosine of the direction of the rotating axis in the n-series.

From the definition of quaternion, $q_1^2 + q_2^2 + q_3^2 + q_4^2 = 1$, The components of the quaternion are not independent of each other, and only three independent quaternion components are needed to describe the rotation of the coordinate axes. However, there is usually a calculation error, defined as $\Delta = 1 - \left(q_1^2 + q_2^2 + q_3^2 + q_4^2\right)$. To correct for this error, the quaternion Q expressed in vector form should be updated with the following formula after each calculation:

$$\hat{Q} = \frac{Q}{\sqrt{1-\Delta}} \approx Q\left(1 + \frac{\Delta}{2}\right) \tag{3.53}$$

Describe the time domain variation of quaternions by using first-order differential equations,

$$\dot{Q} = \frac{1}{2}\Omega(\omega)Q \tag{3.54}$$

where $\Omega(\omega)$ is an anti-symmetric matrix, expressed as follows:

$$\Omega(\omega) = \begin{bmatrix} 0 & \omega_z & -\omega_y & \omega_x \\ -\omega_z & 0 & \omega_x & \omega_y \\ \omega_y & -\omega_x & 0 & \omega_z \\ -\omega_x & -\omega_y & -\omega_z & 0 \end{bmatrix} \tag{3.55}$$

where ω_x, ω_y, ω_z are the carrier rotation angular rate.

To solve the first order differential equation, Determine Q_{k+1} at time t_{k+1} according to Q_k at time t_k, As shown below:

$$Q_{k+1} = Q_k + \left(\frac{1}{2}Q(\omega_k)Q_k\right)\Delta t \tag{3.56}$$

where $\Delta t = t_{k+1} - t_k$. When at time t_k, the quaternion is determined, R_b^n can be directly determined by the following formula:

$$\begin{aligned} R_b^n &= \begin{bmatrix} R_{11} & R_{12} & R_{13} \\ R_{21} & R_{22} & R_{23} \\ R_{31} & R_{32} & R_{33} \end{bmatrix} \\ &= \begin{bmatrix} q_1^2 - q_2^2 - q_3^2 + q_4^2 & 2(q_1q_2 - q_3q_4) & 2(q_1q_2 + q_3q_4) \\ 2(q_1q_2 + q_3q_4) & -q_1^2 + q_2^2 - q_3^2 + q_4^2 & 2(q_2q_3 - q_1q_4) \\ 2(q_1q_3 - q_2q_4) & 2(q_2q_3 + q_1q_4) & -q_1^2 - q_2^2 + q_3^2 + q_4^2 \end{bmatrix} \end{aligned} \tag{3.57}$$

Therefore, as shown in the formula (3.57), the update equation for the quaternion can be obtained as:

$$\begin{bmatrix} q_1(t_{k+1}) \\ q_2(t_{k+1}) \\ q_3(t_{k+1}) \\ q_4(t_{k+1}) \end{bmatrix} = \begin{bmatrix} q_1(t_k) \\ q_2(t_k) \\ q_3(t_k) \\ q_4(t_k) \end{bmatrix} + \frac{1}{2} \begin{bmatrix} 0 & \Delta\theta_z & -\Delta\theta_y & \Delta\theta_x \\ -\Delta\theta_z & 0 & \Delta\theta_x & \Delta\theta_y \\ \Delta\theta_y & -\Delta\theta_x & 0 & \Delta\theta_z \\ -\Delta\theta_x & -\Delta\theta_y & -\Delta\theta_z & 0 \end{bmatrix} \begin{bmatrix} q_1(t_k) \\ q_2(t_k) \\ q_3(t_k) \\ q_4(t_k) \end{bmatrix} \tag{3.58}$$

Get BHA inclination θ、 tool face ϕ and azimuth ψ are as follows:

$$\theta = \arctan\left(\frac{R_{32}}{\sqrt{R_{12}^2 + R_{22}^2}}\right) \tag{3.59}$$

$$\phi = \arctan\left(\frac{-R_{31}}{R_{33}}\right) \tag{3.60}$$

$$\psi = \arctan\left(\frac{-R_{12}}{R_{22}}\right) \tag{3.61}$$

where the wellbore inclination is $I = 90 - \theta$.

In the Strapdown navigation system, due to GPS alignment, the Kalman filtering method can be used to obtain accurate carrier attitude and position information after establishing the error model. However, in the measurement while drilling, GPS signal correction cannot be used, and the Kalman filter model needs to be re-established. Regarding the Kalman filter model of the gyro-while-drilling measurement system, since the inertial gyro and the acceleration sensor are used, the error model is consistent with the Kalman filter error model of the strapdown navigation in the aerospace field. From the perspective of reliability, it is currently in the field of drilling measurement, magnetic-based measurement systems have more application advantages than gyro systems. However, due to magnetic interference and other factors, the magnetic base system cannot be completely perfect. The future development direction must be the composite mode measurement system, that is, the measurement while drilling system combined with the magnetic sensor and the gyro system.

3.5 Calibration Model for Installation Error

In practical cases, even if carefully try to make the three-axis accelerometers and three-axis fluxgates orthogonal, it is impossible to guarantee the coordinate axis are orthogonal and installation is centered exactly. This will bring about the final solution error regardless of whatever the solution method. Therefore, if want to improve the accuracy measurement, we must develop a compensation algorithm to make sensors centered and mutually orthogonal, that is, from the mathematical model of the system, design the corresponding algorithm to solve out the installed error, and to accurately calculate the drilling string attitude.

3.5.1 Mathematical Model

We try to establish the algorithm model of error compensation since the installation error cannot be avoided and the calibration parameters will be obtained through the laboratory experiments. First of all, assuming that A_x, A_y, A_z, were output voltage of the accelerometer before the establishment of mathematical model, so the relationship between the respective components of the gravity and the output voltage as shown in the following formula.

$$\begin{bmatrix} A_x \\ A_y \\ A_z \end{bmatrix} = \begin{bmatrix} K_{xx} & K_{xy} & K_{xz} \\ K_{yx} & K_{yy} & K_{yz} \\ K_{zx} & K_{zy} & K_{zz} \end{bmatrix} \times \begin{bmatrix} G_x \\ G_y \\ G_z \end{bmatrix} + \begin{bmatrix} A_{x0} \\ A_{y0} \\ A_{z0} \end{bmatrix} \tag{3.62}$$

where $K_{ix}, K_{iy}, K_{iz}(i = x, y, z)$ indicated triaxial accelerometer calibration coefficient.

Then assume that I_{Ax}, I_{Ay}, I_{Az} and T_{Ax}, T_{Ay}, T_{Az} indicated triaxial accelerometers installation angle and the phase of the installation angle. The following formula can be obtained.

$$\begin{bmatrix} A_x \\ A_y \\ A_z \end{bmatrix} = \begin{bmatrix} K_{Ax} \cdot \cos I_{Ax} & K_{Ax} \cdot \sin I_{Ax} \cdot \cos T_{Ax} & K_{Ax} \cdot \sin I_{Ax} \cdot \sin T_{Ax} \\ K_{Ay} \cdot \sin I_{Ay} \cdot \sin T_{Ay} & K_{Ay} \cdot \cos I_{Ay} & K_{Ay} \cdot \sin I_{Ay} \cdot \cos T_{Ay} \\ K_{Az} \cdot \sin I_{Az} \cdot \cos T_{Az} & K_{Az} \cdot \sin I_{Az} \cdot \sin T_{Az} & K_{Az} \cdot \cos I_{Az} \end{bmatrix} \times \begin{bmatrix} G_x \\ G_y \\ G_z \end{bmatrix} + \begin{bmatrix} A_{x0} \\ A_{y0} \\ A_{z0} \end{bmatrix} \tag{3.63}$$

Obviously, $K_{xx} = K_{Ax} \cdot \cos I_{Ax}, \cdots, K_{zz} = K_{Az} \cdot \cos I_{Az}$, So, we can calculate the accelerometers calibration coefficient as follows:

$$\begin{bmatrix} K_{Ax} \\ K_{Ay} \\ K_{Az} \end{bmatrix} = \begin{bmatrix} \sqrt{(K_{xx}G)^2 + (K_{xy}G)^2 + (K_{xz}G)^2} \\ \sqrt{(K_{yx}G)^2 + (K_{yy}G)^2 + (K_{yz}G)^2} \\ \sqrt{(K_{zx}G)^2 + (K_{zy}G)^2 + (K_{zz}G)^2} \end{bmatrix} \times \left[\frac{1}{G}\right] \tag{3.64}$$

The bias of the sensors can be obtained:

$$\begin{bmatrix} Bias_{Ax} \\ Bias_{Ay} \\ Bias_{Az} \end{bmatrix} = \begin{bmatrix} 1/K_{Ax} & 0 & 0 \\ 0 & 1/K_{Ay} & 0 \\ 0 & 0 & 1/K_{Az} \end{bmatrix} \times \begin{bmatrix} A_{x0} \\ A_{y0} \\ A_{z0} \end{bmatrix} \tag{3.65}$$

For the accelerometer installed in the x-axis, defined the $\cos(A_xP_x)$, $\cos(A_xP_y)$ and $\cos(A_xP_z)$ are cosine values of angles between accelerometer sensitive axis and three axes of the instrument coordinate system.

$$\begin{aligned} Ax &= \left[G_x \cdot \cos(AxPx) + G_y \cdot \cos(AxPy) + G_z \cdot \cos(AxPz)\right] \cdot K_{Ax} + Bias_{Ax} \cdot K_{Ax} \\ &= \left(G_x \cdot \cos(AxPx) + G_y \cdot \cos(AxPy) + G_z \cdot \cos(AxPz) + Bias_{Ax}\right) \cdot K_{Ax} \end{aligned} \tag{3.66}$$

Obviously, $G = \sqrt{G_x^2 + G_y^2 + G_z^2}$.

Then we get the calibration mathematical model of accelerometer error as shown in Eq. 3.67.

$$\begin{bmatrix} \mathrm{Ax} \\ \mathrm{Ay} \\ \mathrm{Az} \end{bmatrix} = \begin{bmatrix} \mathrm{K_{Ax}} \times \mathrm{cosAxPx} & \mathrm{K_{Ax}} \times \mathrm{cosAxPy} & \mathrm{K_{Ax}} \times \mathrm{cosAxPz} \\ \mathrm{K_{Ay}} \times \mathrm{cosAyPx} & \mathrm{K_{Ay}} \times \mathrm{cosAyPy} & \mathrm{K_{Ay}} \times \mathrm{cosAyPz} \\ \mathrm{K_{Az}} \times \mathrm{cosAzPx} & \mathrm{K_{Az}} \times \mathrm{cosAzPy} & \mathrm{K_{Az}} \times \mathrm{cosAzPz} \end{bmatrix} \times \begin{bmatrix} \mathrm{Gx} \\ \mathrm{Gy} \\ \mathrm{Gz} \end{bmatrix} + \begin{bmatrix} \mathrm{K_{Ax}} \times \mathrm{Bias_{Ax}} \\ \mathrm{K_{Ay}} \times \mathrm{Bias_{Ay}} \\ \mathrm{K_{Az}} \times \mathrm{Bias_{Az}} \end{bmatrix} \quad (3.67)$$

Similarly, we get the calibration mathematical model of fluxgate error as shown in Eq. 3.68.

$$\begin{bmatrix} \mathrm{Fx} \\ \mathrm{Fy} \\ \mathrm{Fz} \end{bmatrix} = \begin{bmatrix} \mathrm{L_{Fx}} \times \mathrm{cosFxPx} & L_{Fx} \times \cos FxPy & \mathrm{L_{Fx}} \times \mathrm{cosFxPz} \\ \mathrm{L_{Fy}} \times \mathrm{cosFyPx} & L_{Fy} \times \cos FyPy & \mathrm{L_{Fy}} \times \mathrm{cosFyPz} \\ \mathrm{L_{Fz}} \times \mathrm{cosFzPx} & L_{Fz} \times \cos FzPy & \mathrm{L_{Fz}} \times \mathrm{cosFzPz} \end{bmatrix} \times \begin{bmatrix} \mathrm{Bx} \\ \mathrm{By} \\ \mathrm{Bz} \end{bmatrix} + \begin{bmatrix} \mathrm{L_{Fx}} \times \mathrm{Bias_{Fx}} \\ \mathrm{L_{Fy}} \times \mathrm{Bias_{Fy}} \\ \mathrm{L_{Fz}} \times \mathrm{Bias_{Fz}} \end{bmatrix} \quad (3.68)$$

wherein, $B = \sqrt{B_x^2 + B_y^2 + B_z^2}$

Then the calibration parameters K_{Ai}, $Bias_{Ai}$, $\cos(A_iP_j)$, or L_{Fi}, $Bias_{Fi}$, $\cos(F_iP_j)$ $\mathrm{L_{Fx}}$ will be used in the algorithm, where $i = x, y, z; j = x, y, z$.

3.5.2 *Experimental Methods*

Designed the experimental instrument that can be put in arbitrary position in the three-dimensional space (as shown in Fig. 3.12 on the left side), and using non-magnetic materials to ensure that the fluxgate sensors will not be disturbed.

First using the orthogonal method to calibrate the installation error. Determine the 24 positions as shown in Table 3.2, correction parameters can be obtained when the inclination and azimuth values for each point have been calculated. As show in Fig. 3.12 on the right side, point A represents the number 2 in Table 1. As an example of A_x, obtain the following formula by Eq. (3.67) and Table 1.

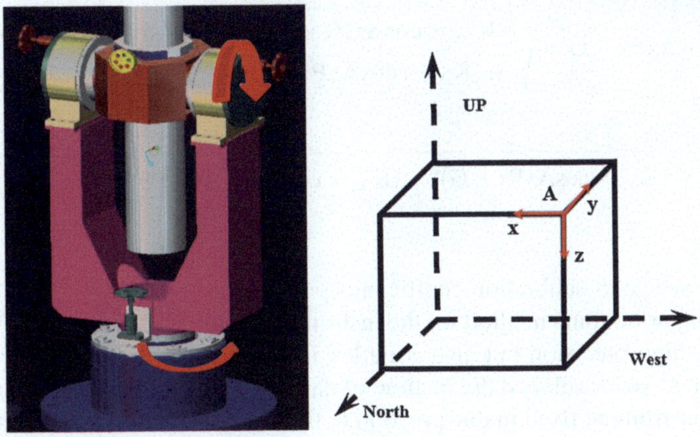

Fig. 3.12 Experimental equipment and location coordinates

Table 3.2 Orthogonal position

No.	Sensor shaft			No.	Sensor shaft		
	X	Y	Z		X	Y	Z
1	north	east	down	13	up	west	south
2(A)	east	south	down	14	west	down	south
3	south	west	down	15	down	east	south
4	west	north	down	16	east	up	south
5	north	west	up	17	up	north	west
6	east	north	up	18	north	down	west
7	south	east	up	19	down	south	west
8	west	south	up	20	south	up	west
9	up	east	north	21	up	south	east
10	east	down	north	22	south	down	east
11	down	west	north	23	down	north	east
12	west	up	north	24	north	up	east

$$K_{Ax} \times \cos AxPx \times Gx = \frac{1}{8}(Ax_1 + Ax_5 + Ax_{18} + Ax_{24} - Ax_3 - Ax_7 - Ax_{20} - Ax_{22}) \tag{3.69}$$

$$K_{Ax} \times \cos AxPx \times Gy = \frac{1}{8}(Ax_2 + Ax_6 + Ax_{10} + Ax_{16} - Ax_4 - Ax_8 - Ax_{12} - Ax_{14}) \tag{3.70}$$

$$K_{Ax} \times \cos AxPx \times Gz = \frac{1}{8}(Ax_{11} + Ax_{15} + Ax_{19} + Ax_{23} - Ax_9 - Ax_{13} - Ax_{17} - Ax_{21}) \tag{3.71}$$

And because of $G = \sqrt{G_x^2 + G_y^2 + G_z^2}$, so we get the following:

$$K_{Ax} \times \cos AxPx \times G = \sqrt{\begin{array}{l}(K_{Ax} \times \cos AxPx \times Ghc)^2 + (K_{Ax} \times \cos AxPx \times Ghs)^2 \\ + (K_{Ax} \times \cos AxPx \times Gv)^2\end{array}} \tag{3.72}$$

$$K_{Ax} \times G = \sqrt{(K_{Ax} \times \cos AxPx \times G)^2 + (K_{Ax} \times \cos AxPy \times G)^2 + (K_{Ax} \times \cos AxPz \times G)^2} \tag{3.73}$$

Finally, we get 5 calibration coefficients.

But using orthogonal method for the instrument required of the calibration system is not only high precision but also complex in structure. It is difficult for practical application, so we developed the method of data fitting. Specific steps are as follows.

Let the instrument fixed in one position (a fixed well inclination and azimuth) and rotated 360°. Sample one data when the instrument rotated 45° (error:±1°) and will get 8 sampling data when the instrument rotated 360°. Using numerical fitting theory which based on the orthogonal trigonometric we can get the sensors output voltage curve when the instrument rotated 360°. Then we can calculate the calibration coefficients for each sensor.

Calculation methods are described with Ax and *Fx* as an example. According to Eqs. 3.67 and 3.68, A_x and F_x are presented as:

$$Ax = (Gx \cdot \cos AxPx + Gy \cdot \cos AxPy + Gz \cdot \cos AxPz + Bias_{Ax}) \cdot K_{Ax} \tag{3.74}$$

$$Fx = (Bx \cdot \cos FxPx + By \cdot \cos FxPy + Bz \cdot \cos FxPz + Bias_{Fx}) \cdot L_{Fx} \tag{3.75}$$

Then,

$$\begin{array}{l} Ax = K_{Ax} \cdot G \cdot \sin I \cdot \cos AxPx \cdot \cos T - K_{Ax} \cdot G \cdot \sin I \cdot \cos AxPy \cdot \sin T \\ \quad + K_{Ax} \cdot (-G \cdot \cos I \cdot \cos AxPz + Bias_{Ax}) \end{array} \tag{3.76}$$

Assume that:

$$M = K_{Ax} \cdot G \cdot \sin I . \cos AxPx \tag{3.77}$$

$$N = -K_{Ax} \cdot G \cdot \sin I \cdot \cos AxPy \tag{3.78}$$

$$P = K_{Ax} \cdot (-G \cdot \cos I \cdot \cos AxPz + Bias_{Ax}) \tag{3.79}$$

If the inclination is unchanged, *M*, *N*, *P* are constants. Brought into Eq. 3.67, A_x and F_x can be presented as:

$$Ax = M \cdot \cos r + N \cdot \sin r + P \tag{3.80}$$

$$Fx = m \cdot \cos r + n \cdot \sin r + p \tag{3.81}$$

$$\text{wherein}: m = L_{Fx} \cdot [(B_h \cdot \cos A \cdot \cos I - B_v \cdot \sin I) \cdot \cos FxPx - B_h \cdot \sin A.\cos FxPy]$$
$$n = L_{Fx} \cdot [-B_h \cdot \sin A \cdot \cos FxPx + (-B_h \cdot \cos A \cdot \cos I + B_v \cdot \sin I) \cdot \cos FxPy]$$
$$p = L_{Fx} \cdot [(B_h \cdot \cos A \cdot \sin I + B_v \cdot \cos I) \cdot \cos FxPz + Bias_{Fx}]$$

Eqs. 3.80 and 3.81 are the output of accelerometer and fluxgate mathematical model, so, in order to achieve higher fitting precision, selected the orthogonal trigonometric as basic functions to fit the curve of the output of each sensor.

Taken A_x as an example, assuming $A_m = \alpha_0$, $I = d_1$, A_x output is:

$$Ax_1 = M_1 \cdot \cos r + N_1 \cdot \sin r + P_1 \tag{3.82}$$

Assuming $A_m = \alpha_0$, $I = d_2 = d_1\ @+ 90°$, A_x output is:

$$Ax_2 = M_2 \cdot \cos r + N_2 \cdot \sin r + P_2 \tag{3.83}$$

Then,

$$Bias_{ax} \times \begin{bmatrix} G \cdot \sin d_1 \cdot \cos AxPx \\ -G \cdot \sin d_1 \cdot \cos AxPy \\ -G \cdot \cos d_1 \cdot \cos AxPz + Bias_{Ax} \\ G \cdot \sin d_2 \cdot \cos AxPx \\ -G \cdot \sin d_2 \cdot \cos AxPy \\ -G \cdot \cos d_2 \cdot \cos AxPz + Bias_{Ax} \end{bmatrix} = \begin{bmatrix} M_1 \\ N_1 \\ P_1 \\ M_2 \\ N_2 \\ P_2 \end{bmatrix} \tag{3.84}$$

We can calculate the d_1, d_2, K_{Ax}, $Bias_{Ax}$, cosAxPx, cosAxPy, cosAxPz as follows:

$$d_1 = \arctan(M_1/M_2) \tag{3.85}$$
$$d_2 = d_1 + 90 \tag{3.86}$$

$$\cos AxPx = 1 \cdot (\tan d_1 + 1) \cdot M_1/(M_1^2 \cdot \tan^2 d_1 + 2 \cdot N_1^2 \cdot \tan d_1 + 2 \cdot M_1^2 \cdot \tan d_1 + M_1^2 + N_1^2 \cdot \tan^2 d_1 + P_1^2 \cdot \tan^2 d_1 + N_1^2 + P_2^2 \cdot \tan^2 d_1 - 2 \cdot \tan^2 d_1 \cdot P_1 \cdot P_2)^{1/2} \tag{3.87}$$

$$\cos AxPy = -\frac{N_1}{M_1} \cdot \cos AxPx \tag{3.88}$$

$$\cos AxPz = \frac{(P_2 - P_1) \cdot \sin d_1}{M_1 \cdot (\sin d_1 + \cos d_1)} \tag{3.89}$$

$$K_{Ax} = \frac{M_1}{G \cdot \sin d_1 \cdot \cos AxPx} \tag{3.90}$$

$$\mathrm{Bias}_{\mathrm{Ax}} = \frac{P_1}{SF_{ax}} + G \cdot \mathrm{cosd}_1 \cdot \mathrm{cosAxPz} \tag{3.91}$$

The fluxgate calculation method is similar as above.

3.5.3 Performance Results after Calibration

Using orthogonal method and the numerical fitting calibration method calculated the calibration coefficients respectively. The results obtained for comparison in the Table 3.3. Two calibration methods have a little difference in calculate the coefficients. Using these coefficients to calculate the borehole inclination and azimuth as shown in Fig. 3.13.

It can be seen that each error range is within the scope of the engineering allows. So, the two calibration methods all achieving the purpose of the system calibration. But the numerical fitting method is easy to operate, and the calibration instrument

Table 3.3 Compared calibration factor of two methods calculated

Coefficients	Numerical fitting	Orthogonal design	Coefficients	Numerical fitting	Orthogonal design
K_{Ax}(V/G)	1.1060	1.1063	L_{Fx}(V/uT)	0.0546	0.0544
Bias_{Ax}(G)	0.0587	0.0567	Bias_{Fx}(uT)	2.2123	2.2975
cosAxPx	0.9965	0.9965	cosFxPx	0.9920	0.9960
cosAxPy	0.0824	0.0832	cosFxPy	0.0880	0.0883
cosAxPz	−0.001	−0.0008	cosFxPz	−0.0155	−0.0156

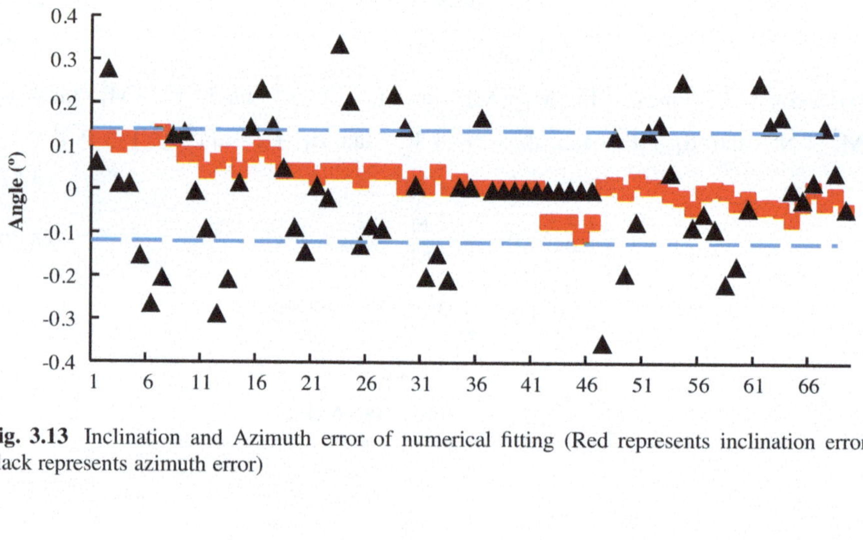

Fig. 3.13 Inclination and Azimuth error of numerical fitting (Red represents inclination error, black represents azimuth error)

has simple structure, even if the calibration instrument has lower precision than before, we can also obtain the very precise calculation coefficients like orthogonal method. Despite the error of the final result still exists, it has been greatly improved that achieve the requirements of field application.

1. In Rotary steerable system, must establish a measurement system equipped with a triaxial fluxgate and triaxial accelerometer, but the installation error cannot be avoided and must be calibrated.
2. Developed the calibration model which can well meet the requirements of field application. The final errors of the measurement of Inclination and Azimuth are small.
3. Orthogonal method and curve-fitting method has little difference of calculating calibration coefficients, however, the curve fitting method is easy to operate and the calibration instrument has simple structure, even if the calibration instrument has lower precision than before, we can also obtain the very precise calculation coefficients like orthogonal method, more suitable for engineering applications.

3.6 Dynamic Algorithm for Stick-Slip Motion (DAS)

In the actual drilling process, the drillstring will appear torsional vibration and stick-slip phenomenon but not always presents uniform rotating. As shown in the Fig. 3.14a, extracting a period of underground data, the rotary table speed maintained at 100 r/min, the drill bit speed has fluctuated between 0–200 r/min, stick-slip phenomenon is very serious. In this situation, the theoretical models based on the drillstring uniform rotating will not be practicable [31].

We first analyzed the movement of the drillstring, actually figure out how often the stick-slip phenomenon emerges throughout the drilling process. The rotary steerable system we have developed has conducted a number of field tests [32], the measurement data show that stick-slip vibration will occur in drilling frequently. We randomly selected speed data of 2 h utilizing the reservoir sampling principle [33], and estimated the overall characteristics from the characteristics of the sample. First select the top of data points of 1 h; suppose there are k data points; From the $k + 1$ data point to the last data point is reached; Select the i-th data point in the probability of $1/i$ ($i = k + 1, k + 2, \ldots, N$), and randomly replace a previously selected elements. This traversal time can guarantee data points of 1 h to be completely randomly selected. As shown in Fig. 3.14b, speed at vicinity of zero represents the emergence of stick-slip. Stick-slip vibration is always existent throughout the drilling process, so the application of stick-slip vibration method to improve measurement accuracy is feasible.

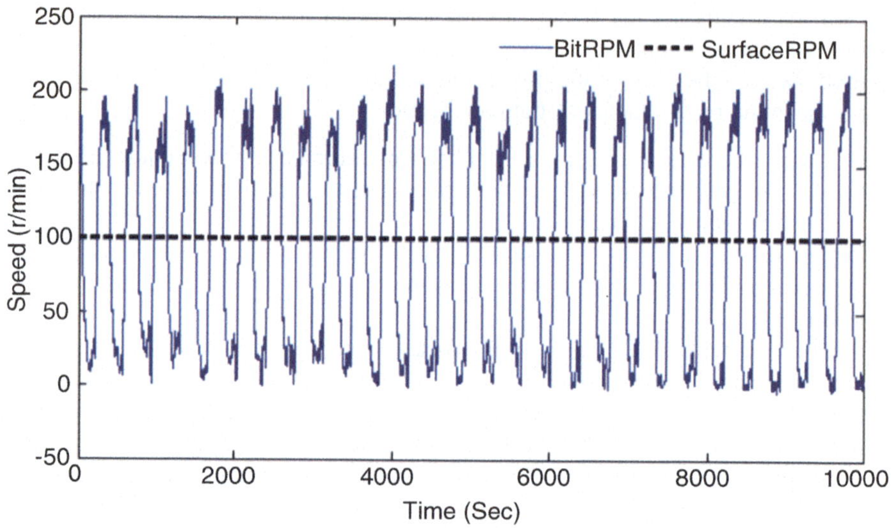

(a) SurfaceRPM and BitRPM

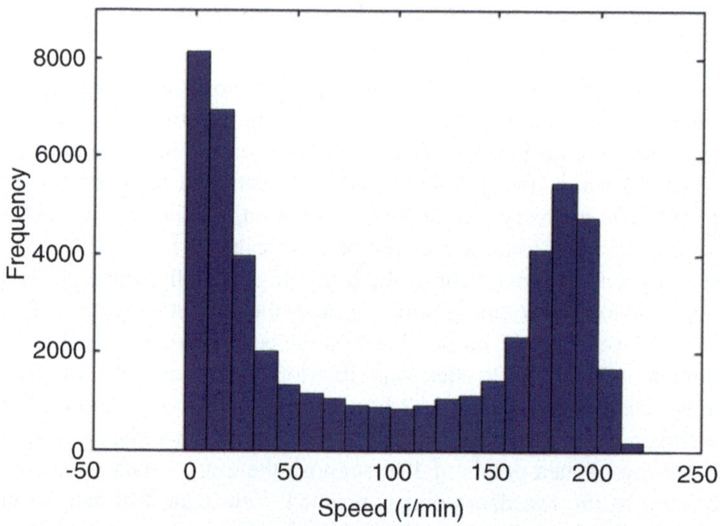

(b) Speed ststistical histogram

Fig. 3.14 Analysis of stick-slip vibration phenomena, as show in the part a, when the surface top drive rotary speed is 100 r/min, the speed of drill bit has fluctuated between 0–200 r/min, stick-slip phenomenon is very serious. Form the part b, Stick-slip vibration is always existent throughout the drilling process

3.6.1 Experiment Data

Using the formula (3.30), (3.31) and the designed filter, establish the DSP program. Put the entire system to the laboratory bench for accuracy testing of the measuring system. Figures 3.15 and 3.16 shows the accelerometer (x, y, z axis) and fluxgate (x, y, z axis) measurement data. It can be seen the measurement noise of accelerometer is relatively larger than that of fluxgate; the main reason is due to accelerometers' sensitivity to the drill string vibrations. Through the filter can be obtained to a certain extent on the noise elimination. Finally, inclination and azimuth values in the test are shown in Fig. 3.17. Wherein X region is the calculated results in the case of vibration noise caused by rotary drillstring.

The design of the measuring algorithm can used for the strap-down rotary steerable drilling system, while the drill string rotary, dynamic algorithm be used

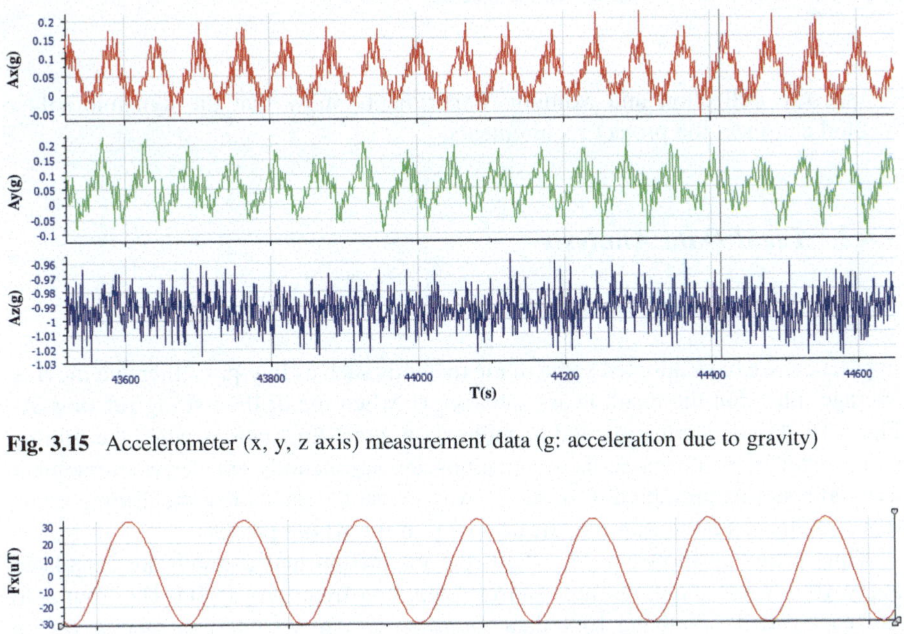

Fig. 3.15 Accelerometer (x, y, z axis) measurement data (g: acceleration due to gravity)

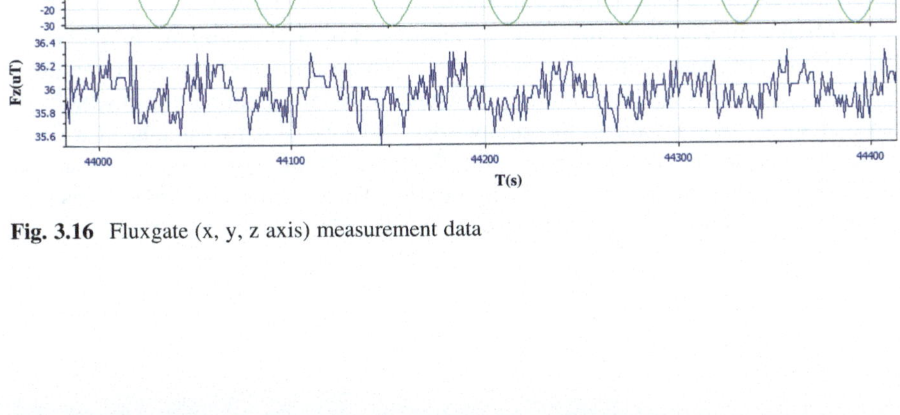

Fig. 3.16 Fluxgate (x, y, z axis) measurement data

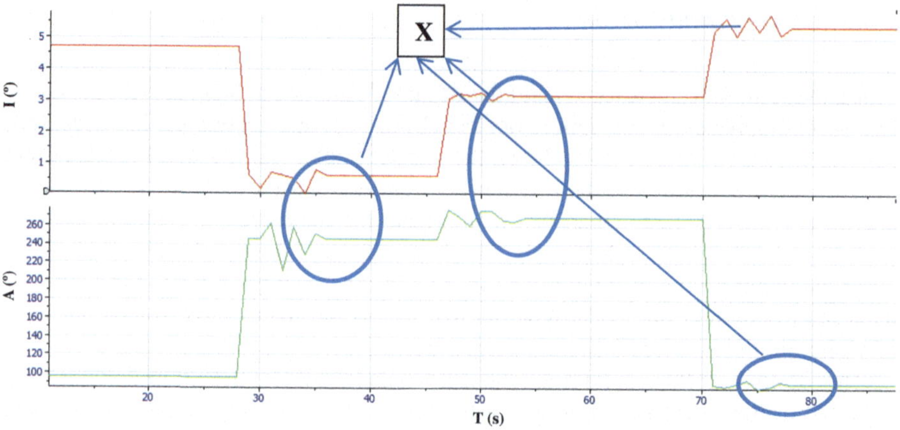

Fig. 3.17 Inclination and azimuth values in the test

to measure deflection and azimuth. Experiments show that the dynamic solver method can meet the project requirements.

3.6.2 *Field Data Analysis*

3.6.2.1 Measurement Data Performance

In practice, we have adopted some of the methods such as low-pass filter and moving average filter, but the result is not satisfactory when the drilling string rotating. As Fig. 3.18 shows, inclination and azimuth show great fluctuations when the drilling string is rotating, otherwise the fluctuations are significantly smaller when momentarily at rest, this prompted us to seek a way to use the data when the drilling string momentarily at rest to improve the accuracy of the whole process.

Time series as shown in Fig. 3.18, the inclination and azimuth are calculated every 10 s, while the down hole sensors sampling frequency is 100 Hz. That is to say, each calculation using 1000 data points, so we can see that non-rotating region in Fig. 3.18 lasted about 400 s. That is the result of the stop artificially in the drilling process. Actually, the down-hole drill string also could automatically stop when on the ground seems to be in continuous rotation.

To observe in a smaller time scale as shown in Fig. 3.19, sampling frequency is 100 Hz and 500 data points representing 5 s. These data are even not enough to calculate a point in Fig. 3.18, however, we have found the stationary area which we can call sticky monument. In the sticky area, fluxgate signal maintains at a fixed value, accelerometer signal shows some fluctuations but much smaller than the performance when the drilling string is rotating. Through the installation structure of Fig. 3.1, it can be inferred that the y-axis signal and the x-axis are similar, and the

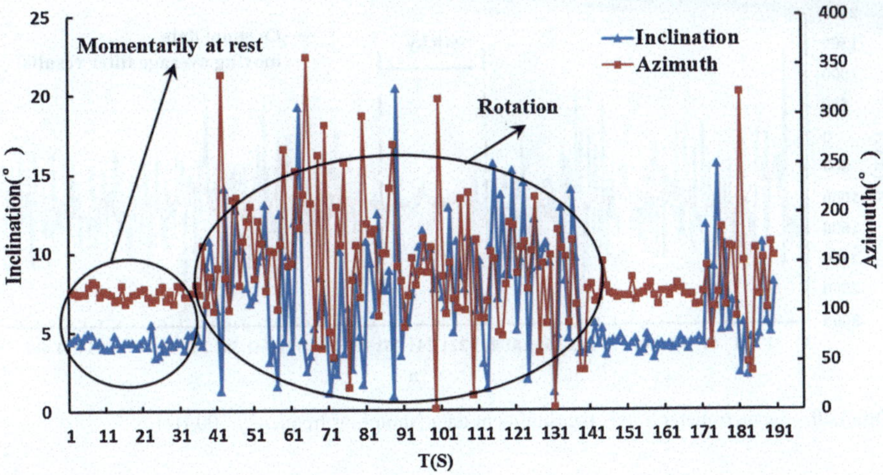

Fig. 3.18 The result of the calculation of the Formula (7), (8), the field tests playback data

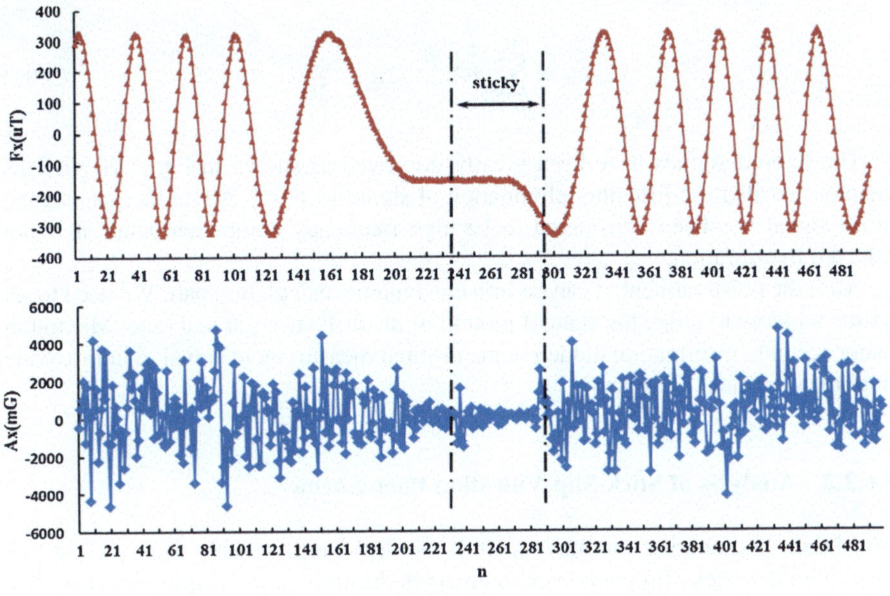

Fig. 3.19 Accelerometer and fluxgate x-axis measurement data (sampling frequency 100 Hz)

phase difference is of 90°, so there is not shown y-axis signal in the figure. Obviously, accelerometer signal should also be presented sine wave like fluxgate signal when in the rotating monument, but we do not see this result because of the strong vibration of the drillstring.

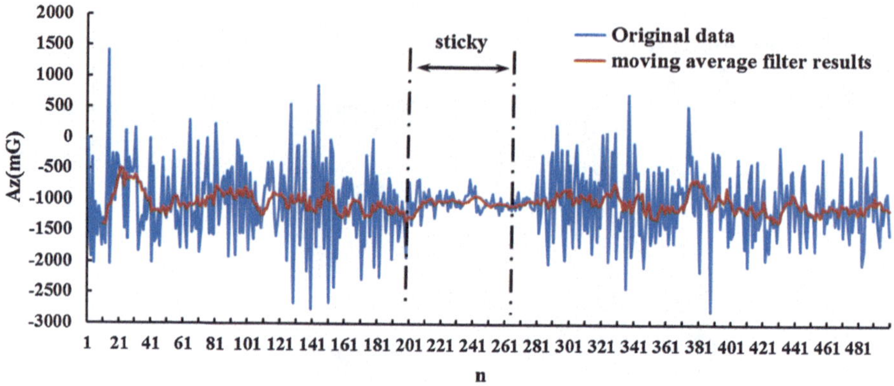

Fig. 3.20 Accelerometer z-axis measurement data (sampling frequency 100 Hz)

Accelerometer signal of z-axis is also subjected to the influence of vibration which is relatively smaller. We use moving average filter for further processing, As in Eq. (3.92).

$$A_n = \sum_{i=1}^{10} a_i A_{n+i-9}, a_i = \frac{1}{10} \tag{3.92}$$

The filtered signals have been greatly improved as shown in Fig. 3.20. In This chapter, we also use FIR filter elimination of signal-to-noise. About fluxgate signal, noise signal has been considered to be high-frequency component since it is not affected by vibration.

After the pre-treatment, it can go into the dynamic calculation part. We need to set a time window to judge the state of motion of the drill string in real time. Maximum using accurate information hidden in the original measurement signal to improve the final measurement accuracy.

3.6.2.2 Analysis of Stick-Slip Vibration Phenomena

We should first analyze the movement of the drillstring, actually to figure out that how often the stick-slip phenomenon emerges throughout the drilling process. The rotary steerable system developed has conducted a number of on-site tests, test data playback show that stick-slip vibration frequency of occurrence is very high. Analysis of the entire measurement data is time-consuming and labor-intensive due to the large amount of data.

So, we randomly selected 2 h of speed data utilizing the reservoir sampling principle [34], in order to estimate the overall characteristics from those of the sample. First select the top of 1 h of data points, suppose there are *k* data points, From the k + 1 data point to the last data point is reached, Select the *i*-th data point in

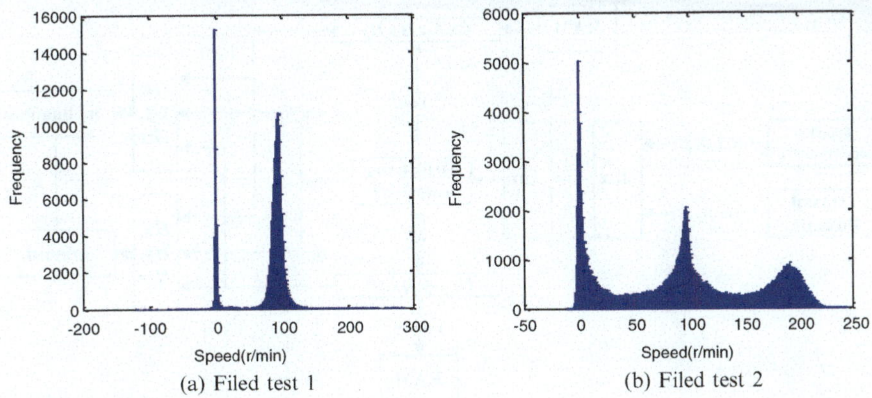

Fig. 3.21 Speed statistical histogram

the probability of $1/i$ ($i = k + 1, k + 2,\ldots,N$), and randomly replace a previously selected elements. This traversal time to get 1 h of data points can be guaranteed to be completely randomly selected.

As can be seen from Fig. 3.21, Speed at vicinity of zero represents the emergence of stick-slip. Stick-slip vibration is always there throughout the drilling process, so the application of stick-slip vibration method to improve measurement accuracy is feasible.

3.6.3 Utilizing Stick-Slip Vibration to Improve Measurement Accuracy

3.6.3.1 Application Methodology of Stick-Slip Vibration

Dynamic solution approach to the bottom of the rotating drill string attitude was proposed in Fig. 3.22, that is, with non-rotating string, filtered real time signals on three-axis were all used for calculation, at the same time, filtered signals on x-axis, y-axis were stored; on condition of rotation string, real time filtered signals on z-axis and stored signals of x-axis, y-axis with non-rotating string were adopted.

In addition, stick-slip state of the down-hole drilling tool was seen as a non-rotating "stationary" state. Real-time judgment method of string rotation state based on down-hole survey data was proposed. Drillstring rotational speed may be used to determine whether the string is rotating or not, it is a practicable method, but the reliability is not high. By analysis of above, the standard deviation statistical methods to determine the drilling string movement will be better, because it reflects the degree of dispersion among the individuals within the group. Using 50 data points as a time window, assumed to be $x_1, x_2,\ldots, x_{49}, x_{50}$, will get the standard deviation σ. As shown in Fig. 3.23, the drill string moment can be considered static when the standard deviation is close to zero.

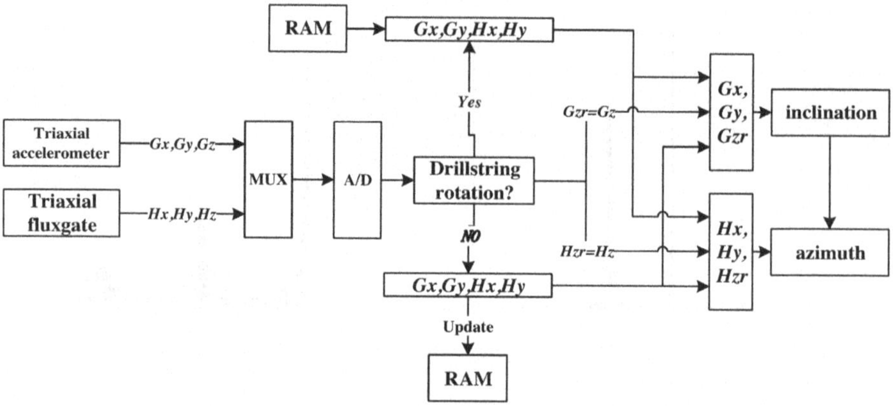

Fig. 3.22 Dynamic measurement algorithm flowchart

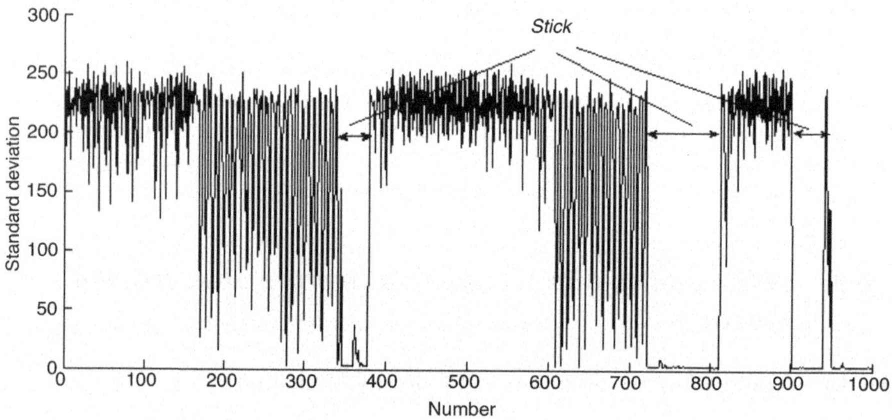

Fig. 3.23 Standard deviation distribution map

3.6.3.2 Kalman Filter Approach

Kalman filter is an optimal recursive data processing algorithm. It is optimal in the sense that it uses all available information to minimize the errors in the state variables of a given system. It is recursive, because it does not require all previous data to be kept in storage. It is also a data processing algorithm since it is not an electrical filter, but rather a computer program [35]. In order to implement a discrete Kalman filter, the error models have to be given in state space form:

$$x_k = F_{k,k-1} + G_{k-1}w_{k-1} \tag{3.93}$$

$$y_k = H_k x_k + v_k \tag{3.94}$$

Equation (3.93) is called the dynamics equation and Eq. (3.94) is the observation or update equation [36]. Here x_k is the process state vector, $F_{k,\,k-1}$ is a square matrix relating x_k to x_{k-1} being called the state transition matrix, and w_{k-1} is a random function considered white noise with G_{k-1} as its coefficient vector. In Eq. (12), y_k is the measurement vector at the kth moment, H_k is the design matrix giving the ideal noiseless relationship between the observations vector and the state vector, and v_k is the observation random noise. It is assumed that v_k has no correlation with w_{k-1}. The Kalman filter model using in This chapter is shown in the Fig. 6.

When defined tool face as φ_G, inclination as θ, azimuth as ψ. Transformation matrix R_b^n can be defined as following:

$$R_b^n = \begin{bmatrix} \cos\psi\cos\phi + \sin\psi\sin\theta\sin\phi & \sin\psi\cos\theta & \cos\psi\sin\phi - \sin\psi\sin\theta\cos\phi \\ -\sin\psi\cos\phi + \cos\psi\sin\theta\sin\phi & \cos\psi\cos\theta & -\sin\psi\sin\phi - \cos\psi\sin\theta\cos\phi \\ -\cos\theta\sin\phi & \sin\theta & \cos\theta\cos\phi \end{bmatrix} \tag{3.95}$$

Assuming that $G^b = [G_x\ G_y\ G_Z]^T$, the rotary speed defined as ω_x, ω_y, ω_z on the xyz axis respectively, $\omega_{ib}^b = \left[\omega_x\ \omega_y\ \omega_z\right]^T$.Then the rotary angle will be obtained.

$$\theta_{ib}^b = \begin{bmatrix} \Delta\theta_x \\ \Delta\theta_y \\ \Delta\theta_z \end{bmatrix}_{ib}^b = \begin{bmatrix} \omega_x \\ \omega_y \\ \omega_z \end{bmatrix} \Delta t \tag{3.96}$$

Using quaternion $Q = [q_1\ q_2\, q_3\ q_4]^T$ to express the coordinate system transformation, then R_b^n can be redefined as following:

$$R_b^n = \begin{bmatrix} R_{11} & R_{12} & R_{13} \\ R_{21} & R_{22} & R_{23} \\ R_{31} & R_{32} & R_{33} \end{bmatrix} = \begin{bmatrix} q_1^2 - q_2^2 - q_3^2 + q_4^2 & 2(q_1q_2 - q_3q_4) & 2(q_1q_2 + q_3q_4) \\ 2(q_1q_2 + q_3q_4) & -q_1^2 + q_2^2 - q_3^2 + q_4^2 & 2(q_2q_3 - q_1q_4) \\ 2(q_1q_3 - q_2q_4) & 2(q_2q_3 + q_1q_4) & -q_1^2 - q_2^2 + q_3^2 + q_4^2 \end{bmatrix} \tag{3.97}$$

Then the dynamics equation in the inclination and azimuth solving will be obtained.

$$Q_{k+1} = \begin{bmatrix} 1 & -\frac{1}{2}\omega_x dt & -\frac{1}{2}\omega_y dt & -\frac{1}{2}\omega_z dt \\ \frac{1}{2}\omega_x dt & 1 & \frac{1}{2}\omega_z dt & -\frac{1}{2}\omega_y dt \\ \frac{1}{2}\omega_y dt & -\frac{1}{2}\omega_z dt & 1 & \frac{1}{2}\omega_x dt \\ \frac{1}{2}\omega_z dt & \frac{1}{2}\omega_y dt & -\frac{1}{2}\omega_x dt & 1 \end{bmatrix} Q_k \tag{3.98}$$

When the drillstring appears torsional or stick-slip vibration in the actual drilling process, we use the algorithm defined as "DAS" method.

3.6.4 Simulation Results

We extracted 4000 data points under the state of the rotating drill string, simulation results in Matlab are shown in Fig. 3.24. It can be seen, using stick-slip phenomenon has greatly improved the borehole inclination and azimuth calculation accuracy when the drilling string is in dynamic rotation.

In directional drilling and rotary steerable drilling technology and application, to accurately measure the spatial attitude (inclination, azimuth, tool face) of the bottom drilling tool in real time with the drill string rotating is a challenging problem.

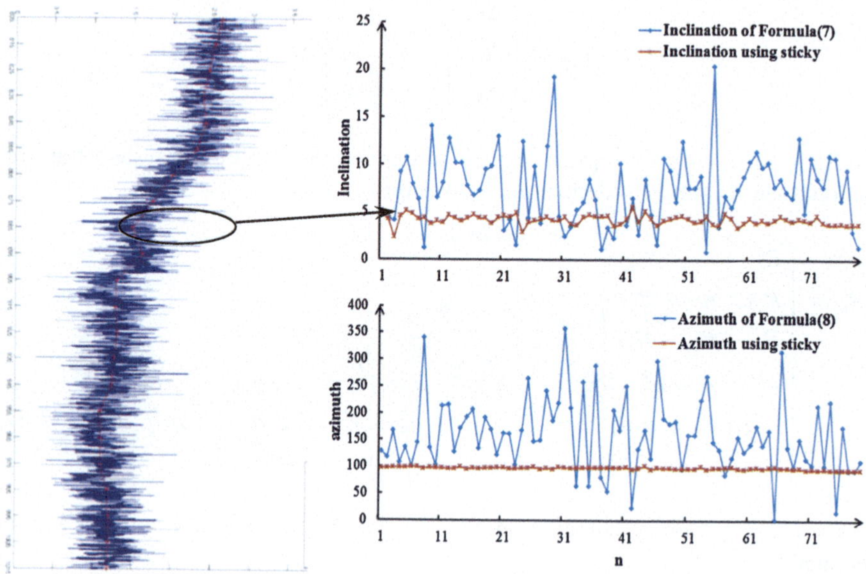

(a) The whole drilling test (b) 4000 data points simulation results

Fig. 3.24 Contrast inclination and azimuth dynamic measurement results with the former

Simulation and experiments show that the dynamic solver methods developed in this chapter can meet the project requirements. But the drilling string vibration seriously affected the dynamic solver accuracy. Through statistical analysis of field data, we found that the stick-slip phenomenon is widespread, so we use the stick-slip state to develop a dynamic algorithm, in order to improve the borehole inclination and azimuth solver accuracy. The simulation and experimental evaluation results show that the designed algorithm has good practicability. That has reference significance for the directional drilling and rotary steerable technology development.

3.7 New Kalman Filter Approach

The application of KF [23, 24, 37] requires that both the system and the measurement models of the underlying process be linear. A discrete-time linear state space system is described by:

$$x_k = \Phi_{k,k-1}x_{k-1} + G_{k-1}w_{k-1} \tag{3.99}$$

where $\Phi_{k,k-1}$ is the state transition matrix, x_k is the state vector, G_{k-1} is the noise distribution matrix, w_{k-1} is the process noise vector, and k is the measurement epoch.

The measurement equation of the system is given by

$$z_k = H_k x_k + \eta_k \tag{3.100}$$

where z_k is the measurement vector of the system output, H_k is the observation or design, and η_k is the measurement noise. The system noise w_k and the measurement noise η_k are unassociated zero-mean white-noise processes with determined autocovariance functions.

We developed an algorithm with new state-space models by analyzing drillstring dynamics. In the solution system, we define the input vector of KF-1 as $X = \begin{bmatrix} a_x & a_y & a_z \\ m_x & m_y & m_z \end{bmatrix}$, which is measured by the three-axis magnetometers and three-axis accelerometers. The input vector of KF-2 is the inclination and azimuth, defined as $M = \begin{bmatrix} I \\ A \end{bmatrix}$, where I is the inclination of the borehole, and A is the azimuth of the borehole. Using the solving process as shown in Fig. 3.25, we develop two

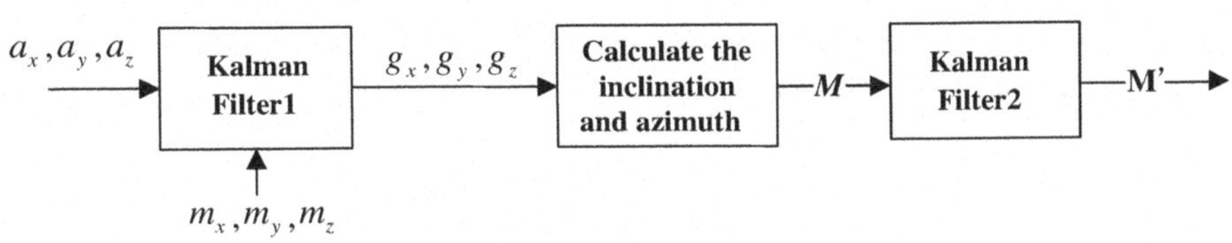

Fig. 3.25 Solving process for the system

KFs for the entire drilling process. After KF-1, we can obtain the more precise signals of gravity acceleration g_x, g_y and g_z, that defined as the output of KF-1. Using the gravity accelerations, we can obtain the inclination and azimuth by the equation developed when the drillstring rotates. KF-2 is then used to further smooth the drilling trajectory. The output of KF-2 defined as $M' = [I'\ A']^T$, which is more precisely.

3.7.1 State-Space Model for KF-1

The sensors installed in the center of the drillstring, the measurement signals of x and y axis will exhibit a sine wave during rotation. Theoretically, accelerometer and magnetometer signals have the same rule. In the actual drilling process, the vibration of the drillstring affects less on the magnetometer signals. That is, the magnetometer signals are used to calibrate the accelerometer signals. From the laboratory testing we can conclude that the changes of signals of fluxgate in line with signals of gravity acceleration, as show in the Fig. 3.25 which will provide in the following.

Assume the angular velocity is $\omega_{x,\ y,\ z}$, sampling interval is Δt, then,

$$\frac{m_{x,y}(k)}{m_{x,y}(k-1)} = \frac{M \sin \omega_{x,y} t}{M \sin \omega_{x,y}(t - \Delta t)} = \frac{G \sin \omega_{x,y} t}{G \sin \omega_{x,y}(t - \Delta t)} = \frac{g_{x,y}(k)}{g_{x,y}(k-1)} \tag{3.101}$$

In the KF-1, we define the state vector as $x_g(k) = \begin{bmatrix} g_x(k) \\ g_y(k) \\ g_z(k) \end{bmatrix}$, as show in Fig. 3.25, from the vibration signals of a_x, a_y, a_z we can obtain the signals of gravity acceleration $x_g(k)$. Thus, the measurement vector of the system output as $z(k) = \begin{bmatrix} a_x \\ a_y \\ a_z \end{bmatrix}$, when the drillstring rotates, the measurement signals of the x- and y-axis will exhibit a sine wave except the signal of z-axis. So, the transformation matrix defined as $H_k = \begin{bmatrix} \frac{m_x(k)}{m_x(k-1)} & 0 & 0 \\ 0 & \frac{m_y(k)}{m_y(k-1)} & 0 \\ 0 & 0 & 1 \end{bmatrix}$. The system noise w_k and the measurement noise η_k are uncorrelated zero-mean white-noise processes. Therefore, we obtain the state-space model of the KF-1 as follows:

$$x_g(k) = \begin{bmatrix} \frac{m_x(k)}{m_x(k-1)} & 0 & 0 \\ 0 & \frac{m_y(k)}{m_y(k-1)} & 0 \\ 0 & 0 & 1 \end{bmatrix} x_g(k-1) + \begin{bmatrix} w_x(k-1) \\ w_y(k-1) \\ w_z(k-1) \end{bmatrix} \tag{3.102}$$

$$z(k) = H_k x_g(k) + \eta(k) \tag{3.103}$$

3.7.2 Calculating the Inclination and Azimuth

When the drillstring rotates, the above equations are not applicable. The sensors installed in the center of the drillstring, the measurement signals of the x- and y-axis, will exhibit a sine wave during rotation.

Through KF-1, we obtain g_x, g_y and g_z which are defined as survey signals of gravity acceleration on the x, y, and z axes, respectively. Then define the input vector of the system as $X' = \begin{bmatrix} g_x & g_y & g_z \\ m_x & m_y & m_z \end{bmatrix}$.

Define the rotational speed as R. If $R = 0$, Eqs. (3.16) and (3.17) are used to calculate the inclination and azimuth. $R \neq 0$ indicates that the drillstring is rotating, and computational formulas of inclination I and azimuth A are given as follows:

$$I = tg^{-1} \frac{\sqrt{\left(\int_{T_M=0}^{2Np} g_x dT_M\right)^2 + \left(\int_{T_M=0}^{2Np} g_y dT_M\right)^2}}{-\int_{T_M=0}^{2Np} g_z dT_M} \tag{3.104}$$

$$A = tg^{-1}\left(\frac{\int_{T_M=0}^{2Np} G(m_y g_x - m_x g_y) dT_M}{\int_{T_M=0}^{2Np} m_Z\left(g_x^2 + g_y^2\right) dT_M - \int_{T_M=0}^{2Np} g_z(m_x g_x + m_y g_y) dT_M}\right) \tag{3.105}$$

where T_M is the magnetic tool-face angle: $T_M = \text{tg}^{-1}\left(\frac{-m_y}{m_x}\right)$. Tool face angle used for near-vertical wells. Magnetic tool face is the angle, or azimuth, of the borehole survey instrument within the wellbore measured clockwise relative to magnetic north and in the plane perpendicular to the wellbore axis; the north, east, south and west directions have magnetic tool face angles of 0°, 90°, 180° and 270°, respectively. Magnetic tool face may be corrected to reference either grid north or true north.

Although the drillstring rotational speed is a way to determine whether the string is rotating or not, it reliability of this approach is not high. Instead, using the standard-deviation statistical methods to determine the drillstring movement is more effective because it reflects the degree of dispersion among the individuals within the group. Using 50 data points as a time window, assumed them to be x_1, x_2, …, x_{49}, x_{50}, we get the standard deviation $\sigma = \sqrt{\frac{1}{N}\sum_{i=1}^{N}(x_i - \overline{x})^2}$. The drillstring can be considered static when the standard deviation σ is close to zero.

3.7.3 State-Space Model for KF-2

In the drilling process, the movement states that σ=0 or $\sigma \neq 0$ will appear alternately. When σ=0, the solving results are more accurate because the vibration is slight when the drillstring is not rotating. We develop another Kalman Filter (KF-2) to smooth the trajectory of the drilling. (KF-2 was divided into KF-2.1 and KF2.2 respectively as show in the Fig. 3.27.)

In a normal drilling process, the drilling operation has to stop at measurement stations frequently in order to measure the inclination and azimuth. The well trajectory is then computed between the two surveying stations based on mathematical assumptions. For instance, it may be assumed that the drilled distance is a straight line, smooth arc, or polygonal line; each requires a different calculating method. Assuming that the three-dimensional coordinates of the measuring N-th point of the actual drilling trajectory are (x_N, y_N, z_N), the measuring (N + 1)-th point is (x_{N+1}, y_{N+1}, z_{N+1}), and the well depth, vertical depth, inclination, and azimuth are L_N, H_N, θ_N, ψ_N, and $L_{N+1}, H_{N+1}, \theta_{N+1}, \psi_{N+1}$, respectively. The well trajectory between two points can be then defined as follows:

$$\begin{cases} L = L_{N+1} - L_N \\ H = H_{N+1} - H_N \\ \theta = (\theta_N + \theta_{N+1})/2 \\ \psi = (\psi_N + \psi_{N+1})/2 \end{cases} \tag{3.106}$$

If the vertical depth H is a known, the three-dimensional coordinates of the measuring (N + 1)-th point can be defined as

$$\begin{cases} x_{N+1} = x_N + H \tan\theta \cos\psi \\ y_{N+1} = y_N + H \tan\theta \cos\psi \\ z_{N+1} = z_N + H \end{cases} \tag{3.107}$$

We can obtain the spatial coordinates of each point by recursive calculation, thereby obtaining the entire drilling trajectory.

As shown in Fig. 3.26, we can use the wellbore trajectory extrapolation method to establish a recursive relationship between two adjacent measurement points. Assume

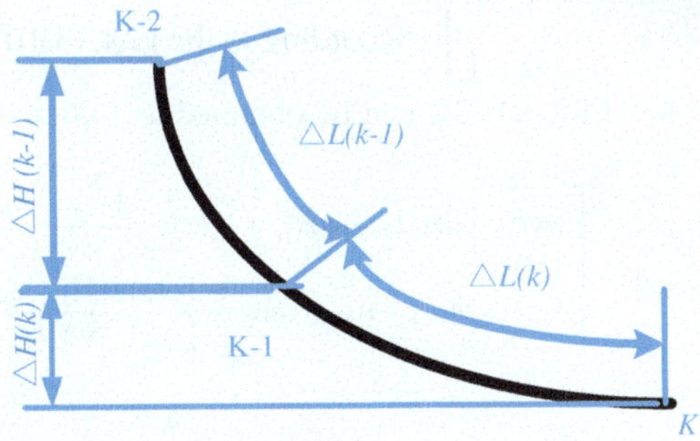

Fig. 3.26 Trajectory prediction using method of inclined plane cirque

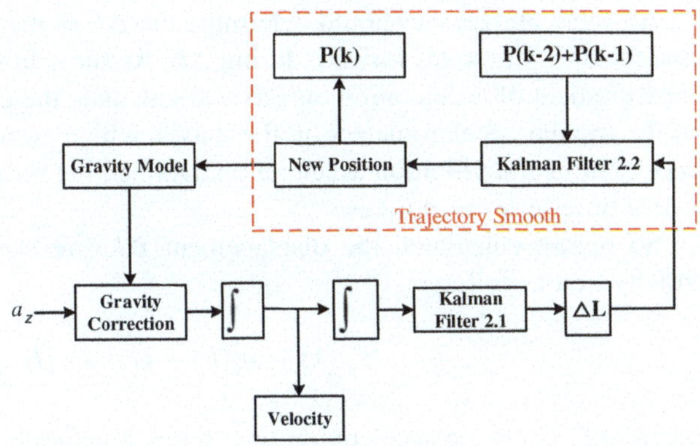

Fig. 3.27 Solving process for KF-2

L is the depth of the drilling, γ is the angle of the drilling trajectory and its tangent. Then,

$$L(k) = L(k-1) + \Delta L(k) \tag{3.108}$$

$$\gamma = \arccos\left[\cos(I_{k-2})\cos(I_{k-1}) + \sin(I_{k-2})\sin(I_{k-1})\cos(A_{k-1} - A_{k-2})\right] \tag{3.109}$$

$$\gamma(k) = \frac{\gamma}{\Delta L(k-1)}\Delta L(k) \tag{3.110}$$

Form the Eq. (3.108) to Eq. (3.110), we can use two points to estimate the next point, so the well trajectory can be smoothed. As shown in Fig. 5, we can calibrate the drilling trajectory using Kalman Filter 2.2, wherein defined the *k-th* measurement point as P(k) = [I_k A_k]. The system input is P(k-2) and P(k-1), KF-2.2 estimated P(K) combined with measured values and the theoretical calculation values.

KF-2.2 can smooth the well trajectory by using inclination and azimuth as inputs. Assuming the state vector as $x(k) = \begin{bmatrix} I_k \\ A_k \end{bmatrix}$, the measurement vector of the system output as $y(k) = \begin{bmatrix} I_k \\ A_k \end{bmatrix}_m$, which is calculated from Eqs.(8) and (9). With

$H(k) = \begin{bmatrix} 1 & 0 \\ 0 & 1 \end{bmatrix}$. According to the Eqs. (3.107), (3.108) and (3.109), the state-space model for KF-2.2 can be obtained as follows:

$$\begin{bmatrix} I_k \\ A_k \end{bmatrix} = \begin{bmatrix} \arccos\left[\cos(I_{k-2})\cos(\gamma+\gamma(k)) - \frac{\sin(\gamma+\gamma(k))}{\sin(\gamma)}(\cos(I_{k-2})\cos\gamma - \cos(I_{k-1}))\right] \\ A_{k-1} + (A_{k-1} - A_{k-2}) \cdot \arccos\left[\frac{\cos(\gamma(k)) - \cos(I_{k-1})\cos(I_k)}{\sin(I_{k-1})\sin(I_k)}\right] \end{bmatrix} + \begin{bmatrix} w_I(k) \\ w_A(k) \end{bmatrix} \tag{3.111}$$

$$y(k) = H(k)x(k) + v(k) \tag{3.112}$$

As show above, we should determine the ΔL as the KF-2.2 input. (The distance that the drill bit moves forward during Δt). As show in the Fig. 3.26, we can use the measurement of acceleration on z axis to calculate the displacement. a_z is the signal of the triaxial accelerometers on the z axis, which combined with the gravitational acceleration and vibration acceleration. Defined the measurement of acceleration on z axis time series as $a_z(k)$.

So before calculated the displacement ΔL, we should exclude the impact of gravity first as follows:

$$f_{g_z}(k) = a_z(k) - G \cdot \cos(I_{k-1}) \tag{3.113}$$

wherein $f_{g_z}(k)$ is the acceleration time series function by remove the acceleration of g_z, can be calculated from $a_z(k)$ and inclination I_{k-1} corresponding to the same time.

Then we can calculate the depth (ΔL) of drilling using acceleration on z axis time series $f_{g_z}(k)$. Define the state vector as $\Delta L(k) = \begin{bmatrix} \Delta L(k) \\ \Delta\dot{L}(k) \\ \Delta\ddot{L}(k) \end{bmatrix}$, the measurement vector of the system output as $z(k) = f_{g_z}(k)$, we develop a state-space model for KF-2.1 as follows:

$$\begin{bmatrix} \Delta L(k) \\ \Delta\dot{L}(k) \\ \Delta\ddot{L}(k) \end{bmatrix} = \begin{bmatrix} 1 & t & \frac{1}{2}t^2 \\ 0 & 1 & t \\ 0 & 0 & 1 \end{bmatrix} \begin{bmatrix} \Delta L(k-1) \\ \Delta\dot{L}(k-1) \\ \Delta\ddot{L}(k-1) \end{bmatrix} + \begin{bmatrix} \varepsilon(k) \\ \dot{\varepsilon}(k) \\ \ddot{\varepsilon}(k) \end{bmatrix} \tag{3.114}$$

$$f_{g_z}(k) = [0\ 0\ 1] \begin{bmatrix} \Delta L(k) \\ \Delta\dot{L}(k) \\ \Delta\ddot{L}(k) \end{bmatrix} + \eta(k) \tag{3.115}$$

By preprocessing the measured signals, a dynamic azimuth and inclination solving algorithm is established based on dynamical analysis of the bottom rotating drilling tool. Based on the theoretical model, we develop a Kalman filter to improve the solver accuracy as well; the state space equations used for the Kalman filter have been established based on the drilling trajectory predicted. The dynamic measurement algorithm with the Kalman filter is a new model that can greatly reduce the solution errors.

3.7.4 *Experimental Results*

Dynamic measurement algorithms developed were tested through laboratory bench and field measurements data respectively [37].

3.7.4.1 Laboratory Testing

The combined measurement system was tested first in a laboratory environment, as shown in Fig. 3.28. Measurement data were obtained under different inclination and

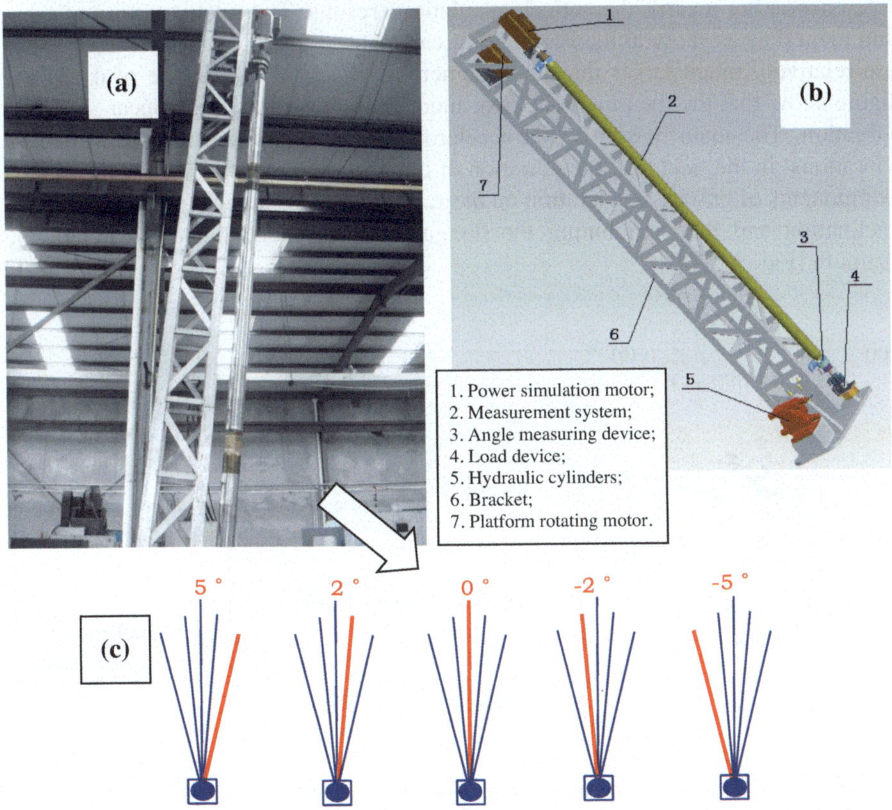

Fig. 3.28 (**a**) Laboratory bench, (**b**) Three-dimensional simulation model of the test bench, (**c**) Experimental equipment at inclinations of −5°, −2°, 0°, 2°, and 5° (Shown by red line)

rotating speed conditions. In the experimental system, we use the encoder measure the drillstring rotational speed and positioned the inclination mechanically.

As show in Fig. 3.28 (b), part 1 (power simulation motor) will work to provide electricity, although this time the drillstring not continuously rotates, the power simulation motor will generate vibration making the static solver results as above are not very accurate. Part 7 (Platform rotating motor) working makes the drillstring continuously rotates, we could use Eqs. (3.16) and (3.17) to dynamic solve the inclination and azimuth. In the experiment we set the lab test at different inclination and different rotate speed, as show in Fig. 3.28(c).

When the drillstring rotates continuously with a constant speed and is located at a particular borehole inclination and azimuth, accelerometer (x, y, z axis) and fluxgate (x, y, z axis) measurement data are obtained as shown in Figs. 3.29(b) and 3.29(c), respectively.

We put the entire system on a laboratory bench [Fig. 3.28(a)] to conduct an accuracy test of the measurement system. Figure 3.29(b) and 3.29(c) show the measurement data for the magnetometers (*x*, *y*, *z* axes) and accelerometers (*x*, *y*, *z* axes). Here, we can calculate the signals of the magnetometers and accelerometers if we know the inclination according to the Eqs. (3.104) and (3.105). Defined the reference is the reverse theoretical calculations results, The relative errors of measurement data as show in the Fig. 3.30, it can be seen that the accelerometer noises are relatively much larger than the magnetometer signal noises, the peaks in the figure show that the measurement error increases when drillstring appear stick-slip vibration. The main reason is that accelerometers are hypersensitive to drillstring vibrations. In the field test, the vibration of the drillstring is more violent. Using the components of gravity acceleration on the *x*, *y*, and *z* axes can help us calculating the inclination and then combining the fluxgate measurement signals to obtain the azimuth [Fig. 3.29(d)].

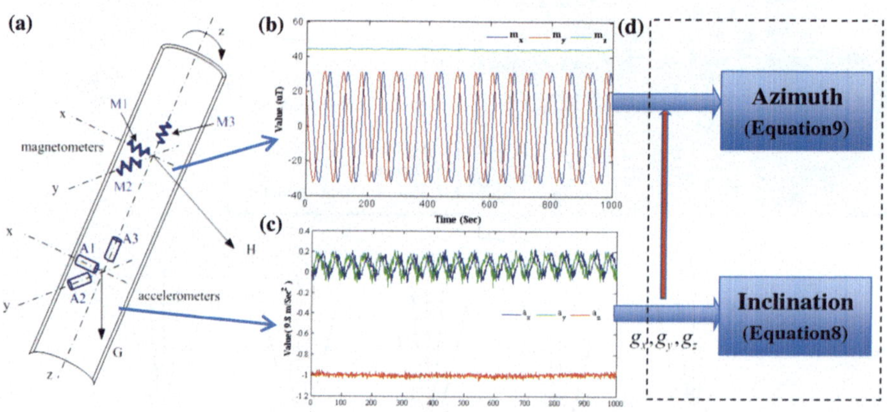

Fig. 3.29 Measurement system. (**a**), construction of measurement system; (**b**), laboratory survey signals of triaxial magnetometers on the xyz axis; (**c**), laboratory survey signals of triaxial accelerometers on the xyz axis; (**d**), Continuous survey method, g_x, g_y, g_z are the gravity acceleration on the *xyz* axis, in the Eqs. 8 and 9, we only concerned about the acceleration of gravity

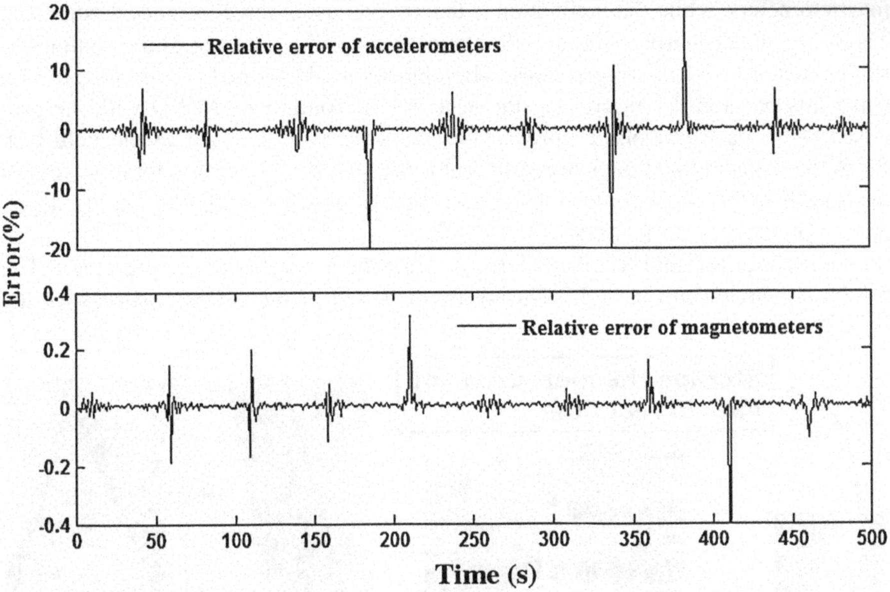

Fig. 3.30 The relative errors of accelerometers and magnetometers measurement data

Table 3.4 Error of inclination

Inclination /(°)	The error of results	
	Max/(°)	Min/(°)
5	1.58	0.02
2	0.53	0.03
0	1.2	0
−2	0.61	0.01
−5	0.27	0.01

In engineering applications, the inclination error of 0.1 will suffice. As shown in Fig. 3.28(c), when the experimental equipment is at inclinations of −5°, − 2°, 0°, 2°, and 5°, the test results are shown in Table 3.4. Through laboratory tests, the theoretical models can be verified as entirely feasible when the drillstring rotates. Experiments show that the dynamic solving methods in this chapter meet engineering requirements.

3.7.4.2 Field-Drilling Testing

In the field that some methods such as a low-pass filter and moving average filter are considered, but the results are not satisfactory when the drillstring rotates. The accelerometer signals of the field tests are completely different from those obtained via laboratory survey. The azimuth is determined by using three-axis magnetometers, while the inclination is determined using three-axis accelerometers. Therefore, the drillstring vibration dramatically amplifies the error of the continuous survey when the drillstring rotates. The algorithms developed in this chaper will solve this problem by improving the accuracy of continuous MWD with the proposed state-space models. Figure 3.31 is the schematic diagram of the field test. From the measurement data stored in real time, we can obtain the vibration signals measured by the accelerometer (x, y, and z axes), which contain the gravity acceleration on the x, y, and z axes.

We use the real field-test data to demonstrate the feasibility of our algorithm. The results are shown in Fig. 3.32. During the first step of the solving process for the

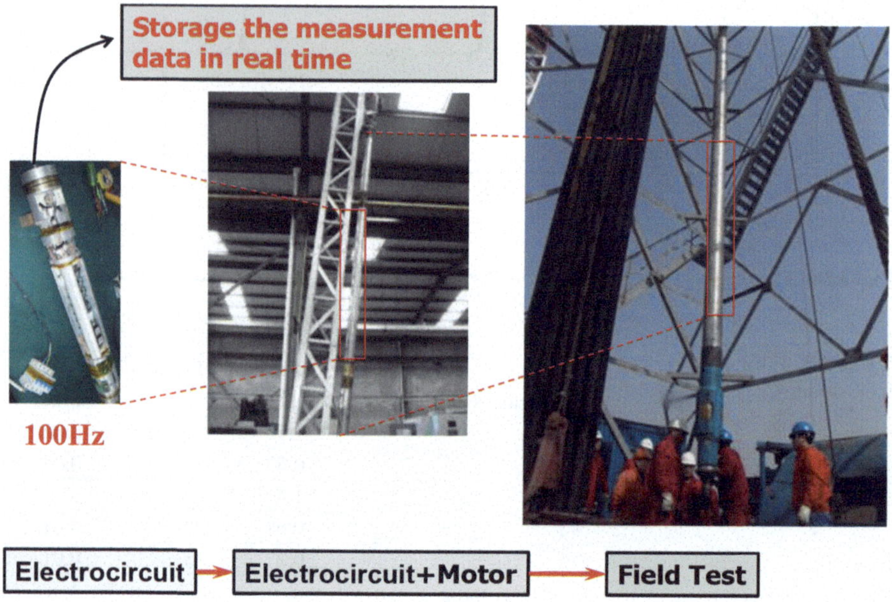

Fig. 3.31 Schematic diagram of field test. The electro circuit is installed on a compressive cylinder and placed in the axis of the drill collar, the sampling frequency of measurement and the control system is 100 Hz, and the measurement data can be stored in real time

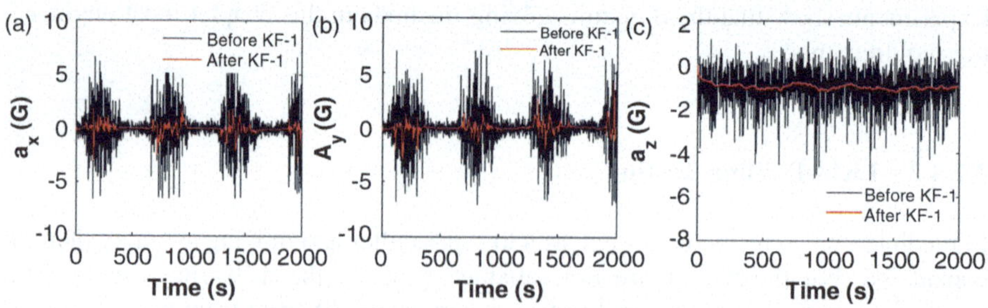

Fig. 3.32 Signals of triaxial accelerometers on the x, y, and z axes, before KF-1 and after KF-1

system (Fig. 3.25), the signals of the triaxial accelerometers (a_x, a_y, a_z) going though KF-1 get the gravity acceleration (g_x, g_y, g_z).

The results of KF-1 are shown in Fig. 3.32. The red line is the output of KF-1, that is, the gravity acceleration. Because the accelerometer on the z-axis is not influenced by the rotation of the drillstring, the curve tends to be a straight line when the inclination does not change [Fig. 3.32(c)]. We can not estimate the error of this process, because the data is came from the field-test, impossible to know the exact measurements.

In the actual drilling process, we need to find a way to measure how good the measurement accuracy is. For continuous drilling trajectory, there is no method that could be used to accurately measure the wellbore trajectory. So we could only use the MWD static measuring points for comparison. It should be feasible to evaluate the advantages of the dynamic measurement algorithm. Tests of more than 10 wells were conducted throughout China. The dynamic measurement system was deployed at multiple formations. The field tests were carried out on wells using automatic vertical or rotary steerable drilling. Three and four ribs actuators were adopted. A lot of raw measurement data were accumulated. These data were input into our algorithm for authentication. Conclusions on bit movement rules were gained through analyzing and summarizing large amounts of data.

Two wells in China (Xuanye1 and Anshun1) are selected here to evaluate the algorithm. The results for the inclination and azimuth are shown in Figs. 3.33 and 3.34. The basic parameters of field tests are shown in Table 3.5. The proposed Kalman filter method can significantly improve the precision of the survey and reduce vibration interference in the solution results.

We can also use the periodic static measurements to improve the dynamic solver accuracy. Dynamic solution approach to the bottom of the rotating drillstring attitude is proposed in [9, 37]. With non-rotating string, filtered real time signals on three-axis are all used for calculation, and at the same time, filtered signals on x-axis, y-axis were stored; on condition of rotation string, real time filtered signals on z-axis and stored signals of x-axis, y-axis with non-rotating string are adopted; In addition, stick-slip state of the down-hole drilling tool is considered as a non-rotating "stationary" state.

Inclination and azimuth show great fluctuations when the drilling string is rotates. Using the method we developed, the results are shown again in Figs. 3.33 and 3.34. Comparing static measurement points, it can be seen that the dynamic measurement inclination error is less than 1°, and the azimuth error is relatively large, between 5°~20°, because it is calculated under small inclination angles. When the angle of inclination is small, accelerometer signals of G_x and G_y are small as well, which lead to the impact of noise increased. Moreover, the error accumulated in the formula (3.17) causes the azimuth error become larger [37].

1. Drilling at Xuanye1

This drilling field test was conducted on May 19, 2010. We use the static measurement data to evaluate the continuous solving results. From Figs. 3.33 and 3.34, we can see that the inclination error is less than 1°, additionally, the error is

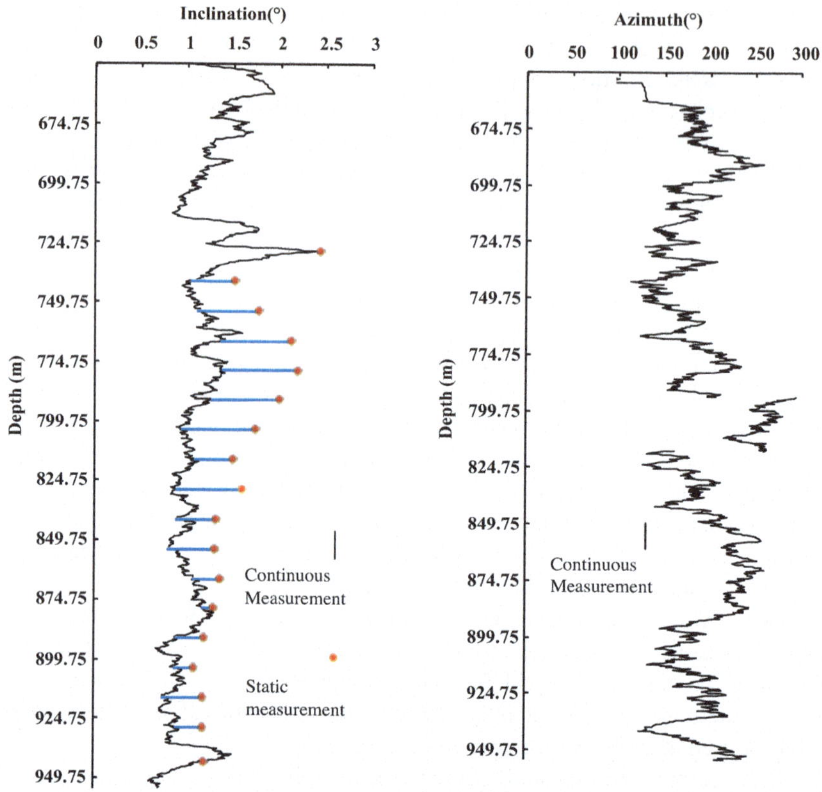

Fig. 3.33 The solving results of inclination and azimuth at XuanYe 1

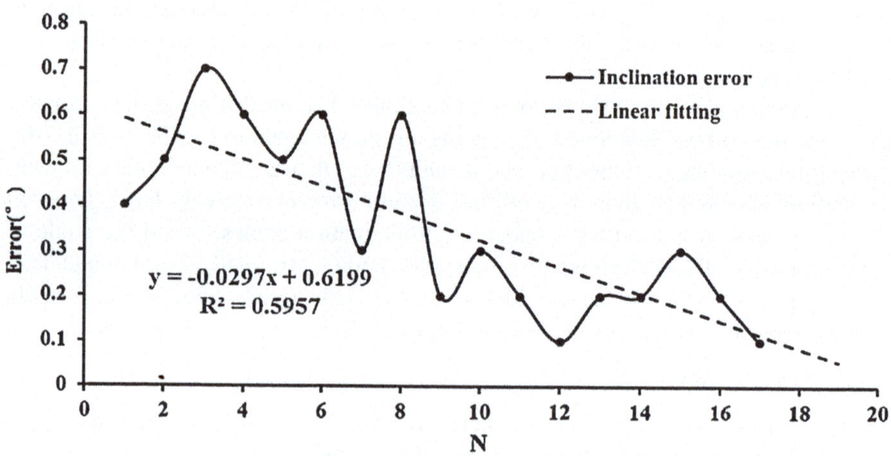

Fig. 3.34 The inclination error of solving results

Table 3.5 Basic parameters of field test

Location	Segment/m	Length/m	During/h	Diameter/mm
Xuanye1	650.03–791.75	141.72	32	φ228.6
Anshun1	2436–2518.37	82.37	81	φ300

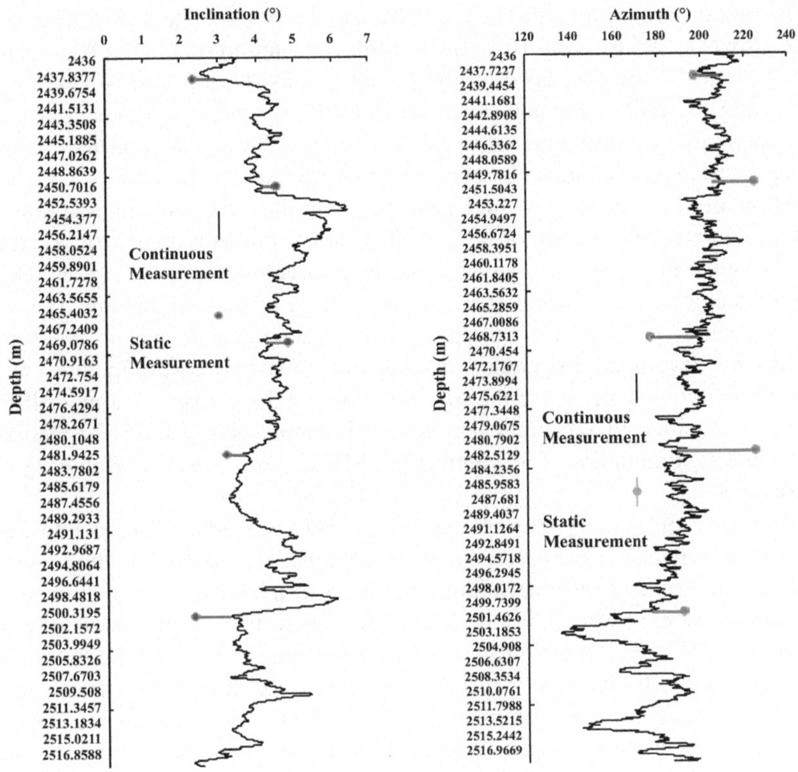

Fig. 3.35 Solving results of inclination and azimuth at Anshun1

gradually reduced. Unfortunately, we did not get the static field measurement data of Azimuth.

2. Drilling at Anshun1

This drilling field test was conducted on January 23, 2012. We use the static measurement data to evaluate the continuous solving results (Fig. 3.35). From Fig. 3.35, we can see that the inclination error is less than 1°, and that the azimuth error is between 5°~20°.

3.8 Conclusion

In directional drilling and rotary-steerable drilling technology, it is a challenge to measure the spatial attitude (inclination, azimuth, and tool face) of the bottom-drilling tool accurately in real time while the drillstring is rotating. The drillstring vibration seriously affects the accuracy of a dynamic solution. In drilling engineering, the bottom drillstring attitude is usually measured when the drillstring does not rotate. However, with recent progress in drilling technology, continuous measurement of the well trajectory becomes increasingly important. It also becomes necessary in rotary-steerable and automatic vertical drilling systems. In recent years, the requirement for a continuous survey that captures the actual trajectory between stationary surveying stations has become an urgent need. This allows for a better estimation of the casing pipe and cementing of the borehole. In addition, this provides an actual estimation of the curvature along the well trajectory. Therefore, in this chapter the trajectory between the two surveying stations is continuously surveyed using a triad of accelerometers and a triad of magnetometers. The calculation algorithm is based on a strap-down computation mechanism and Kalman filtering. We developed downhole measurement signal-processing methods and an algorithm to dynamically solve the spatial attitude of the bottom rotating drillstring. Initially we established a theoretical system of dynamic solving and laid a theoretical and technical foundation for continuous MWD and rotary-steerable systems development.

As the core of a rotary-steerable system, MWD urgently needs to improve its accuracy and reliability. However, due to the complexity of BHA vibration, dynamic real-time measurement, precise solution has become a key technology affecting the rapid development of MWD. For the downhole signal filtering of sensors, a specific algorithm needs to be proposed by analyzing the motion state of the bottom-hole drillstring. After filtering, the measurement signals are input into a space attitude dynamic measurement algorithm model that is based on motion analysis. The use of stick-slip vibration phenomena can improve the accuracy of the solution. Using the dynamic measurement theory model and a trajectory prediction, a new space-state equation can be established. A more effective Kalman filter model was developed. This further reduces the errors of the results of a dynamic solution. The dynamic measuring algorithm proposed is analyzed through downhole measurement instance data. Results show that the algorithm can effectively reduce inclination and azimuth solving errors. Field measurement data analysis shows that the algorithm can effectively solve the dynamic inclination and azimuth when the inclination is large.

References

1. Brzezowski S, Fagan J. Analysis of alternate borehole survey systems, Proceeding of 39th Annual Meeting Institute of Navigation, Houston, 1983, p. 71–78.
2. Joshi SD, Ding W. The cost benefits of horizontal drilling, Proceeding of American Gas Association, Arlington, 1991, p. 679–684.

3. Rehm WA, Garcia A, Cia SA. Horizontal drilling in mature oil fields, Proceeding of SPE/IADC Drilling Conference, New Orleans, 1989, p. 755–764.
4. Thorogood JL, Knott DR. Surveying techniques with a solid state magnetic multi-shot device. SPE Drill Eng. 1990;5(3):209–14.
5. Warren T. Rotary steerable technology conclusion: implementation issues concern operators. Oil Gas J. Dec. 1998;96(12):23–4.
6. T. Yonezawa, E. J. Cargill, et al., Robotic controlled drilling: a new rotary steerable drilling system for the oil and gas industry, Proceeding of IADC/SPE Drilling Conference, Dallas, 2002, p. 744–758.
7. Poli S, Donaco F, Oppelt J, Ragnitz D. Advanced tools for advanced wells: rotary closed-loop drilling system-results of prototype field testing. SPE Drill Complet. 1998;13(2):67–72.
8. Sugiura J, Bowler A, Hawkins R, Jones S, Hornblower P. Downhole steering automation and new survey measurement method significantly improves high-dogleg rotary-steerable system performance, SPE Annual Technical Conference and Exhibition, New Orleans, 2013.
9. Xue Qilong, Wang Ruihe, Sun Feng, et al. Continuous measurement-while-drilling utilizing strap-down multimodel surveying system, IEEE Trans Instrum Meas, vol. 63, no. 3, pp. 650–657, Mar. 2014.
10. Xue Q, Wang R, Huang L, Sun F. Dynamic solution approach to the inclination and Amizuth of bottom rotating drill string, SPE Western Regional Meeting, Monterey, 2013.
11. Wang R, Xue Q, Han L, et al. Torsional vibration analysis of push-the-bit rotary steerable drilling system. Meccanica. Jul. 2014;49(7):1601–15.
12. Xue Q, Wang R, Feng S. Study on lateral vibration of rotary steerable drilling system. J Vibroeng. 2014;16(6):2702–12.
13. ElGizawy M, Noureldin A, Georgy J, Iqbal U, El-Sheimy N. Wellbore surveying while drilling based on Kalman filtering. Am J Eng Appl Sci. 2010;3(2):240–59.
14. Elgizawy M, Noureldin A, El-Sheimy N. Continuous wellbore surveying while drilling utilizing MEMS gyroscopes based on Kalman filtering, Proceeding of SPE Annual Technical Conference Exhibit, Florence, 2010, pp. 5416–5428.
15. Jurkov AS, Cloutier J, Pecht E, Mintchev MP. Experimental feasibility of the in-drilling alignment method for inertial navigation in measurement-while-drilling. IEEE Trans Instrum Meas. Mar. 2011;60(3):1080–90.
16. Noureldin A, Irvine-Halliday D, Mintchev MP. Accuracy limitations of FOG-based continuous measurement-while-drilling surveying instruments for horizontal wells. IEEE Trans Instrum Means. Jun. 2002;51(6):1177–91.
17. A. Noureldin, H. Tabler, D. Irvine-Halliday, and M. Mintchev, Testing the applicability of fiber optic gyroscopes for azimuth monitoring for measurement-while-drilling processes in the oil industry, Proceeding IEEE Position Location Navigation Symposium, San Diego, 2000, p. 291–298.
18. Pecht E, Mintchev MP. Observability analysis for INS alignment in horizontal drilling. IEEE Trans Instrum Meas. May, 2007;56(5):1935–45.
19. Zhang Y, Wang S, Fang J. Measurement-while-drilling instrument based on predigested inertial measurement unit. IEEE Trans Instrum Meas. Dec. 2012;61(12):3295–302.
20. Chen C, Zhang Y, Li C. Surveying method of measurement while drilling based on the inertial sensor, Proceeding of 1st International Conference on PCSPA, Harbin, 2010, p. 1192–1195.
21. Walters PH. Method of determining the orientation of a surveying instrument in a borehole, US Patent 4709486, 1986.
22. Russell MK, Russell AW. Surveying of boreholes. US Patent 4163324, 1978.
23. Noureldin A, et al. Fundamentals of inertial navigation, satellite-based positioning and their integration. Berlin: Springer; 2013.
24. Minkler G, Minkler J. Theory and application of kalman filtering. Palm Bay: Magellan Book Co; 1993.
25. Ding JJ. Time frequency analysis and wavelet transform class note. Taipei: Department of Electrical Engineering, National Taiwan University; 2007.

26. Russel MK, Russel AW. Surveying of boreholes, U.S. Patent 4163324, 1979.
27. DiPersio RD, Cobern ME. Method for measurement of azimuth of a borehole while drilling: US4813274, 1989.
28. Salychev O. Inertial systems in navigation and geophysics. Moscow: Bauman MSTU press; 1998.
29. Schwarz KP, Wei M. A framework for modelling kinematic measurements in gravity field applications. Billeting Geodesique. 1990;64(4):331–46.
30. Salychev O. Inertial systems in navigation and geophysics. Moscow: Bauman MSTU press; 1998.
31. Qilong X, Ruihe W, Feng S, Leilei H. Continuous measurement-while-drilling utilizing strap-down multi-model surveying system. IEEE Trans Instrum Meas. 2014;63(3):650–7.
32. Qilong X, Ruihe W, Baolin L, Leilei H. Dynamic measurement of spatial attitude at bottom rotating drillstring: simulation, experimental, and field test. J Energy Resour Technol Trans ASME. 2016;138(2):1–9.
33. Vitter JS. Random sampling with a reservoir. ACM Trans Math Softw. 1985;11(1):37–57.
34. Vitter JS. Random sampling with a reservoir. ACM Trans Math Softw. 1985;11(1):37–57.
35. Maybeck PS. Stochastic models, estimation and control. New York: Academic Press Inc; 1979.
36. Vanicek P, Omerbasic M. Does a navigation algorithm have to use a Kalman filter? Can Aeronaut Space J. 1999;45(3):292–6.
37. Qilong X, Henry L, Ruihe W, Baolin L. Continuous real-time measurement of drilling trajectory with new state-space models of Kalman filter. IEEE Trans Instrum Meas. 2016;65(1):144–54.

Chapter 4
Analysis of Vibration Noise on the Fiber-Optic Gyroscope

Lu Wang

Abstract The random noise of the sensor is the main cause of the error in navigation accuracy. In addition, strong downhole vibration will generate greater random noise of fiber-optic gyroscope and accelerometer during MWD operation. The random noise in vibration features concerning time series mutation, slowness and periodicity in its varying. The results show a wide internal noise band and the changing of the noise over time. Considering all these features, this section explores with the dynamic Allan variance method the dynamic characteristics of the random noise produced by fiber-optic gyroscope and accelerometer in vibration, to offer theoretical guidance for improving the environmental adaptability of sensors in vibration and offer theoretical support for noise modeling.

Keywords Allan variance · Navigation accuracy · Fiber-optic gyroscope · Dynamic characteristics

4.1 Allan Variance

In statistics, mean value and variance value are two classical parameters in describing random variables. Traditional variance method is adopted in quantitative characterization of frequency stability in early days. While in 1966, when the scholar D.W. Allan analyzed the frequency stability of cesium atom frequency standard, called Allan Variance (AVAR, Allan Variance) [1]. Allan Variance is recommended by IEEE (IEEE, Institute of Electrical and Electronics Engineers) as a time-domain method for frequency stability analysis. Also, Allan Variance is widely used in random error modeling of those inertial devices with the characteristic of the oscillators. In 1995, Allan Variance is also adopted in the modeling analysis of laser gyroscope in *IEEE Standard 647-1995 specification format guide and test procedure for single-Axis Laser Gyros*.

Contributions by Lu Wang and Qilong Xue, China University of Geosciences (Beijing).

Q. Xue, *Data Analytics for Drilling Engineering*, Information Fusion and Data Science, https://doi.org/10.1007/978-3-030-34035-3_4

At present, the method of Allan Variance has been a crucial means to observe and evaluate various errors and noise characteristic of the inertial devices. Allan Variance is also used to identify various random errors and statistical characteristics, and to calculate the coefficients in the characterization of various random errors. Noise sources in data can thus be identified.

4.1.1 Allan Variance Theory

Suppose sample data *N*, and the sampling interval τ_0. Group the sample data according to the time interval τ_0, $2\tau_0$, … $k\tau_0$ (k < N/2). Data are grouped due to different time intervals. For each group, quadratic sum of the difference between the mean value of the adjacent data set is called Allan Variance. Different groups have its different Allan Variance. Thus, Allan Variance is the function of group time interval $k\tau_0$. Allan Variance can be expressed as follows:

$$\sigma_\omega^2(\tau) = \frac{1}{2}\left\langle \left(\overline{\omega}(t+\tau) - \overline{\omega}(t)\right)^2 \right\rangle \tag{4.1}$$

Where τ is the varying group time intervals, also called observation time. $\langle\rangle$ is total time averaging. $\overline{\omega}(t)$ is the mean value of group at time t. It can be given by:

$$\overline{\omega}(t) = \frac{1}{\tau}\int_t^{t+\tau} \omega(u)du \tag{4.2}$$

u is integral variable. The formula (3.1) is the definition of Allan Variance for continuous data. The formula of Allan Variance for discrete data is expressed in (3.3).

$$\sigma_\omega^2(k) = \frac{1}{2k^2\tau_0^2}\frac{1}{N-2k}\sum_{n=0}^{N-2k-1}\left(\overline{\omega}[n+k] - \overline{\omega}[n]\right)^2) \tag{4.3a}$$

$$\overline{\omega}[n] = \frac{1}{k}\sum_{i=n}^{n+k\text{-}1}\omega[i] \tag{4.3b}$$

In the reference Ng, L.C. [2], the study shows that Allan Variance is related with power spectral density (PSD, power spectral density) of the noise in primary data. The relationship between Allan Variance and PSD $S_\Omega(f)$ can be expressed as:

$$\sigma^2(\tau) = 4\int_0^{+\infty} S_\Omega(f)\frac{\sin^4(\pi f\tau)}{(\pi f\tau)^2}df \tag{4.4}$$

Here, $S_\Omega(f)$ is the PSD of random process $\Omega(t)$, is the integral time

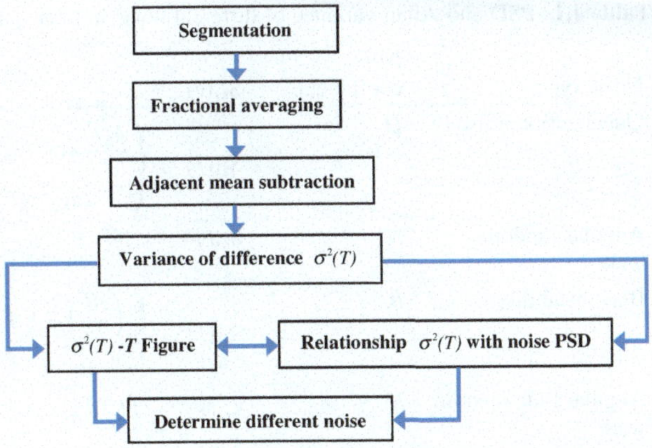

Fig. 4.1 Flow chart of allan variance calculation

Equation (4.4) indicates that Allan variance [3] of the random process can be obtained through integral of PSD in random process. It should be noted that Allan variance is proportional to the noise energy in the random process when passing through the filter, transfer function can be expressed as $\frac{\sin^4 u}{u^2}$. This specific transfer function is obtained through the method of array generation and operation. That is, the frequency band of filter is determined by τ. That is to say, different types of random process can be verified by adjusting the frequency band of the filter, or rather to say, different τ can be used to verify different types of random process [4] (Fig. 4.1)

Analysis of Allan variance for random process is inputting the random signal into a series of filter with different frequency bandwidth τ, then the output of a group of random noise can be obtained. It is Allan variance for the noise of a certain frequency band $\sigma^2(\tau)$. Thus, the different noise in random process can be identified and expressed quantitatively by quantitative Eq. (4.4) of the relation between Allan variance and noise PSD. Generally, log-log curve $\sigma^2(\tau) - \tau$ should be drawn for visually display.

Generally, Noise of fiber optic gyroscope can be divided into five types: the quantization noise-Q, angular random walk-N, bias instability-B, rate random walk-K, the rate slope-R. PSD corresponding to different noise is shown in Table 4.1. Different noise will display in different observation time.

If the various noise items are mutually independent of each other, its Allan variance can be expressed as Allan variance sum of all types of noise, as shown in the Eq. (4.5).

$$\begin{aligned}\sigma^2(\tau) &= \sigma_Q^2(\tau) + \sigma_N^2(\tau) + \sigma_B^2(\tau) + \sigma_K^2(\tau) + \sigma_R^2(\tau) \\ &= \frac{3Q^2}{\tau^2} + \frac{N^2}{\tau} + \frac{2B^2}{\pi}\ln 2 + \frac{K^2\tau}{3} + \frac{R^2\tau^2}{2}\end{aligned} \tag{4.5}$$

Table 4.1 PSD and Allan variance of different noise in fiber optic

Noise type	Noise coefficient	$S_\Omega(v)$	$\sigma^2(\tau)$	Slope
Quantization noise	Q	$S_\Omega(f)=\begin{cases}\frac{4Q^2}{\tau}\sin^2(\pi f\tau) & f\geq\frac{1}{2\tau}\\(2\pi f)^2\tau Q^2 & f<\frac{1}{2\tau}\end{cases}$	$\frac{3Q^2}{\tau^2}$	−1
Angular random walk	N	$S_\Omega(f) = N^2$	$\frac{N^2}{\tau}$	−1/2
Bias instability	B	$S_\Omega(f)=\begin{cases}\left(\frac{B^2}{2\pi}\right)\frac{1}{f} & f\leq f_0\\0 & f>f_0\end{cases}$	$\frac{2B^2}{\pi}\ln 2$	0
Angular Rate random walk	K	$S_\Omega(f) = \left(\frac{K^2}{2\pi}\right)\frac{1}{f^2}$	$\frac{K^2\tau}{3}$	1/2
Rate slope	R	$S_\Omega(f) = \frac{R^2}{(2\pi f)^3}$	$\frac{R^2\tau^2}{2}$	1

Where, $\sigma_Q^2(\tau), \sigma_N^2(\tau), \sigma_B^2(\tau), \sigma_K^2(\tau), \sigma_R^2(\tau)$ are variances of the five different noise items. When the analysis of random noise produced by fiber optic gyroscope and accelerometer is carried out, the unit of data for fiber optic gyroscope is (°/h).

4.1.2 *Gyroscope Noise Analysis*

1. Angle (Velocity) Random Walk

The noise source is caused by unstable excitation source or external interference, such as electron-thermal noise, shot noise and mechanical vibration, resulting from integral of broad-band noise. Other high frequency noise with less correlation time than the sampling time can also be produced by angle (velocity) random walk, which always exists in the output of the light source. But most of these noise source can be eliminated by designing. The angular random walk noise of gyroscope (or accelerometer rate) increases slowly with time with its bandwidth less than 10 Hz. Therefore, if angle (velocity) random walk cannot be accurately determined within the bandwidth range of the most navigation system, it will become the main noise limiting the performance of the navigation system. When the noise amplitude is N, the noise power spectral density is [15]:

$$S_\Omega(f) = N^2 \tag{4.6}$$

By substituting the above equation into the Eq. (4.4), Allan Variance of angle (velocity) random walk is:

$$\sigma_N{}^2(\tau) = \frac{N^2}{\tau} \tag{4.7}$$

Meaning: the slope of angle (velocity) random walk coefficient in $\sigma_N(\tau)$ for log-log curve τ is $-1/2$. The value of angle (velocity) random walk coefficient N can be estimated when $\tau = 1$ [15].

2. Bias Instability

Bias instability mainly refers to low-frequency null-shift jitter, mainly reflected in $1/f$ noise, which is common in resonators. $1/f$ noise is related not only to the random fluctuation of the conductance of the contact points between electronic components, but also to the quality factor Q of the resonator. A mass of experiments demonstrated that the power spectrum of noise mainly comes from discharge components, plasma discharge, circuit noise, environmental noise or other random flicker sensitive items. The rate power spectral density of the noise is [5]:

$$S_\Omega(f) = \begin{cases} \left(\frac{B^2}{2\pi}\right)\frac{1}{f}, f \le f_0 \\ 0, f > f_0 \end{cases} \tag{4.8}$$

In the equation, B is bias instability coefficient, f_0 is cut-off frequency. By substituting the above equation into the Eq. (4.4), it can be obtained:

$$\sigma_B{}^2(\tau) = \frac{2B^2}{\pi}\left[\ln 2 - \frac{\sin^3 x}{2x^2}(\sin x + 4x\cos x) + Ci(2x) - Ci(4x)\right] \tag{4.9}$$

When $x = \pi f_0 \tau$, in the $\sigma_B(\tau)$ for log-log curve τ, the bias instability coefficient B can be estimated from the straight part of the curve [14].

3. Rate (Acceleration) Random Walk

Rate (acceleration) random walk is a result of the integral of the power spectral density for the broadband angular acceleration rate, along with long-term effect of the resonator. Still the source of a random process has not been determined yet, which may be due to the limit case of exponential correlation noise with long correlation time, or to the aging effect of crystal oscillator. The power spectral density of burst noise at high frequency can approximate to rate (acceleration) random walk. The power spectral density of the noise is [14]:

$$S_\Omega(f) = \left(\frac{K}{2\pi}\right)^2 \frac{1}{f^2} \tag{4.10}$$

In the above equation, K is the coefficient of rate (acceleration) random walk. By substituting the above equation into the Eq. (4.4) and integrating, it can be obtained:

$$\sigma_K{}^2(\tau) = \frac{K^2\tau}{3} \tag{4.11}$$

The above equation shows that, in the $\sigma_K(\tau)$ for log-log curve τ, the slope of rate (acceleration) random walk noise is 1/2, and the coefficient of rate (acceleration) random walk K can be estimated when τ=3.

4. Rate (Acceleration) Drift Ramp

For extremely long but limited time intervals, the noise is more like a deterministic noise than a random noise. The noise in the data will show a rather slow monotonic change in the intensity of the light source for long period. Such change may also be due to a minimal acceleration of the platform in the same direction over a long period of time. The rate power spectral density of the noise is [14]:

$$S_\Omega(f) = \frac{R^2}{(2\pi f)^3} \tag{4.12}$$

In above equation, R is coefficient of the rate (acceleration) drift ramp. By substituting the above equation into the Eq. (4.4) and integrating, it can be obtained:

$$\sigma_R{}^2(\tau) = \frac{R^2\tau^2}{2} \tag{4.13}$$

In the equation, in $\sigma_R(\tau)$ for log-log curve τ, the slope of rate (acceleration) drift ramp noise is 1, and the coefficient of rate (acceleration) drift ramp R can be estimated when $\tau=\sqrt{2}$ [14]

5. Quantization Noise

Quantization noise is caused by quantization and digital output characteristic of the sensor during the output, which reflects the minimum resolution of the sensor. It is caused by A/D conversion during sampling. The size of quantization noise depends on the data collection system accuracy and the initial sampling time. Due to a very short correlation time (equals to a very wide bandwidth), quantization noise will cause a great impact in the environment application where fast sampling speed is required. And measures shall be taken to suppress. In order to reduce the noise, the accuracy of the collection system can be improved and the initial sampling time can be shortened. Also, a filter can be used in the system for filtering. The rate power spectral density is [15]:

$$S_\Omega(f) = \begin{cases} \dfrac{4Q^2}{\tau_0}\sin^2(\pi f\tau_0) \\ \approx (2\pi f)^2\tau_0 Q^2, f < \dfrac{1}{2\tau_0} \end{cases} \tag{4.14}$$

In the equation, Q is coefficient of quantization noise. By substituting the equation into the Eq. (4.4), it can be obtained:

$$\sigma_Q{}^2(\tau) = \frac{3Q^2}{\tau^2} \tag{4.15}$$

The equation shows that: in $\sigma_Q(\tau)$ for log-log curve τ, the slope of quantization noise is -1, and the coefficient of quantization noise Q can be estimated when $\tau = \sqrt{3}$ [15].

By Allan variance, the effectiveness of time series filtering method can be quantitatively analyzed by analyzing and comparing the five noise source coefficients given above before and after filtering.

4.2 Dynamic Allan Variance Method

Allan variance method is the main tool for static test data analysis, which can effectively identify and analyze the noise characteristics of the static data. However, Allan variance method is usually based on stationary signals and cannot characterize the noise under non-stationary signals. While the noise of fiber optic gyroscope and accelerometer operated in dynamic environment is often unstable. Therefore, the dynamic Allan variance method is proposed to characterize the noise in dynamic data.

In 2003, dynamic Allan variance (DAVAR) was proposed by L.Galleani and P. Tavella on the basis of Allan variance to analyze the time-domain stability of atomic clock signal [6]. It is a new method of tracking and describing dynamic characteristics of time series, which can be used to reveal the stability changes of atomic clocks affected by humidity, temperature, radiation, sudden failure, aging and other factors [7–9]. In 2007, DAVAR was first used in Galileo's test satellites, GIOVE-A and GIOVE-B [10–12]. In 2009, L.Galleani studied the fast algorithm of DAVAR application in atomic clock signal analysis [13, 14]. In 2010–2011, he made another study of DAVAR confidence and detection sensitivity [15, 16], and then applied DAVAR in the evaluation of the time-domain stability under the case of atomic clock data loss in 2013 [17]. While in China, Wei Guo and other people applied DAVAR in the evaluation of random error for mechanically dithered RLG in 2008 [18]. While Li Ying and other people applied it into the analysis of static test data of fiber optic gyroscope [19]. Li Xuyou, Zhang Na et al. made an analysis of signal swing of marine fiber optic gyroscope, and studied the impact of window function on DAVAR [20, 21]. However, there is still no complete analysis of DAVAR correctness, effectivity and practicability, and no quantitative two-dimensional display of the coefficient dynamic characteristics for FOG noise.

This chapter gives a brief introduction of DAVAR algorithm and fast DAVAR algorithm is also mentioned. Simulation data is used to verify the algorithm. DAVAR not only quantitatively reflects the non-stationary characteristics of the output data of fiber optic gyroscope and accelerometer, but also presents quantitative two-dimensional display of the dynamic characteristics for each noise coefficient.

4.2.1 Basic Principle of Dynamic Allan Variance

Allan variance can describe the characteristics of various errors and noise by signal processing, but it cannot capture various non-stationary factors in signals. Dynamic Allan variance is an improvement and extension of Allan variance. Allan variance of signals can be repeatedly estimated in various time periods by integrating the information of time and Allan variance. Finally, it is expressed in the form of three-dimensional graph, from which the change of statistical characteristics over time for signal noise can be seen clearly, thus making up for the deficiency in Allan variance method.

DAVAR analyzed the signals as follows [22]:

1. Fixed-analyze time point t_1;
2. Definition-the window function $P_L(t)$ with a window length of L, use $P_L(t)$ to cut off the signal $\omega(t)$ on the time axis. t_1 is set as the center point of the window and if the condition $t - L/2 \leq t_1 \leq t + L/2$ is satisfied, the signal $y(t, t_1)$ obtained through cut-off of the window function can be expressed as follows:

$$y(t, t_1) = \omega(t)\mathrm{P_L}(t - t_1) \tag{4.16}$$

3. Assuming that the sum of processing sample is N, sample time is $y(t, t_1)$, composed of N_w continuous data of t_0. Then make Allan variance analysis using cut-off data N_w as samples.
4. Choose another time point t_2 (the cut-off data of window function at the time of t_2 should overlap the cut-off data at the time of t_1), and repeat the steps (2) and (3) to get $\overline{\Omega}_p(\tau) = \frac{1}{\tau}\int_{t_p}^{t_{p+\tau}} \Omega(t)dt$, and so on, $\overline{\Omega}_p(\tau) = \frac{1}{k}\sum_{i=p}^{p+k} \Omega_i(t_0)$ is thus obtained (where $m = 1, 2,\ldots K, K \leq M - m + 1$). Finally, DAVAR is obtained by drawing in the same three-dimensional graph in chronological order. In general, dynamic Allan deviation (dynamic Allan deviation, DADEV) $\sigma(t_m, \tau)$ is used to represent in the three-dimensional graph.

The above procedure of signal processing with Allan variance, it can be obtained [23]:

$$\sigma_\omega^2(t, \tau) = \frac{1}{2\tau^2(N_w - 2\tau)} \int_{t-\frac{N_w}{2}+\tau}^{t+\frac{N_w}{2}-\tau} (\varpi(u + \tau) - \varpi(u))^2 du \tag{4.17}$$

Where N_w is cut-off window length, t is analysis time point, τ is observation time interval. $\sigma_\omega^2(t, \tau)$ is the value of DAVAR. u is integral variable. In the above equation, t and τ has time-continuous relation. The discrete time series are:

$$u = m\tau_0, \left(m = \frac{N_W}{2}, 2, \cdots N - \frac{N_W}{2}\right), \tau = k\tau_0, \left(k = 1, 2, \cdots \frac{N_W}{2}\right) \tag{4.18}$$

By substituting the Eq. (4.18) in the (4.17), DAVAR of the discrete time series can be obtained:

$$\sigma_\omega^2[n,k] = \frac{1}{2k^2\tau_0^2}\frac{1}{N_w - 2k} \times \sum_{m=n-N_w/2}^{n+N_w/2-2k-1} (\varpi[m+k] - \varpi[k])^2 \tag{4.19}$$

τ_0 is sample interval, $k\tau_0$ is observation interval, N is the total sample, N_w is the cut-off window length (Assuming N_w is even).

Summing up the above analysis process, by moving time analysis points with certain principles, and based on those points, appropriate window functions are adopted to cut-off the signals. Then calculate the Allan variance or standard deviation, and draw it into the same three-dimensional graph in chronological order. By observing the change of Allan variance curve or standard deviation curve with analysis time point, the dynamic characteristic of the signal can be quantitatively reflected.

4.2.2 *Two-Dimensional Display of the Noise Based on Dynamic Allan Variance*

Ng, L.C. [24] shows a uniquely determined relation between Allan variance $\sigma_\omega^2(\tau)$ and power spectral density (power spectral density, PSD) of fixed noise, as shown in the following equation:

$$\sigma_\omega^2(\tau) = 4\int_0^\infty S_\omega(f)\frac{\sin^4(\pi f\tau)}{(\pi f\tau)^2}du \tag{4.20}$$

Where, $S_\omega(f)$ is PSD of random process $\omega(t)$, any random process of the physical meanings can be substituted into the above equation, thus obtaining the related function between Allan variance of random process $\sigma_\omega^2(\tau)$ and the time τ [25]. As we know, there are five types of noise in a gyroscope: the quantization noise (*Q*), angular random walk (*N*), bias instability (*B*), rate random walk (*K*), the rate slope (*R*). Equation (4.4) shows that Allan variance is proportional to the total power of the random process after passing the filter with the transfer function of $\frac{\sin^4(\pi f\tau)}{(\pi f\tau)^2}$. It also means that different types of random noise can be filtered by changing τ. PSD of five types of noise and its corresponding Allan variance are given in Table 5.

Different noise will appear in different time zones so as to separate different noise in gyro data. Supposing that each noise is statistically independent of each other,

Allan variance can be expressed as the sum of Allan variance of different noise, as follows:

$$\begin{aligned} \sigma^2(\tau) &= \sigma_Q^2(\tau_Q) + \sigma_N^2(\tau_N) + \sigma_B^2(\tau_B) + \sigma_K^2(\tau_K) + \sigma_R^2(\tau_R) \\ &= \frac{3Q^2}{\tau_Q{}^2} + \frac{N^2}{\tau_N} + \frac{2B^2}{\pi}\ln 2 + \frac{K^2\tau_K}{3} + \frac{R^2\tau_R{}^2}{2} \end{aligned} \tag{4.21}$$

Where, represent Allan variance of five types of noise respectively. Then the noise coefficients of the five types of noise can be obtained through the method of curve fitting. When the gyro data unit is °/h and the noise coefficient is converted to common unit, the coefficients of the five types of noise can be expressed as follows:

$$\begin{aligned} N &= \frac{\sqrt{C_{-1}}}{60}\left({}^\circ/h^{\frac{1}{2}}\right) \\ K &= 60\sqrt{3C_1}\left({}^\circ/h^{\frac{3}{2}}\right) \\ B &= \frac{\sqrt{C_0}}{0.664}\left({}^\circ/h\right) \\ Q &= \frac{10^6\pi\sqrt{C_{-2}}}{180\times 3600\times\sqrt{3}}('') \\ R &= 3600\sqrt{2C_2}\left({}^\circ/h^2\right) \end{aligned} \tag{4.22}$$

DAVAR is a slip Allan variance analysis method. At each analysis time point, Allan variance can be obtained. Least square fitting of this Allan variance, and noise characteristics of cut-off data corresponding to each analysis time point can be obtained. Hence, at each time point t, it can be obtained as follows:

$$\begin{aligned} \sigma^2(t) &= \sigma_Q^2(t,\tau_Q) + \sigma_N^2(t,\tau_N) + \sigma_B^2(t) + \sigma_K^2(t,\tau_K) + \sigma_R^2(t,\tau_R) \\ &= \frac{3Q^2}{\tau_Q^2} + \frac{N^2}{\tau_N^2} + \frac{2B^2}{\pi}\ln 2 + \frac{K^2\tau_K}{3} + \frac{R^2\tau_R^2}{2} \end{aligned} \tag{4.23}$$

By presenting the two-dimensional display in chronological order, the change of each noise coefficient over time can be obtained, thus quantitatively reflecting the dynamic characteristics of the signal noise over time. Therefore, dynamic Allan variance method can be applied to achieve the quantitative two-dimensional description of random noise. The algorithm flow chart of dynamic Allan variance is shown below (Fig. 4.2).

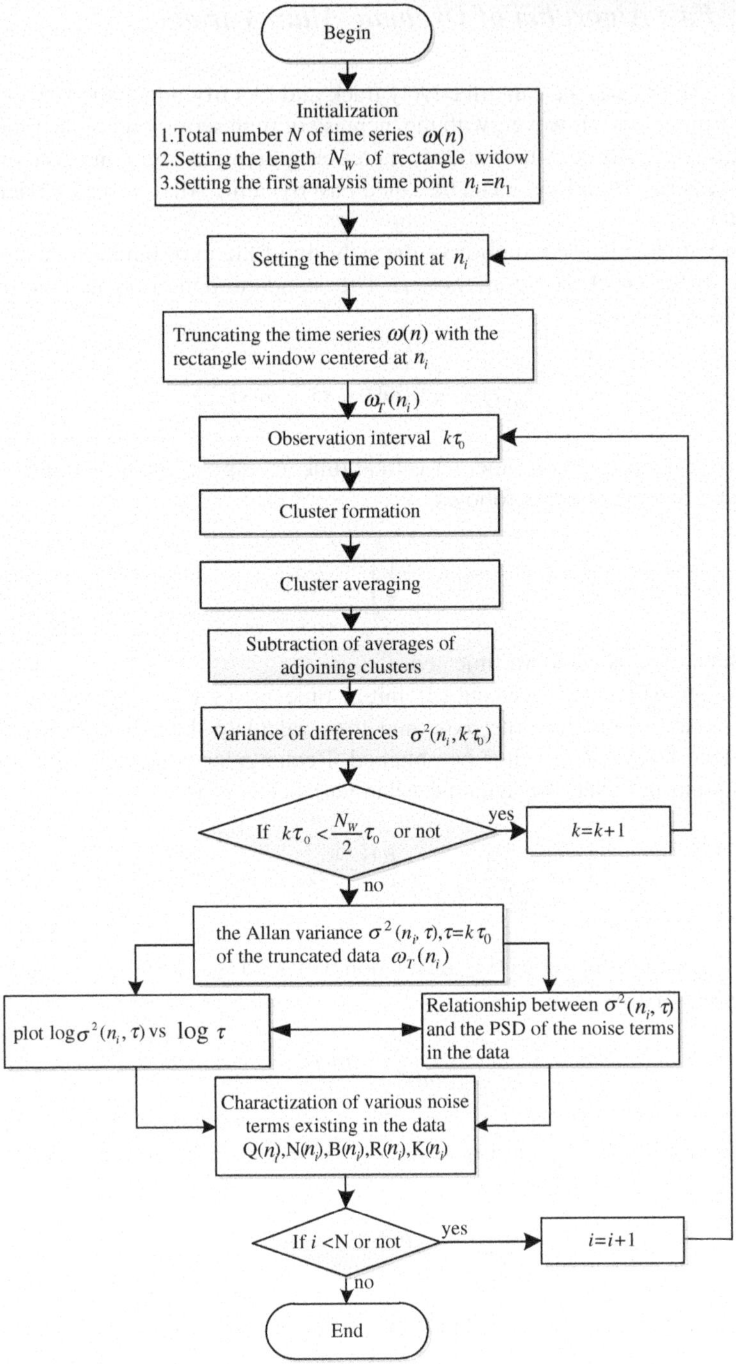

Fig. 4.2 The algorithm flow chart of DAVAR

4.2.3 Fast Algorithm of Dynamic Allan Variance

Dynamic Allan variance can effectively track and identify non-stationary characteristics in time series. However, with the increase of time series length, the time for the calculation of dynamic Allan variance increases greatly, resulting in waste of time in data processing. Therefore, it can be called Fast dynamic Allan variance briefly (Fast DAVAR).

Allan variance is used to observe the stability of the gyro output time series $w(t)$ through different observation time τ. The standard definition equation can be expressed as follows:

$$\sigma_{\omega}^{2}(\tau) = \frac{1}{2}\left\langle (\overline{\omega}(t+\tau) - \overline{\omega}(t))^{2} \right\rangle \tag{4.24}$$

Where, τ is observation time, $\langle\ \rangle$ is total time averaging, the total time averaging of $\omega(t)$ can be expressed as follows:

$$\overline{\omega}(t) = \frac{1}{\tau}\int_{t}^{t+\tau} \omega(t')dt' \tag{4.25}$$

In which, ω(t) is standard time series.

Based on Allan variance, cut-off initial time series by using slip rectangular window function on the time axis, and then calculate the Allan variance of the cut-off data, DAVAR can thus be obtained. Rectangular window length is T, time analysis point is t, then the dynamic Allan variance is $\sigma_{\omega}^{2}(t,\tau)$.

$$\sigma_{\omega}^{2}(t,\tau) = \frac{1}{2(T-2\tau)} \times \int_{t-\frac{T}{2}}^{t+\frac{T}{2}-2\tau} (\overline{\omega}(t'+\tau) - \overline{\omega}(t'))^{2} dt' \tag{4.26}$$

Define another time series $\theta(t)$, its relation with $\omega(t)$ is as follows, $\theta(t)$ and $\omega(t)$ are both random process:

$$\omega(t) = \frac{d\theta(t)}{dt} \tag{4.27}$$

Combining Eq. (4.26) and Eq. (4.27), DAVAR can be rewritten as follows:

$$\sigma_{\omega}^{2}(t,\tau) = \frac{1}{2(T-2\tau)\tau^{2}} \times \int_{t-\frac{T}{2}}^{t+\frac{T}{2}-2\tau} (\theta(t'+2\tau) - 2\theta(t'+\tau) + \theta(t'))^{2} dt' \tag{4.28}$$

For discrete time series $\theta(n)$, DAVAR can be expressed as the Eq. (4.28).

$$\sigma_{\omega}^{2}[n,k]=\frac{1}{2k^{2}\tau_{0}^{2}}\frac{1}{N_{w}-2k}\times\sum_{m=n-N_{w}/2}^{n+N_{w}/2-2k-1}(\theta[m+2k]-2\theta[m+k]+\theta[m])^{2} \quad (4.29)$$

Wherein, τ_0 is sampling interval, $k = \tau/\tau_0$ is observation interval, N is the total sample of time series $\theta(n)$, N_w is analyzed width in rectangular window. At the time of t, when $k = 1, 2, \cdots, \frac{N}{2} - 1$, dynamic Allan variance of the discrete data can be obtained. Suppose N is the even number.

Through a further study on the calculation process of DAVAR, it can be found that DAVAR has recursive nature. With the unique characteristic, fast calculation of DAVAR can thus be obtained. In other words, DAVAR at the time of $n + 1$ can be obtained by the recursive nature of DAVAR at the time of n. Based on the Eq. (4.29), a fast algorithm of DAVAR is deduced.

First, define the second-order difference $\Delta_k[m]$

$$\Delta_k[m]=\theta[m+2k]-2\theta[m+k]+\theta[m] \quad (4.30)$$

Thus, the Eq. (4.29) can also be written as the Eq. (4.31) as follows:

$$\sigma_{\omega}^{2}[n,k]=\frac{1}{2k^{2}\tau_{0}^{2}}\frac{1}{N_{w}-2k}\times\sum_{m=n-N_{w}/2}^{n+N_{w}/2-2k-1}\Delta_k^2[m] \quad (4.31)$$

In the same way, DAVAR can be obtained at the time of $n + 1$:

$$\sigma_{\omega}^{2}[n+1,k]=\frac{1}{2k^{2}\tau_{0}^{2}}\frac{1}{N_{w}-2k}\times\sum_{m=n+1-N_{w}/2}^{n+N_{w}/2-2k}\Delta_k^2[m] \quad (4.32)$$

Expand and recombine the above sum, it can be obtained:

$$\begin{aligned}\sigma_{\omega}^{2}[n+1,k]&=\frac{1}{2k^{2}\tau_{0}^{2}}\frac{1}{N_{w}-2k}\times\sum_{m=n-N_{w}/2}^{n+N_{w}/2-2k-1}\Delta_k^2[m]\\&+\frac{1}{2k^{2}\tau_{0}^{2}}\frac{1}{N_{w}-2k}\times\left(\Delta_k^2[n+N_w/2-2k]-\Delta_k^2[n-N_w/2]\right)\end{aligned} \quad (4.33)$$

Obviously, the first item in the right side of the equation is $\sigma_{\omega}^{2}[n,k]$ at the time of n, so DAVAR at the time of $n + 1$ can be expressed as:

$$\begin{aligned}\sigma_{\omega}^{2}[n+1,k]&=\sigma_{\omega}^{2}[n,k]+\frac{1}{2k^{2}\tau_{0}^{2}}\frac{1}{N_{w}-2k}\\&\times\left(\Delta_k^2[n+N_w/2-2k]-\Delta_k^2[n-N_w/2]\right)\end{aligned} \quad (4.34)$$

Equation (4.34) is the recursive algorithm of DAVAR. So DAVAR at the time of $n + 1$ can be obtained through $\sigma_{\omega}^{2}[n, k]$ at the time of n and calculating two separated second-order difference. That is, when the analysis data increases $\theta[n + N_w/2 - 2k]$, and decreases $\theta[n - N_w/2]$, it needs to increase on the basis of the previous DAVAR, and then increase the second-order difference $\Delta_k^2[n - N_w/2]$, and decrease the second-order difference $\Delta_k^2[n - N_w/2]$.

From the recursive algorithm, an initial value is needed before calculation. If DAVAR is calculated from the time of n_0, the initial value is:

$$\sigma_{\omega}^{2}[n_0, k] = \frac{1}{2k^2\tau_0^2}\frac{1}{N_w - 2k}\sum_{m=n_0-N_w/2}^{n_0+N_w/2-2k-1}\Delta_k^2[\mathrm{m}] \tag{4.35}$$

So when the sample size is N, rectangular window length is N_w, step length is B_w, and observation intervals are $k = 1, 2, \cdots, \frac{N_w}{2} - 1$, using dynamic Allan variance to calculate the times of Allan variance, it can be expressed as S_{DAVAR}:

$$S_{DAVAR} = \frac{N_W}{2} \times \left(\frac{N - N_W}{B_W}\right) \tag{4.36}$$

While dynamic Allan variance based recursive sequence only requires calculating at the initial time Allan variance for $\frac{N_W}{2}$ times and then carrying out addition and subtraction. The flow chart for the calculation of the fast algorithm of dynamic Allan variance is shown as follows (Fig. 4.3):

4.3 Verification of the Fast-Dynamic Allan Variance

Dynamic Allan variance can be used to judge the stability of time series, track the changes of time signal, identify the sudden signal, the signals of periodic changes and slow changes. To prove its effectiveness, two simulated signals with known variation characteristics are generated. Two groups of simulated signals are uncorrelated white noise with zero mean value. The noise model is shown as follows:

$$x[n] = \sigma[n]f[n] \tag{4.37}$$

Where, $f[n]$ is Gaussian white noise and $\sigma[n]$ is the standard deviation of Gaussian white noise. $x[n]$ is the simulation data of fiber optic gyroscope with the unit of °/h. Utilize the above model, two sets of simulation data of fiber optic gyroscope are generated. And $x_1[n]$ is stationary Gaussian white noise with its mean value of 0 and variance of 1. $x_2[n]$ is non-stationary white noise which variance increases with time. The sampling interval of the two groups of simulation data is

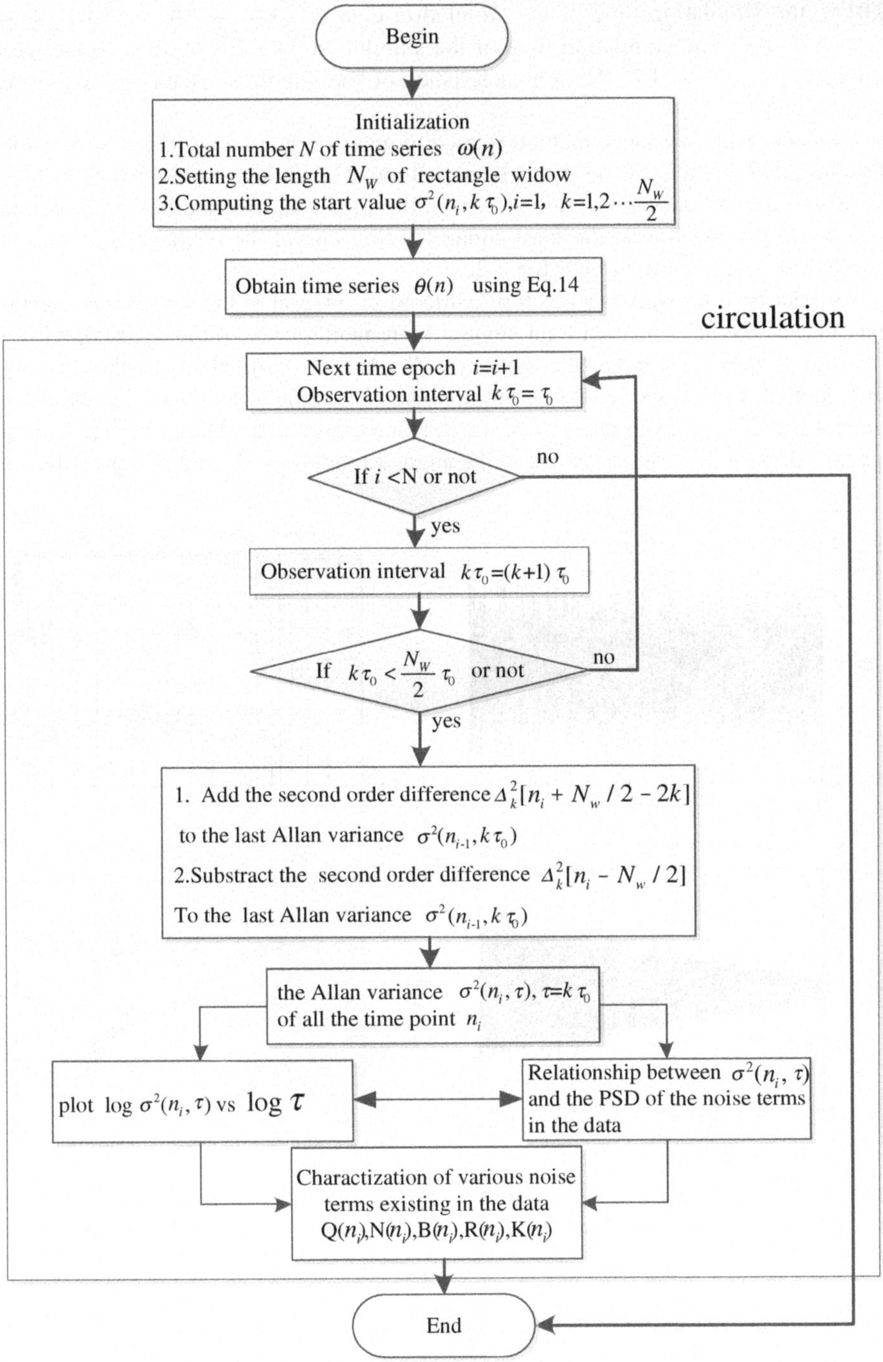

Fig. 4.3 The flow chart of fast algorithm of DAVAR

0.01 s, the simulation time of the simulation data 1 is 60s, so the data length is $L_1 = 6 \times 10^3$. The simulation time of the simulation data 2 is 6000 s, so the data length is $L_2 = 6 \times 10^5$. The characteristics of the simulation data are shown as follows:

Classical Allan variance method is used to analyze the two groups of simulation data, and then fit them by the least square method. The initial simulation data and the corresponding Allan variance are shown in the Fig. 4.4., the blue line is the initial data while the red line is the least squares fitting curve, the characteristics of the simulation data as shown in Table 4.2.

Angular random walk is a result of white noise integral of the broadband angular rate, that is, the accumulated total angular incremental error of the gyroscope from the time of zero is shown as the random walk. And the equivalent angular velocity error at each time is expressed as white noise. The simulation data 1 is stationary white noise. The analysis chart of Allan variance presents the Allan characteristic of specific slope − 1/2 corresponding to the angular random walk, and also presents the

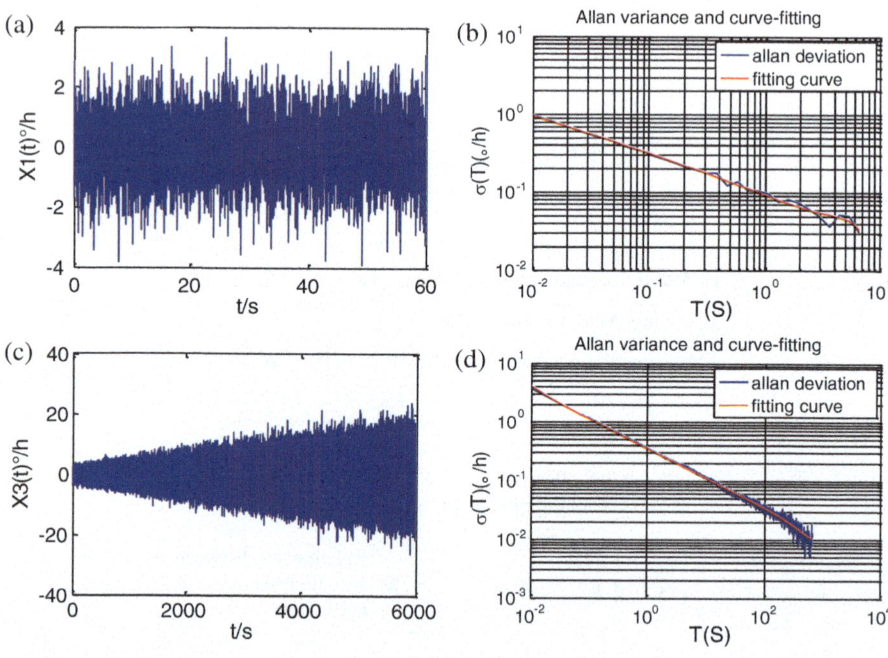

Fig. 4.4 Three types of simulation data and the corresponding Allan variance (**a**) Gaussian white noise $x_1[n]$ (**b**) Allan variance of $x_1[n]$ Gaussian white noise with linear growth $x_2[n]$ (**d**) Allan of $x_2[n]$

Table 4.2 The characteristics of the simulation data

Simulation data	Variance of the noise	Simulation time	Data length
$x_1[n]$	$\sigma[n]_{=1}$	60s	$L_1 = 6 \times 10^3$
$x_2[n]$	$\sigma[n] = 1 + \frac{10}{L_2} \times n$	6000s	$L_2 = 6 \times 10^5$

characterization of the simulation data $x_1[n]$ as white noise. The angular random walk coefficient N of the stationary white noise $x_1[n]$ is $0.00214814^{\circ}/\sqrt{h}$. Allan variance of the simulation data $x_2[n]$ also presents the particular slope of angular random walk, $-1/2$, with the characterization of simulation data as white noise. The angular random walk coefficient N of the stationary white noise $x_2[n]$ is $0.01141631^{\circ}/\sqrt{h}$. The simulation data $x_2[n]$ is not stationary white noise, but it has the same shape with Allan variance chart of the stationary white noise $x_1[n]$, the slope is both $-1/2$. From the analysis of Allan variance, we can only get the noise coefficient of the angular random walk caused by white noise rather than reflect the dynamic changes of the noise.

DAVAR algorithm and fast DAVAR algorithm are used to analyze the above two kinds of simulation data. The analysis results are shown in the following figure. Choose rectangular window as truncation window. The length of the rectangular window for the first group simulation data is 1000 with the step length of 100. The length of the rectangular window for the second group simulation data is 2000 with the step length of 300. The computer model (Intel(R) Core(TM) i7–3770, CPU@3.4GHz), matlab2010a is used. Figure 4.5 (a)(c) are DAVAR results in the analysis of two sets of the simulation data, (b)(d) are fast DAVAR results in the analysis of two sets of the simulation data.

As is shown in Fig. 4.5 (a)(b), DAVAR changes stationary with time. The slope of Allan variance curve at each time point is $-1/2$, indicating of the data as stationary

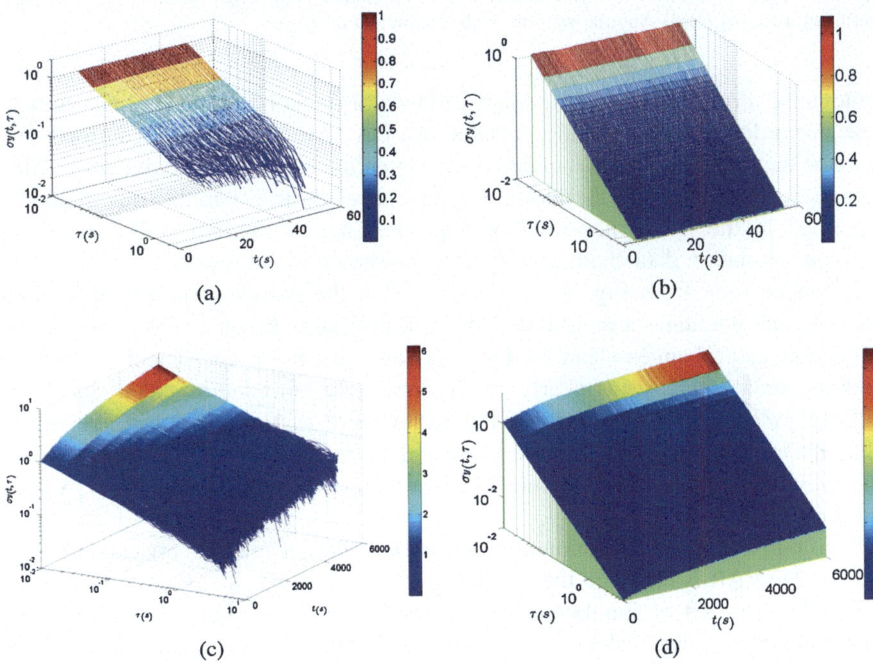

Fig. 4.5 DAVAR of the simulation data and fast DAVAR **(a)** DAVAR of $x_1[n]$ **(b)** Fast DAVAR of $x_1[n]$ **(c)** DAVAR of $x_2[n]$ **(d)** Fast DAVAR of $x_2[n]$

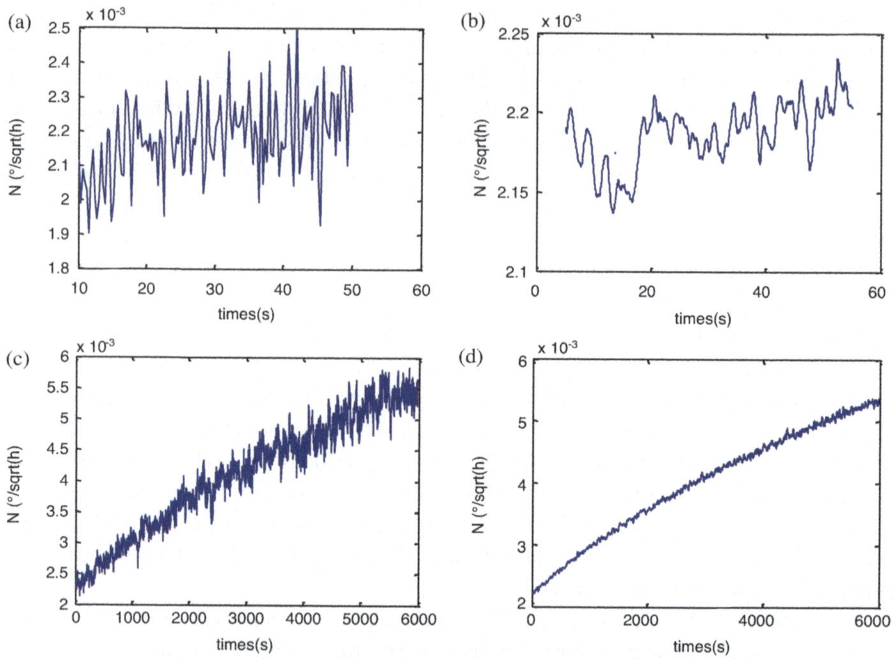

Fig. 4.6 Angular random walk coefficient of the simulation data **(a)(b)** Angular random walk coefficient of $x_1[n]$ **(c)(d)** Angular random walk coefficient of $x_2[n]$

white noise. From Figure (c)(d), Allan variance of the noise increases linearly with time. According to the simulation results, fast DAVAR and DABAR can track and describe the non-stationary changes of the signal over time. By the dynamic Allan variance, the change of the noise coefficient with time can be quantitatively expressed in the two-dimensional graph. The angular random walk coefficient N of the simulation data changes with time, as shown in the figure.

It can be seen from Fig. 4.6 (a) and (b) that the noise coefficient of angular random walk fluctuates around $0.00215^{\circ}/\sqrt{h}$, indicating that the white noise in the simulation data changes steadily. From (c) and (d), noise coefficient of angular random walk increases linearly with time, the minimum value of it is $0.00215^{\circ}/\sqrt{h}$, and the variance of the white noise is 1. As the variance of the white noise increases gradually, the angular random walk coefficient also increases gradually, it is not a constant value. The correctness of DAVAR and fast DAVAR can thus be proved.

The comparison of calculation time and the calculation amount between DAVAR and fast DAVAR is drawn in the following table.

For the same set of simulation data, by using the same window length and the same calculation. The calculation of fast DAVAR takes far less time than the calculation of DAVAR. When the data size is relatively small $6{*}10^3$, DAVAR and fast DAVAR both takes a rather short time for calculation. Hence, fast DAVAR

Table 4.3 Comparison of the calculation amount for DAVAR and fast DAVAR

				Calculation time		Times of Allan variance calculation	
Data	Length	N_W	Step length	DAVAR	Fast DAVAR	DAVAR	Fast DAVAR
$x_1[n]$	6×10^3	1000	30	3.656864	0.770367	166	1
$x_2[n]$	6×10^5	2000	300	960.422362	27.92727	1993	1

saves 78.93% of the time. When the large data size $6*10^5$, DAVAR takes about 16 min while fast DAVAR takes only 27.92727 s, saving about 97.09% of calculation time. The last column in the table shows the times of Allan variance calculated by DAVAR and fast DAVAR. As the increase of the data amount, DAVAR needs more and more times of Allan variance calculation, which takes a lot of calculation time. But fast DAVAR needs to be calculated only once, that is, the initial value of Allan variance. Then DAVAR at each time point can be obtained by performing addition and subtraction, which greatly saves not only the calculation amount but also the calculation time (Table 4.3).

4.4 Downhole Vibration Simulation Test Verification

The fiber optic gyroscope inertial navigation system was places on a vibration table with the vibration frequency of 25 Hz, vibration acceleration of 1 g, vibration time of 2 h and vibration direction of Y axis. Static data were collected for 5 min before and after the vibration.

Taking the data of Y-axis fiber optic gyro and Y-axis accelerometer as an example, fast DAVAR is used for the analysis of the vibration data. The results are shown in the Fig. 4.7 (a) and (b). From the figure, noise change before, during and after the vibration can be clearly drawn. At each moment before vibration, Allan variance of the noise is relatively low, indicating that the noise in the pre-vibration data is relatively low. During vibration, Allan variance of the noise becomes strong, indicating the the noise in the sensor data during vibration gets bigger. After vibration, Allan variance of the noise returns to the pre-vibration level. As can be drawn from DAVAR figure of fiber optic gyroscope, before vibration, the slope of Allan variance is -1, indicating that the quantization noise is the main noise before vibration. During the vibration, the slope of Allan variance is -1 and $-1/2$, indicating that quantization noise and angular random walk are main noise during the vibration. And as can be drawn from DAVAR figure of the accelerometer, before vibration, the slope of Allan variance is $-1/2$, indicating that the rate random walk is the main noise of the accelerometer before vibration. During the vibration, the slope of Allan variance is -1, indicating that the quantization noise is the main noise. We can see from DAVAR figure that fast DAVAR method can track and characterize the non-stationary characteristics of noise under vibration state, and make quantization analysis of the characteristics of the noise under vibration state.

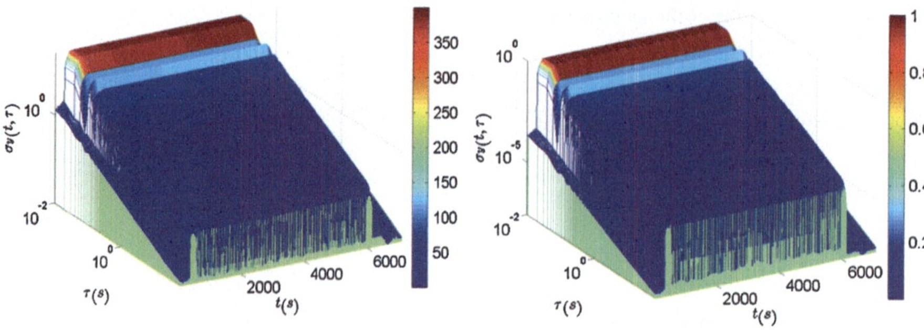

(a) DAVAR of Y-axis fiber optic gyroscope **(b)** DAVAR of Y-axis accelerometer

Fig. 4.7 DAVAR of Y-axis fiber optic and accelerometer

In Fig. 4.7 (a) and (b), quantitative two-dimensional display for each noise coefficient is presented. The value of noise coefficient before, during and after the vibration is listed in the table. It can be seen that five types of noise get stronger during the vibration compared with that before vibration. The quantization noise coefficient of fiber optic gyroscope is 201.4074 times of that before vibration, and angular random walk coefficient is 23.5417 times of that before vibration. Other noise of fiber optic gyroscope increased by about three or four times. The quantization noise of the accelerometer is 70.1951 times of that before vibration. Angular random walk coefficient is 53.9532 times of that before vibration. Other noise of the accelerometer increased by about 14–18 times. The vibration excites the quantization noise and angular random walk in fiber optic gyroscope and the accelerometer. Through the analysis of noise source, the quantization noise and angular random walk noise show high frequency characteristics, which can be filtered by filtering technology. Due to the fixed vibration of the test, the noise coefficient is also in a relatively stable state under the vibration. But the coefficient fluctuates greatly under the state of motion, such as at the beginning or at the end of the vibration. When the motion changes to stable state, the noise coefficient returns to the stable state. So dynamic Allan variance can track and characterize the dynamic characteristics of the noise. The angular random noise in the fiber optic gyroscope comes from the white noise generated by light source, photoelectric detector and the noise from electronic devices. Therefore, it can be concluded that the performance of these optical devices is partially deteriorated in the vibration, such as to reduce the vibration white noise, the anti-vibration performance of optical devices and electronic devices in fiber optic gyroscope needs to be improved.

As can be seen from Fig. 4.8, the five types of noise of gyroscope and the accelerometer changes dramatically when the state of motion changes. When it returns to a stable state, the noise of the gyroscope and the accelerometer becomes stable. When the fiber optic inertial navigation is in a stable fixed vibration of 25hz, the vibration is in a stable motion state. It can be seen from the figure that the three types of the noise of the gyroscope, bias instability noise, angular random walk noise

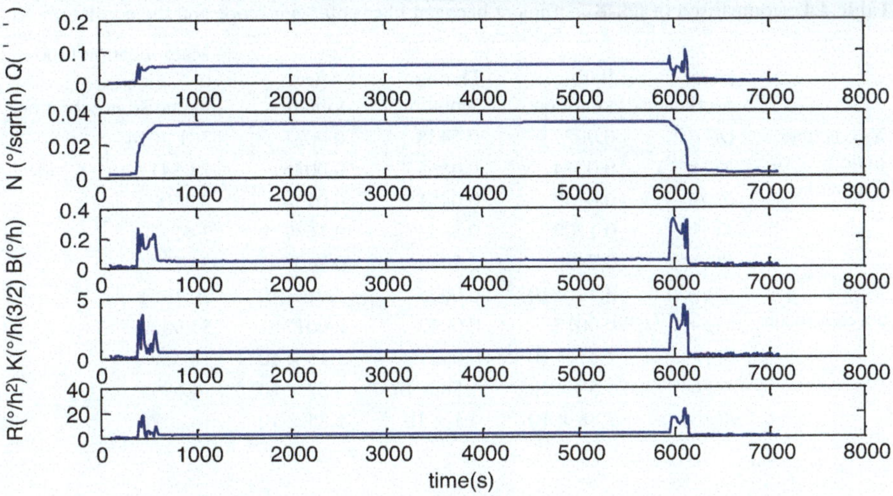

(a) Variation diagram of the noise coefficient in Y-axis fiber optic gyroscope

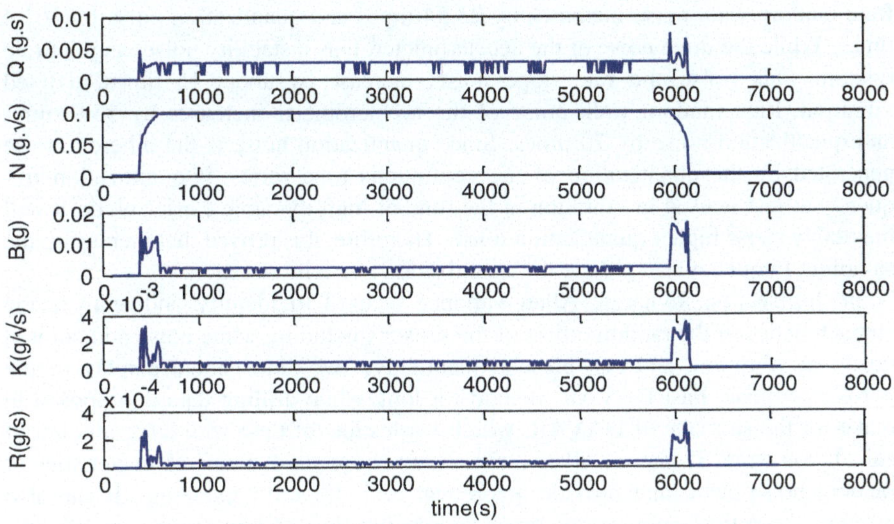

(b) Variation diagram of noise coefficient in Y-axis accelerometer

Fig. 4.8 Variation diagram of noise coefficient

and rate slope noise, are relatively small and stable in fixed vibration. These three-noise items of the accelerometer are also small and stable under stable fixed vibration. But the quantization noise and the angular random walk noise of the gyroscope and the accelerometer will get bigger in fixed vibration. It shows that although the fixed vibration is in a stable state, it can excite quantization noise and white noise of the sensor. Table 8 shows the comparison of noise coefficient between the fiber optic gyroscope and the accelerometer under different states. From the Table 4.4, three noise of the gyroscope, bias instability noise, angular rate random

Table 4.4 Comparison of noise coefficient between fiber optic gyroscope and the accelerometer

	Noise coefficient	Before vibration	During vibration	After vibration	Magnification times of the noise in vibration
Y-axis fiber optic gyroscope	$Q('')$	0.0027	0.5438	0.0031	201.4074
	$N(°/h^{1/2})$	0.0024	0.0565	0.0026	23.5417
	$B(°/h)$	0.0148	0.0481	0.0158	3.2500
	$K(°/h^{3/2})$	0.1708	0.583	0.1637	3.4133
	$R(°/h^2)$	0.7436	2.731	0.7401	3.6727
Y-axis accelerometer	$Q(g \cdot s)$	3.52×10^{-5}	0.0025	1.91×10^{-5}	70.1951
	$N(g \cdot s^{1/2})$	0.0017	0.0943	0.0017	53.9532
	$B(g)$	1.47×10^{-4}	0.0027	1.15×10^{-4}	18.369
	$K(g/s^{1/2})$	3.57×10^{-5}	5.58×10^{-4}	3.57×10^{-5}	14.6998
	$R(g/s)$	2.88×10^{-6}	4.1×10^{-5}	1.85×10^{-6}	14.2578

walk noise and rate slope noise, increase by more than 3 times under fixed vibration. Rate random walk noise increases by 23.54 times and quantization noise by 201.4 times. While the three noise of the accelerometer, bias instability noise, angular rate random walk noise and rate slope noise, increase by about 15 times in fixed vibration. Rate random walk noise of the accelerometer increases by 53.9 times and quantization noise by 70 times. Since quantization noise is the inherent noise generated by the quantization of the continuous time series, if quantization frequency at rest is used in vibration at the time of high dynamic carrier work, it will inevitably cause higher quantization noise. Therefore, it is proved that increasing the sampling frequency can reduce the quantization noise in vibration.

In this section, dynamic Allan variance is used to identify and analyze the characteristics of the random noise of the sensor (including white noise and colored noise) in vibration, and can present quantitative two-dimensional display of the gyroscope noise. Fast DAVAR method for long while drilling data is proposed to make for the shortage of DAVAR, which wastes lots of time with large amount of calculation work. Dynamic Allan variance can show the dynamic characteristics of random noise error, and provide theoretical basis for error modeling. It can also provide theoretical support for error source locating and improvement of anti-vibration performance of the inertial devices.

References

1. Allan DW. Statistics of atomic frequency standards. Proc IEEE. 1966;54(2):221–30.
2. Lawrence CN, Pines DJ. Characterization of ring laser gyro performance using the Allan variance method. J Guid Control Dyn. 1997;20:211–4.
3. Tehrani MM. Ring laser gyro data analysis with cluster sampling technique. Proceeding SPIE 1983, 0412, https://doi.org/10.1117/12.935818.

4. IEEE Standard Specification Format Guide and Test Procedure of Single Axis Interferometric Fiber Optic Gyros; IEEE Std 952-1997; The Institute of Electrical and Electronics Engineers, Inc.: New York, 16 September 1997.
5. Blaser JP, Lockerbie NA, Paik HJ, et al. Eotvos, an inertial instrument for testing the equivalence principle[J]. Adv Space Res. 2003;32(7):1433–6.
6. Galleani L, Tavella P. The characterization of clock behavior with the dynamic allan variance, proceedings of the 2003IEEE international frequency control symposium and PDA exhibition jointly with the 17th European frequency and time Forum 2003, pp. 239–244.
7. Galleani L, Tavella P. Tracking nonstationarities in clock noises using the dynamic Allan variance. Proceeding joint FCS-PTTI meeting August 2005, Vancouver, 2005, pp. 392–396.
8. Galleani L, Tavella E. Interpretation of the dynamic Allan variance of nonstationary clock data. IEEE FCS-EFTF 2007, May 29–June 1 2007, Geneva, pp. 992–997.
9. Nunzi E, Galleani L, Tavella P. Detection of anomalies in the behavior of atomic clocks. IEEE Trans Instrum Meas. 2007;56(2):523–8.
10. Sesia I, Galleani L Tavella P. Implementation of the dynamic Allan variance for the Galileo system test bed V2. Frequency control symposium, 2007 joint with the 21st European frequency and time Forum (2007) 946–949.
11. Sesia I, Galleani L, Tavella P. Implementation of the dynamic allan variance for the galileo system test bed V2. 2007, pp. 946–949.
12. Galleani L, Tavella P. The dynamic Allan variance. IEEE Trans Ultrason Ferroelectr Freq Control. 2009;56(3):450–64.
13. Galleani L, Tavella P. Fast computation of the dynamic Allan variance. Proceeding IEEE FCS-EFTF, 2009, pp. 685–687.
14. Galleani L. The dynamic Allan variance II: a fast computational algorithm. IEEE Trans Ultrason Ferroelectr Freq Control. 2009;57(1):182–8.
15. Galleani L. The dynamic Allan variance III: confidence and detection surfaces. IEEE Trans Ultrason Ferroelectr Freq Control. 2011;58(8):1550–8.
16. Sesia I, Galleani L, Tavella P. Application of the dynamic Allan Varinacne for the characterization of space clock behavior. IEEE Trans Aerosp Electron Syst. 2011;47(2):884–95.
17. Galleani L, Tavella P. Characterization of atomic clock anomalies in the dynamic allan variance domain, 2013 Joint UFFC, EFTF and PFM symposium, pp. 654–648.
18. Guo W, Xinwu L. Research on stochastic errors of dithered ring laser gyroscope based on dynamic Allan variance [J]. Chin J Lasers. 2010;37(12):2975–9.
19. Yin L, Xinlin C, Shenming S. Dynamic allan variance analysis for the drift error of fiber optical gyroscope [J]. J Optoelectron Laser. 2008;19(2):183–6.
20. Xuyou L, Na Z. Analysis of dynamic characteristics of a fiber-optic gyroscope based on dynamic Allan variance [J]. J Harbin Eng Univ. 2011;32(2):183–7.
21. Na Z, Xuyou L. Research on theoretical improvement of dynamic allan variance and its application [J]. Acta Opt Sin. 2011;31(11):1–6.
22. Stein SR. The Allan variance-challenges and opportunities. IEEE Trans Ultrason Ferroelectr Freq Control. 2010;57(2):540–7.
23. Wang L, Zhang C, Lin T, Li X, Wang T. Characterization of a Fiber optic gyroscope in a measurement while drilling system with the dynamic Allan variance. Measurement. 2015;75:263–72.
24. El-Sheimy N, Hou H, Niu X. Analysis and modeling of inertial sensors using Allan variance [J]. IEEE Trans Instrum Meas. January 2008;57(1):140–9.
25. IEEE Standard Specification Format Guide and Test Procedure for Single-Axis Interferometric Fiber Optic Gyros, IEEE Standard 952, 1997, pp. 62–73.

Chapter 5
Measurement and Analysis of Drillstring Dynamics

Abstract We shown that drilling dynamics is a crucial problem must be attached great importance to. Our innovation lies in combining actual measurement with theoretical modeling. Two evaluation methods are compared systematically, such as theoretical and measurement methods. In this chapter, our investigations concern the dynamics of drilling, unveiling a chaotic regime and suggesting practical ways of improving current drilling techniques. The reveal of chaos provides a new way to detect early fatigue cracks as weak signals in a noisy environment to reduce engineering cost and the possibility of disaster. Data from all nine fields in China are used in our studies. Proposes a theoretical model for drilling dynamics, and a real drilling system is developed to validate the theoretical dynamical behaviour. The existence of chaos in drilling may open a new concept of drilling chaos in the solid flow mechanics that will benefit to both the physicists and the drilling engineers.

Keywords Drilling dynamics · Chaotic · Drillstring vibration · Fatigue cracks · Detection

5.1 Introduction

Oil spill disaster as in the Gulf of Mexico has caused great impact on lost assets, health, safety and environment. The cement work has agreed to be the main reason relating to the investigation of the disaster. But what are the factors that affect the quality of cementing?

Drilling dynamics [1, 2] is a crucial problem must be attached great importance to. Many dangerous phenomena related to drilling are normally caused by the dynamics of the drillstring and its interactions with the surroundings. Such as deterioration of the borehole quality caused by the drilling dynamics will directly affect the cementing quality. Most of all, the drilling dynamics will accelerate fatigue damage to the downhole tools and drillstring, which is the most common

---

Contributions by Qilong Xue and Jin Wang.

Q. Xue, *Data Analytics for Drilling Engineering*, Information Fusion and Data Science, https://doi.org/10.1007/978-3-030-34035-3_5

phenomenon in the drilling process. Drillstring failure may occur frequently. Not only does it cost billions of dollars per year, but it also puts the drilling engineering at risk. Downhole vibrations can decrease rate of penetration (ROP), interfere with MWD tools and even cause premature fatigue of the components. The axial, lateral (as transverse or bending)) and torsional vibration modes with different destructive properties are the three main modes of drillstring vibration. Around 73% of inspected drillstrings are defective because of fatigue cracks. In order to improve the drilling efficiency and tools reliability, an effective detection [23] of downhole drillstring dynamics should be done in order to detect the early fatigue cracks. Nowadays, the three-dimensional downhole measurements could provide us the vibration signals supporting the study. However, it is a great challenge like financial crisis forecasting because the potential signals are so weak.

The drillstring dynamics such as vibrations and shocks (V&S) can limit the optimization of drilling performance, which is a key problem for trajectory control, wellbore design, increasing drill tools life, rate of penetration, and intelligent drilling. As a whole drilling system, slenderness ratio of the drillstring is large, and the stiffness is very small, Drillstring V&S is divided into three basic forms such as axial mode, torsional mode, transverse mode. Which the Torsional vibrations are caused due to an irregular rotation of the drillstring when rotated from the surface at constant speed.

Torsional vibration [3] in the drilling process has drawn extensive attention because of its huge danger and unavoidable. It can be seen from the downhole measurement data that applying a constant rotational speed from the surface does not result in a stable rotational motion of the drill bit. In fact, during a large part of the drilling time, the down-hole torsional speed may experience large fluctuations due to the flexibility of the drillstring. Based on lots of theoretical and experimental studies, self-excited vibration of drillstring caused by drill negative damping torque [4, 5], and the nonlinear friction formed between the drillstring and the borehole wall lead to drillstring torsional vibration [6, 7]. A study [8] shows that the axial natural frequency of the drillstring can be derived using a partial differential equation for longitudinal rod vibration. The axial vibration of the drillstring can be defined using discrete mass segments and springs [9] and the frequency response function can be used to demonstrate the similarity of the model to the real drillstring response. The axial vibration problem would more obvious during the extreme-depth wells, velocity changes at the top of the drillstring can produce transient axial pulse excitation. Lubinski [10] investigates the dynamics of the drillstring during tripping movement. Aadnoy et al. [11] studies using the static model of the drillstring and the rotation of the drillstring to reduce the drillstring borehole friction during torsional vibration.

Some studies are dedicated to better understanding the full dynamics of rotary drilling systems [12~16]. Sunit [17] develops the global dynamics of coupled axial torsional vibration and the possibility of bit bounce. Huang et al. [18] establishes a generalized quasi-static model of drillstring system. A unified generalized model is further combined by a forward model and an inversion model. Lian et al. [19] establishes a nonlinear dynamical model to investigate the vibration behavior of drillstring in drilling horizontal wells, the finite element model is also established and analysis of the buckling of the drillstring and the contact with the wellbore. However, none of these studies reports the dynamical motion state of the bottom hole assembly (BHA) in automatic vertical drilling system (VDS) [20], The hydraulically driven top drive on the surface drives the entire drillstring from surface to downhole. The VDS has push-the-bit tools like rotary steerable drilling system (RSS) [21]. It is mounted on the outside of the tool to cause the bit to create a force on the opposite side of the high side of the wellbore during drilling, resulting in a change in the direction of the wellbore trajectory. The pads of the actuator in the VDS or RSS are continuously pushed toward the borehole wall to form a non-linear cyclic damping force, which leads to chaotic vibration of the bottom drilling tool [22].

At present, the exploration and analysis of the vibration mechanism of the drillstring mainly adopts two methods, namely theoretical analysis and numerical simulation method [24–29] and vibration signal measurement method [30–36].

The theoretical analysis and numerical simulation methods for the vibration of the drillstring mainly include the analytical method based on the energy method [26, 27] and the numerical solution method based on the finite element method [28, 29].

When using the analytical method and the numerical method to solve the vibration of the drillstring, the drillstring calculation model is inevitably greatly simplified, so the calculation result often differs greatly from the actual vibration state of the downhole drillstring. For this reason, the downhole drillstring Vibration measurement has become another focus of attention. The work in this area mainly includes the development of downhole vibration measurement tools, the analysis of the influence of drilling parameters on the vibration of the drillstring under the well [30–33] and the vibration excitation mechanism [34]. As early as the middle of the twentieth century, the research on the relevant aspects of the drillstring vibration test has been carried out internationally, and great progress has been made. The world-renowned petroleum technology service companies such as Baker Hughes, Schlumberger and Halliburton have independently developed the reliable drillstring vibration measurement and analysis system [35–37]. Through the vibration signal measurement and analysis under different working conditions of the drillstring, the actual working state of the drillstring can be identified in time, thus effectively preventing the occurrence of drill accidents. The development of vibration measurement technology makes up for the shortcomings of numerical methods in studying the vibration mechanism of drill string, and effectively promotes the development of drill string vibration research [38, 39].

According to past theories, indoor experiments, and field studies, the relationship of ten kinds of V&S is listed, which contains basic forms, modes, frequencies, amplitudes response, and tool damage, Stick slip is the highest, most severe form of torsional vibration. Taken from high speed downhole data, the BHA is coming to a full stop for up to 3 s and peaks at 230 rpm, with a stick slip period of 7.3 s while the top drive is at a constant 92 rpm. The stick slip causing the drill string to periodically be torqued up and then spin free is a phenomenon of nonuniform drill string rotation. The whirl is the eccentric rotation of the BHA & bits and is associated with the BHA rolling around the wellbore. Drillstring lateral shock not only causes harm to drillstring itself but also can cause whirl and serious damage to the wellbore quality.

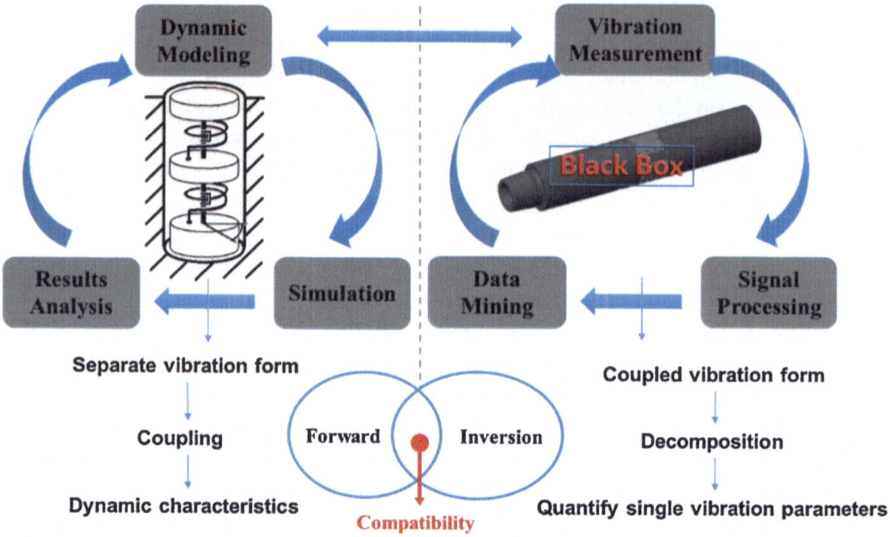

Fig. 5.1 Forward and inversion algorithms of dynamic modeling and vibration measurement data

We should find the reason Why the vibrations are so bad? To drill bit, it will cause Broken cutting structure Failed bearings Uneven wear and so on, to drillstring itself, it will cause Wash outs, Twist offs Tool joint damage Tool failure electrical or mechanical and so on. From the perspective of drilling process, it will Reduce ROP, reduce bit life, reduce efficiency of drilling, also will cause Steering problems Poor hole quality and so on.

In order to clarify the rules of dynamic drillstring in the rotary steerable system, we fully utilize modern computing methods and try to establish a model close to the real drilling environment. According to field measured data correct theory model; complex dynamic problems of drillstring may have a clear understanding. Our innovation lies in combining actual measurement with theoretical modeling. Two evaluation methods are compared systematically, such as theoretical and measurement methods. As show in the Fig. 5.1, in the Dynamic Modeling we should coupling the separate vibration form at the same time decomposition the coupled vibration form in the vibration measurement data. This is like forward and inversion algorithms in geophysics.

5.2 Modeling of Dynamics

As shown in Fig. 5.2 is a typical drilling rig, the torque applied to the surface causes the drillstring to rotate and transfer the rotary motion to the drill bit., it consists of a rotary table at the surface and a series of drillstring which transmits the rotary power

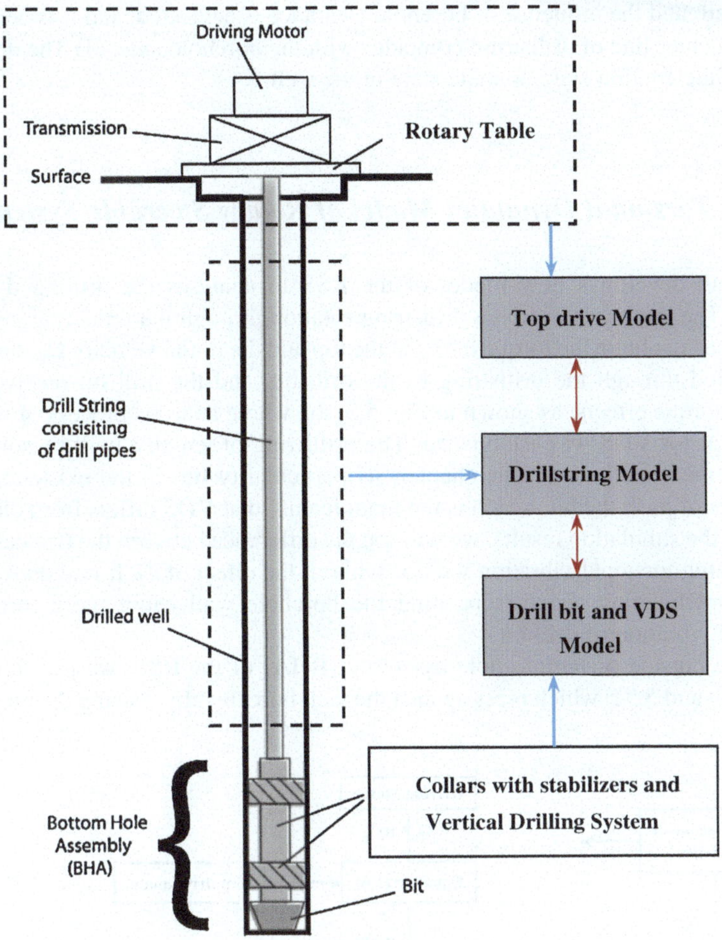

Fig. 5.2 A schematics of a typical drilling rig, contains the most important components: drill bit, drill collars, drillstring, stabilizers, wellbore annulus, lifting system for controlling WOB and electric drive for providing rotary motion to the drill string [17]

required for drilling to the bottom hole assembly (BHA) and drill bit to break the rock.

In the drilling process, the top drive to generate power, relying on the drillstring to transmit torque to the bottom of the drill-bit. The bottom of the drilling tool in the process of rotation generated periodic friction damping in the RSS or VDS, the mathematical model of the dynamics mechanism of VDS can be fully described by the model combination of the entire drilling system, including top drive, dynamics of drillstring, the interaction between the drill bit and the rock being cut, and the depth of cut, excitations produced by push-the-bit in the automatic guidance process. The following assumptions are made: (1) the drillstring is in the range of linear elastic deformation and the cross section is circular; (2) The borehole wall is regarded as a rigid shell, and the influence of borehole clearance is neglected, and it is considered that the center line of drillstring coincides with the borehole axis; (3) The drillstring is in sliding friction state or static state in the well.

5.2.1 Torsional Dynamics Model of Rotary Steerable System

We try to develop a new model of the RSS to analyses the drilling dynamics [2, 40]. The motor rotor driven drillstring rotation through the reduction gear, then the drillstring obtain the torque $n \cdot T_m$ at the top and the initial velocity Ω_0, the torque transmitted through the drillstring to the drill bit, and the drill bit receives three counter-torque effects, as shown in Fig. 5.3, the whole analysis process of torsional oscillation for VDS was established. The drillstring of length l_e will be solving by finite element method. In this chapter, its particularity lies in the existence of the dynamics of push the bit, which is the main feature that VDS differs from other drill tools. In the simulation results, we will see the difference between the two cases, that is drillstring torsional vibration with or without the effect of T_f. It is shown that the VDS implementing agencies pushing the borehole wall cause more serious the torsional vibration of drill bit.

In the outside of bottom hole assembly (BHA) of the RSS, we use three pads (Fig. 5.5b and 5.5c) which press against the well bore thereby causing the bit to press

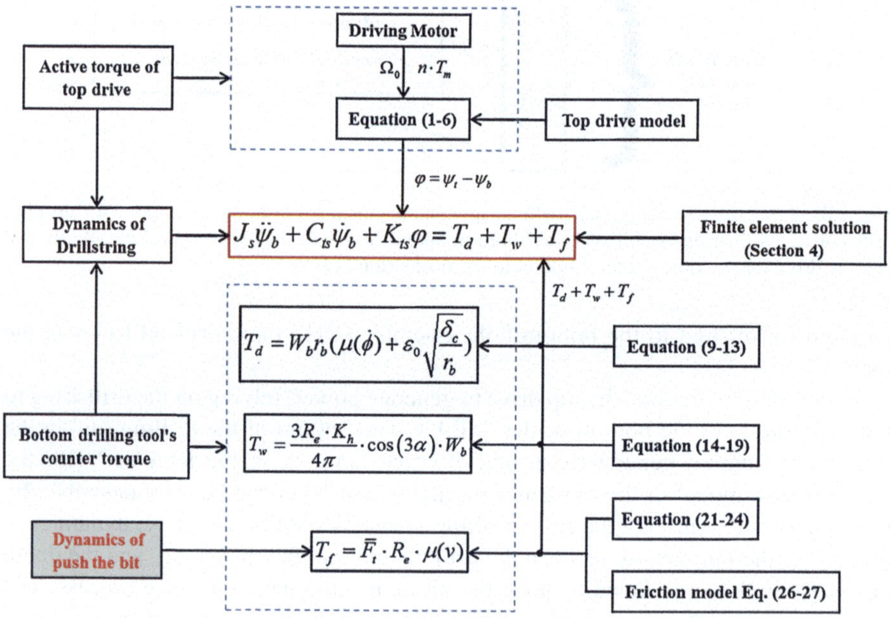

Fig. 5.3 Analysis process of torsional oscillation for VDS or RSS

on the opposite side causing a direction change. The pads of the implementing agency constantly push against the borehole wall, making bottom hole a cycle of nonlinear damping force. The oriented actuator is used to push the pads to steer the drilling trajectory. Power section consists of turbo generator, which is basically driven by drilling fluid. The servo section stabilizes the disk valve at a certain tool face angle, and then drilling fluid moves the pads out by flowing through valves (Fig. 5.5b). Drilling fluid pushes one of the pads to the borehole and produces the steering force (Fig. 5.5c) that can orient the drilling.

5.2.1.1 Modelling of Top Drive

The motor of top drive is connected to a slide rail that is fixed to the derrick, as shown in Fig. 5.4a. The motor can move up and down along the rail but cannot rotate. Therefore, the motor housing can be considered to be stationary for torsional vibration of the drillstring. The motor rotor driven drillstring rotation through the reduction gear. Assume that the gearbox gear ratio is n, J_t is the top drive equivalent moment of inertia, (kg·m^2); θ_t is the rotation angle at the top of the drillstring, (rad); c_2 is equivalent rotational damping of the motor rotor and gear reduction system, (N·m·s/rad); T_m is the electromagnetic torque produced by the rotor, (N·m); T_{st} is the torque acting on the top of the drillstring to the top drive, (N·m);

The following equation can then be obtained:

$$J_t \cdot \ddot{\theta}_t + c_2 \cdot \dot{\theta}_t = n \cdot T_m - T_{st} \tag{5.1}$$

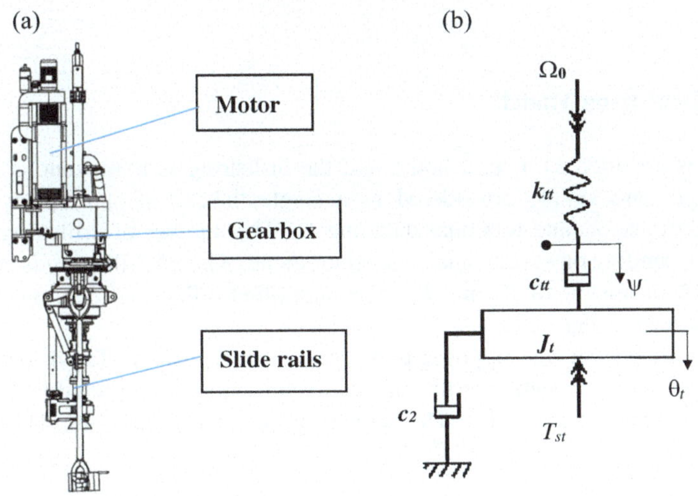

Fig. 5.4 The equivalent mechanical model of top drive

As shown in Fig. 5.4b, suppose that the torque transmitted by the equivalent torsion spring and equivalent damper is $n{\cdot}T_m$, Ω_0 is the rotational speed of motor, $\boldsymbol{K_{tt}}$ is the spring stiffness of top drive, and $\boldsymbol{C_{tt}}$ is the viscous damping coefficient of top drive, the equivalent mechanical model can be explained as:

$$\Omega_0 - \frac{d(n \cdot T_m/k_{tt})}{dt} - \frac{n \cdot T_m}{c_{tt}} = \dot{\theta}_t \tag{5.2}$$

In the equivalent mechanical model, the connection rotation angle of equivalent damper and equivalent torsion spring is assumed to the ψ_t, we have:

$$k_{tt} \cdot (\Omega_0 \cdot t - \psi_t) = c_{tt} \cdot \left(\dot{\psi}_t - \dot{\theta}_t\right) \tag{5.3}$$

$$n \cdot T_m = k_{tt} \cdot (\Omega_0 \cdot t - \psi_t) \tag{5.4}$$

Solving Eq. (5.3) by set $\psi_t(0) = 0$, we have:

$$\psi_t = e^{-\frac{k_{tt}}{c_{tt}}.t} \cdot \int_0^t \left(\dot{\theta}_t + \frac{k_{tt}}{c_{tt}} \cdot \Omega_0 \cdot \tau\right) \cdot e^{\frac{k_{tt}}{c_{tt}}.\tau} \cdot d\tau \tag{5.5}$$

Substitute Eq. (5.5) into Eq. (5.4), we can obtain the torque transmitted by the equivalent torsion spring:

$$n \cdot T_m = k_{tt} \cdot \left[\Omega_0 \cdot t - e^{-\frac{k_{tt}}{c_{tt}}.t} \cdot \int_0^t \left(\dot{\theta}_t + \frac{k_{tt}}{c_{tt}} \cdot \Omega_0 \cdot \tau\right) \cdot e^{\frac{k_{tt}}{c_{tt}}.\tau} \cdot d\tau\right] \tag{5.6}$$

This model is equivalent to giving the boundary conditions at the top of the drillstring.

5.2.1.2 Drillstring Model

The RSS is assumed as a rigid body, and the drillstring is homogenous along its entire length and simply considered as a single linear torsional spring [41] of torsional stiffness K_t and torsional damping C_t. The number of drillstrings can be modified depending on system analysis requirements. The *jth* drillstring is connected to the BHA by means of K_b and C_b. The simplified drillstring torsional vibration model is shown in Fig. 5.5a.

The top drive torque is supposed to be constant and positive. The drillstrings are considered to have the same inertia, and the drilling fluid is simplified by a viscous-type friction element at the hit. The drillstring torsional model [40, 41] takes the following form:

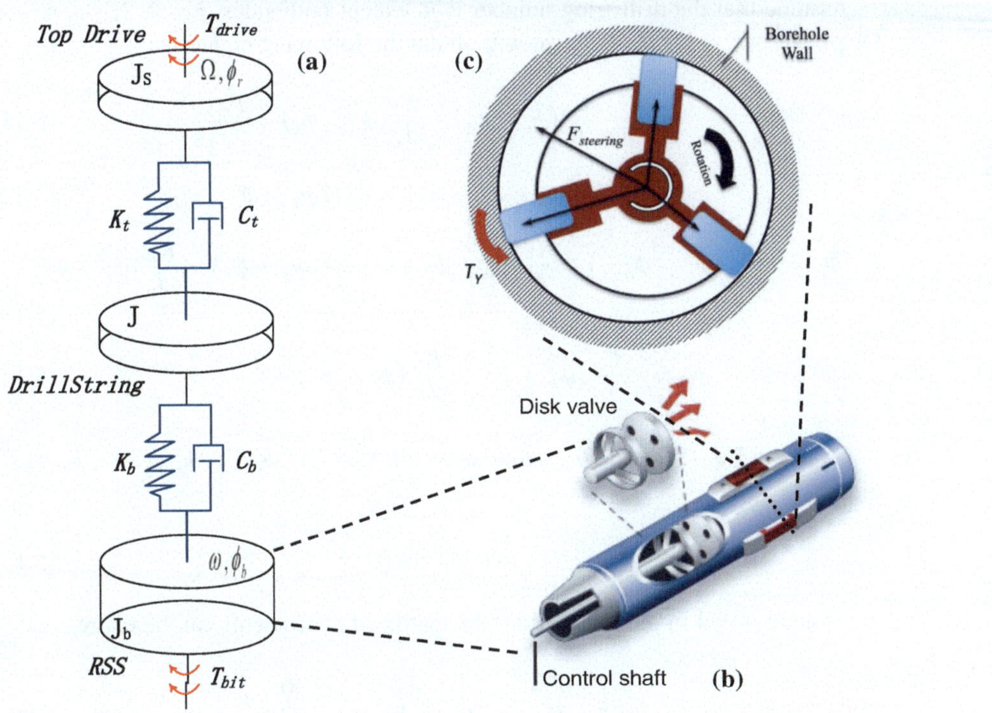

Fig. 5.5 Torsional dynamics model (**a**) drillstring torsional vibration model, there are many drillstrings not indicated in the figure; (**b**) structure of the strap-down rotary steerable drilling system; (**c**) the cross-sectional view of the pads coming from (**b**), if has three pads, and the outside is borehole wall

$$J_s\dot{\Omega} + C_t(\dot{\phi}_r - \dot{\phi}_j) + K_t(\phi_r - \phi_j) = T_{drive} \tag{5.7}$$

$$J\ddot{\phi}_j + C_t(2\dot{\phi}_j - \dot{\phi}_{j+1} - \dot{\phi}_{j-1}) + K_t(2\phi_j - \phi_{j+1} - \phi_{j-1}) = 0 \tag{5.8}$$

$$J_b\dot{\omega} + C_b(\dot{\phi}_b - \dot{\phi}_j) + K_b(\phi_b - \phi_j) = T_Y + T_{bit} \tag{5.9}$$

where ϕ_b is the angular displacement of BHA; ω is the angular speed of BHA; ϕ_r is the angular displacement of the top-drive; Ω is the angular speed of the top drive; ϕ_j is the angular displacement of *j*th drillstring; J_b is the moment of the inertia of BHA; J_s is the moment of the inertia of the rotary table; J is the moment of the inertia of drill string; T_{drive} is top drive input torque and T_{bit} is the torque on the bit.

From the Eq. (5.8), angular acceleration of the *jth* drillstring $\ddot{\phi}_j$ can be expressed as:

$$\ddot{\phi}_j = -\frac{C_t}{J}(2\dot{\phi}_j - \dot{\phi}_{j+1} - \dot{\phi}_{j-1}) - \frac{K_t}{J}(2\phi_j - \phi_{j+1} - \phi_{j-1}) \tag{5.10}$$

Assume that the drillstring number is p, except both sides, $j = 2, 3, \cdots, p - 1$. Obviously, $\dot{\phi}_r = \Omega$, $\dot{\phi}_b = \omega$, we will obtain the following equations:

$$\ddot{\phi}_r = -\frac{C_t}{J_s}(\dot{\phi}_r - \dot{\phi}_1) - \frac{K_t}{J_s}(\phi_r - \phi_1) + \frac{T_{drive}}{J_s} \tag{5.11}$$

$$\ddot{\phi}_1 = -\frac{C_t}{J}(2\dot{\phi}_1 - \dot{\phi}_r - \dot{\phi}_2) - \frac{K_t}{J}(2\phi_1 - \phi_r - \phi_2) \tag{5.12}$$

$$\ddot{\phi}_p = -\frac{C_t}{J}(\dot{\phi}_p - \dot{\phi}_{p-1}) - \frac{K_t}{J}(\phi_p - \phi_{p-1}) - \frac{C_b}{J}(\dot{\phi}_p - \dot{\phi}_b) - \frac{K_b}{J}(\phi_p - \phi_b) \tag{5.13}$$

$$\ddot{\phi}_b = -\frac{C_b}{J_b}(\dot{\phi}_b - \dot{\phi}_p) - \frac{K_b}{J_b}(\phi_b - \phi_p) + \frac{T_Y + T_{bit}}{J_b} \tag{5.14}$$

From the Eq. (5.9)~(5.14), define a vector $X = \left(\phi_r, \dot{\phi}_r, \cdots, \phi_j, \dot{\phi}_j, \cdots, \phi_b, \dot{\phi}_b\right)^T$, then:

$$\dot{X}(t) = AX(t) \tag{5.15}$$

Assume $p = 4$ in our simulation, the matrix of coefficients can be express as:

$$A = \begin{bmatrix} 0 & 1 & 0 & 0 \\ -\frac{K_t}{J_s} & -2\frac{C_t}{J_s} & \frac{K_t}{J_s} & \frac{C_t}{J_s} \\ 0 & 0 & 0 & 1 \\ \frac{K_t}{J_b} & \frac{C_t}{J_b} & -\frac{K_t}{J_b} & -\frac{C_t + C_b}{J_b} \end{bmatrix} \tag{5.16}$$

The parameters in the simulation as show in the Table 5.1.

Table 5.1 System parameter values used to generate numerical results for comparison with experimental results

Variable	Value	Units	Variable	Value	Units
J_s	0.518	$kg \cdot m^2$	J_b	0.0318	$kg \cdot m^2$
J	0.025	$kg \cdot m^2$	C_b	0.01	N · m · s/rad
C_t	0.0001	N · m · s/rad	K_b	0.073	N · m/rad
K_t	0.073	N · m/rad	K_s	1.2×10^{-4}	N · m/rad
M_1	1000	kg	M_2	12.5	kg
R, r_h	0.1555	Mm	e_1	1.57×10^{-5}	–
e_2	−0.658	–	μ_d	0.3	–
μ_s	0.35	–	ζ	0.01	–

5.2.1.3 Deterministic Excitations of Push-the-Bit

In the VDS or RSS system, the frictional force between the pads and borehole wall will make the bottom drill tool instantaneous rotational speed reduce [22]. The VDS pads pushing the borehole wall cause the drill bit torsional vibration more severe. The mechanism of torsional vibration generation between push-the-bit RSS and ordinary drilling systems is completely different. During the rotation of the drillstring, the pads of the VDS is periodically pushed out and acts on the fixed position of the borehole wall for a period of time. This is the main difference between VDS and traditional drilling systems. The friction between the pads and the borehole wall will instantaneously reduce the speed of the drill bit. Assume that the radius of the borehole is R, at the angle θ when the drillstring rotate to a circle, there will be a speed value at the position of ($R*sin\theta$, $R*cos\theta$) in the orthogonal coordinates.

As shown in Fig. 5.6, the pads of the actuator are constantly pushed toward the borehole wall, forming a cyclical nonlinear damping force at the bottom of the hole. The launch of the pads in the actuator is to change the drilling trajectory. The disk valve at a certain tool face angle, and then drilling fluid moves the pads out by flowing through valves. Drilling fluid push one of the pads to the borehole and produce the steering force that orients the drilling.

When the push tool is in operation, the pushing force of the actuator is the resultant force of the three pads. The phase difference between the three round holes on the bottom plate is 120°. When the central angle of the high-pressure hole on the top plate valve is 180°, the thrust force is in the optimal state, and the coverage angle of the thrust force is equal to 60°.

The pushing force F_{ti} can be calculated,

$$\mathrm{F_{ti}} = \Delta \mathrm{p} \pi r_p^2 \ (i = 1, 2, 3) \tag{5.17}$$

where, $\mathrm{F_{ti}}$ is the push force of a single pad pushing against the borehole wall, Δpis the internal and external pressure difference, r_p is the piston radius.

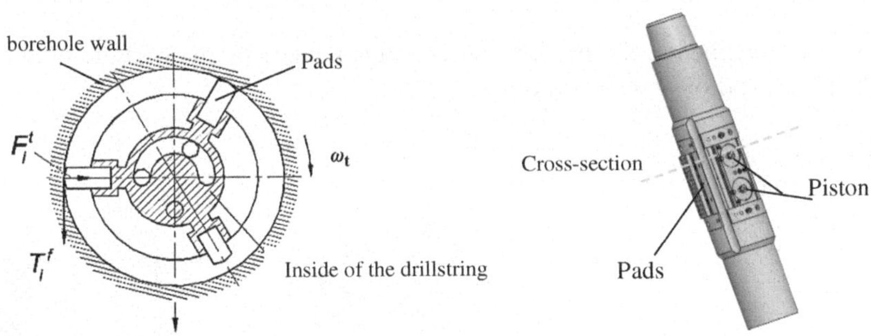

(a) The cross-sectional view of the pads (b) Actuator of push the bit

Fig. 5.6 Actuator's thrust model of VDS

The pushing force of the piston is the pressure difference of the drilling fluid on both sides of the piston chamber (inside and outside of the drilling tool). This pressure difference is mainly caused by the pressure drop caused by the drill nozzle. Δp can be expressed as:

$$\Delta \mathrm{p} = \frac{4\rho_w Q_n{}^2}{5\pi^2 C_{ne}{}^2 d_{ne}^4} \tag{5.18}$$

where ρ_w is the drilling fluid density, g/cm^3; Q_n is the nozzle flow, 1/s; C_{ne} is the nozzle flow coefficient, d_{ne} is the nozzle equivalent diameter, (mm).

When the bottom tool rotates at a constant speed, the concentration effect of the piston's pushing force can be expressed as:

$$\mathrm{F_{ti}t_p} = \mathrm{F_{ti}}\frac{\theta_E - \theta_S}{\omega_t} \tag{5.19}$$

where t_P is the pushing time of one of the pads, θ_S is the initial phase angle of pads pushing, θ_E is the termination phase angle of pads pushing, ω_t is the rotation angular speed of bottom tool.

As shown in Fig. 5.7, due to $\theta_S - \pi/2 = \pi/2 - \theta_E$, $\alpha_f = \theta_E - \theta_S$ is the sweeping angle of the pushing force, comprehensive effect of push force F_{ti} within the time t_p can be expressed as:

$$\overline{\mathrm{F}}_t = \mathrm{F_{ti}}\frac{\sin(\theta_E - \theta_S)}{\theta_E - \theta_S} = \mathrm{F}_{ti}\frac{\sin\alpha_f}{\alpha_f} \tag{5.20}$$

The difference between VDS and ordinary drilling system is the friction torque. As shown in Fig. 5.6 (a), when the pads push against the borehole wall and rotates together with the tool, the wall of the well gives a frictional torque to the drilling tool. Friction torque T_f of the push the bit can be expressed as:

$$T_f = \overline{F}_t \cdot R_e \cdot \mu(v) \tag{5.21}$$

where $\mu(v)$ is defined as the friction coefficient generated by the friction model, v is the relative speed of movement, (m/s).

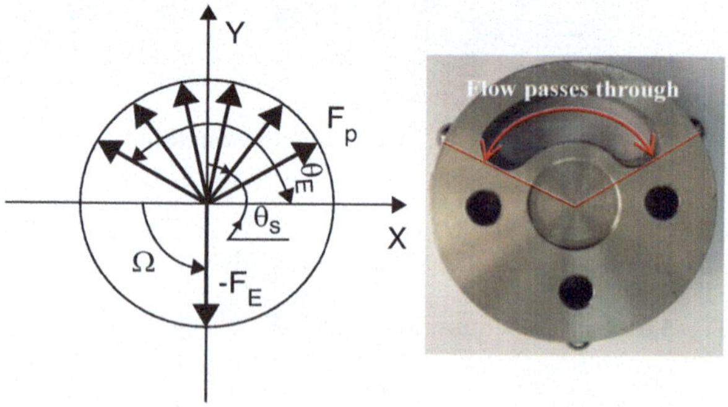

Fig. 5.7 Left is schematic of equivalent directing force, right is the actual photo of disk valve

5.2.1.4 Rock Breaking Model

We use the equation proposed by Spanos et al. [42], assuming that the friction is considered to be evenly distributed on the front face of the bit.

$$T_{bit} = F_w \left[\frac{2}{3} r_h u(\omega) + \zeta \sqrt{r_h \delta_c} \right] \tag{5.22}$$

where F_w is the weight on bit, r_h is the drill bit radius, δ_c is the average cutting depth, and ζ is a dimensionless parameter that characterizes the force necessary to cut the rock. The average cutting depth, δ_c, is obtained from the following relation:

$$\delta_c = \frac{2\pi r_p}{\Omega} \tag{5.23}$$

where r_p is the average rate of penetration, calculated as a function of the applied weight-on-bit, F_w, and the rotary table rotation, Ω, using the following empirical relation:

$$r_p = e_1 F_w \sqrt{\Omega} + e_2 \tag{5.24}$$

where e_1 and e_2 are constants.

5.2.1.5 Friction Model

We define the dry friction as a continuous function to describe both the static and dynamic friction. Using the coulomb friction model, as shown in Fig. 5.8, the constants μ_s and μ_d are the static and dynamic friction coefficients, respectively.

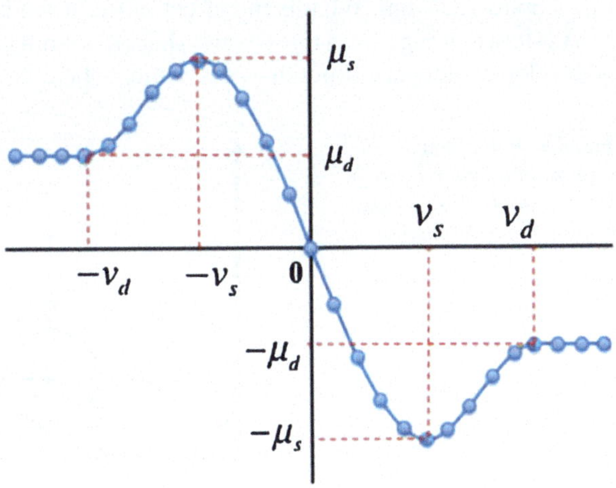

Fig. 5.8 The curve generated from a dry friction of model

Assuming that v is the slip velocity at the friction point, v_s is stiction transition velocity; v_d is the friction transition velocity.

Set $\mu(-v_s) = \mu_s$, $\mu(v_s) = -\mu_s$, $\mu(0) = 0$, $\mu(-v_d) = \mu_d$, $\mu(v_d) = -\mu_d$. Then the function $u(v)$ can be defined as follows:

$$\mu(v) = -sign(v) \cdot \mu_d \mid v \mid > v_d \tag{5.25}$$

$$\mu(v) = -s(|v|, v_d, \mu_d, v_s, \mu_s) \cdot sign(v) v_s \leq \mid v \mid \leq v_d \tag{5.26}$$

$$\mu(v) = s(v - v_s, \mu_s, v_s - \mu_s) \mid v \mid < v_s \tag{5.27}$$

The $s(x, x_0, h_0, x_1, h_1)$ function can be defined as follows, assume x is the independent variable, x_0 is a real variable that specifies the x value at which the s function begins, x_1 is a real variable that specifies the x value at which the s function ends, h_0 is the initial value of the step, h_1 is the final value of the step, and assume $a = h_1 - h_0$, $\Delta = (x - x_0)/(x_1 - x_0)$, then:

$$s(x, x_0, h_0, x_1, h_1) = \begin{cases} h_0 & x \leq x_0 \\ h_0 + a \cdot \Delta^2(3 - 2\Delta) & x_0 < x < x_1 \\ h_1 & x \geq x_1 \end{cases} \tag{5.28}$$

We do not allow $v_s = 0$, the friction transition velocity v_d is greater than the stiction transition velocity v_s by definition.

5.2.1.6 Phase Torque of Drill Bit

In the drilling progress, wells often appear rugged wavy, which is called wave bottom. The bottom hole coring shows that the crests and troughs of the undulating bottom are different depending on the drilling parameters and the rock type, sometimes up to 12.7 mm. What is the effect of the wave bottom on the bit torque?

As shown in Fig. 5.9, a three-wave-shaped bottom surface with a sinusoidal shape is developed along a cutting line of a radius, where the abscissa is the product of the

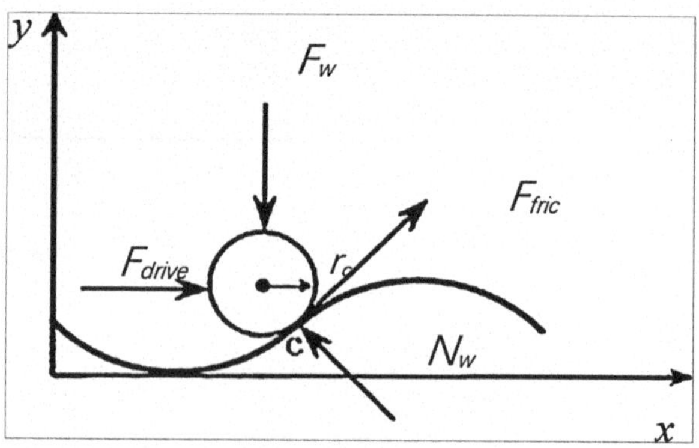

Fig. 5.9 Force analysis of drill bit in rolling bottom, a wave bottom surface along a certain radius cutting line

phase angle of the cone at the bottom of the well and the corresponding radius, Radius on the bottom of the well to the height of the trough.

If y represents the ordinate, x represents the abscissa, and assuming that the height difference between the crest and the trough is proportional to the size of the radius r_c, the ratio coefficient is K_h, then,

$$y = \frac{r_c \cdot K_h}{2} \cdot [1 + \sin(3\alpha)] \tag{5.29}$$

$$\alpha = \frac{x}{2\pi \cdot r_c} \tag{5.30}$$

Where α is the phase angle of the cone in the bottom of the well, (rad).

Due to the close linear contact between the cone and the bottom of the well, the wellbore radius is assumed to be R_e. Fig. 5.9 shows the force analysis diagram of the folded cone bit rolling at the bottom of the corrugated bottom. Point c is the contact point between the cone and the corrugated bottom, F_w is the pressure of the drillstring acting on the roller through the bearing, N_w is the positive pressure acting on the roller cone, F_{drive} is the driving force of the drill string on the cone. Consider that the inertia force of the roller cone when moving downhole is small compared with other forces, it is ignored. Thus, according to the static balance conditions of force and moment, we have:

$$F_w \cdot r_c \cdot \sin\beta - F_{drive} \cdot r_c \cdot \cos\beta = 0 \tag{5.31}$$

$$F_w \cdot \cos\beta + F_{drive} \cdot \sin\beta = N_w \tag{5.32}$$

Where r_c is the corresponding cone diameter at the equivalent radius, $\beta = \arctan\left(\frac{3K_h}{4\pi} \cdot \cos(3\alpha)\right)$.

From Eqs. (5.31) and (5.32), we can obtain:

$$F_{drive} = \frac{3K_h}{4\pi} \cos(3\alpha) \cdot F_w \tag{5.33}$$

$$N_w = \sqrt{1 + \left[\frac{3K_h}{4\pi} \cos(3\alpha)\right]^2} \cdot F_w \tag{5.34}$$

Assume the weight on the bit is evenly distributed on the drill bit, the bit torque caused by the wave bottom can be expressed as:

$$T_w = \frac{3R_e \cdot K_h}{4\pi} \cdot \cos(3\alpha) \cdot W_b \tag{5.35}$$

Where T_w is the drill bit torque caused by the wave bottom hole, (N·m);

Then the total bit torque is $T_d + T_w$, The corrugated hole bottom causes the drill bit to act on the phase moment associated with its angle of rotation. Furthermore, the drillstring with uneven bending rigidity is also affected by the phase moment in the curved wellbore. Therefore, the phase moment is still present in the actual drilling process, but in most cases the value is relatively small.

5.2.2 *Modeling of Lateral Vibration*

The drillstring is assumed as a beam with two cross sections (collars and pipes). A planar wire shape sketch with hollow pipe profiles is used to model the entire drillstring. The "Hermite cubic" beam element is used, which does not account for the shear flexibility, although axial strain is considered. The drillstring–mud interaction effects on the drillstring dynamic analysis are important. The effect of internal and external mud flow is sensitive to the annular space between the drillstring and wellbore. When the fluid flow was considered in the dynamic equations, the lateral dynamic response was a bit larger initially. However, steady-state response was unchanged.

The partial differential equations of drillstring lateral vibration will be obtained according to the drillstring bending theory and D'Alembert's principle [43, 45].

$$\frac{\partial^2 y}{\partial t^2} + a^2 \frac{\partial^4 y}{\partial x^4} = 0 \tag{5.36}$$

wherein, $a^2 = \frac{EIg}{A\rho}$; A is cross-sectional area of drillstring; I is moment of inertia of drillstring; ρ is density of drillstring material; y is the deflection of the drillstring; t is the time。

Study of the fundamental oscillation mode only, the displacement of each point on the axis is the harmonic function with respect to time, then,

$$y = X(A\cos pt + B\sin pt) \tag{5.37}$$

$X$is a function of x, then,

$$\frac{\partial y}{\partial t} = X(-Ap\sin pt + Bp\cos pt) \tag{5.38}$$

$$\frac{\partial^2 y}{\partial t^2} = X\left(-Ap^2\cos pt - Bp^2\sin pt\right) \tag{5.39}$$

$$\frac{\partial y}{\partial X} = (A\cos pt + B\sin pt)\frac{\partial y}{\partial X} \tag{5.40}$$

$$\frac{\partial^4 y}{\partial X^4} = (A\cos pt + B\sin pt)\frac{\partial^4 y}{\partial X^4} \tag{5.41}$$

Simultaneous Eq. (5.38) -(5.41),

$$\frac{\partial^4 y}{\partial X^4} = \frac{p^2}{a^2} X \tag{5.42}$$

The general solution is:

$$\begin{aligned} X = C_1(\cos Kx + \cosh Kx) + C_2(\cos Kx - \cosh Kx) + \\ C_3(\sin Kx + \sinh Kx) + C_4(\sin Kx - \sinh Kx) \end{aligned} \tag{5.43}$$

where in: $K = p^2 A\gamma / EIg$.

5.2.3 Modeling of Longitudinal Vibration

There are more than 5 types of longitudinal free vibration of drillstring model have been developed, in them, three types of most widely used mechanical models are shown in Fig. 5.10 [44]. Compared with Model 1, Model 2 has omitted the quality of the damper and driving system, which includes the quality of the traveling system, Kelly and steel wire rope. In the meantime, Model 3 has ignored the quality of the damper and simplified the quality of drill collar as a focus quality. Due to the best versatility in these mathematical and mechanical models, a lot of special longitudinal vibration can be analyzed by using Model 1, which is currently selected as the longitudinal vibration model of drillstring in this chapter. In Fig. 1, k_1 and m_1 represents the equivalent stiffness coefficient and the equivalent mass of the driving system respectively; k_2 and m_2 respectively express the stiffness coefficient and the mass of damper, which could be approximately considered as the equivalent

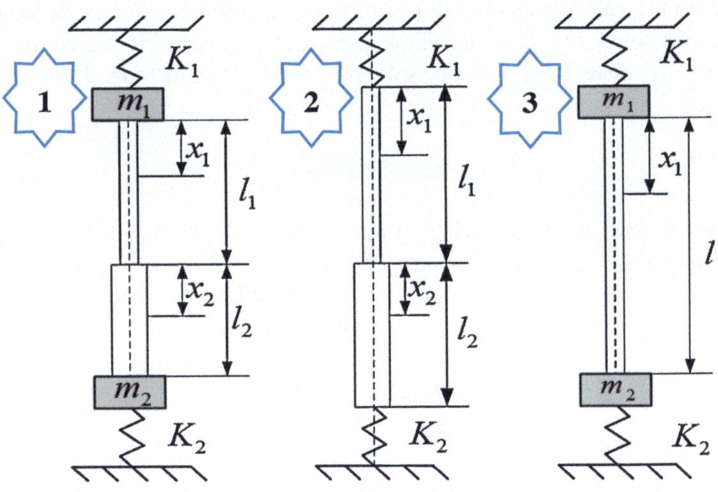

Fig. 5.10 The longitudinal vibration models of the drillstring

integrated structure of BHA; l_1 and l_2 denotes the length of drill pipe and drilling collar, both of them are considered as two elastic links; x_1 and x_2 means the length from the sampled differential element to the top of drill pipe and the top of drilling collar respectively.

Model 1 consists of drill collar, drill stem, joint, stabilizer, damper, and other downhole tools, which could be assembled by any arbitrary combinations and applied for any arbitrary RPM & WOB. During drilling process, the BDTs are inevitably influenced by disturbing factors, such as WOB induced by disturbing forces, the disturbing torsion on stabilizers, the inertial force induced by rotating of drillstring, the damping & viscosity resistance of drilling fluid, et al. Thus, in order to facilitate an efficient analysis, some assumptions must be proposed as follow: (a) With homogeneous density, the drillstring is an elastic straight bar owning a circular cross section, the inner & outer peripheries of which is rigid; (b) The inner wall of borehole is rigid, the annular between borehole and drillstring is a circular ring, the friction between them is neglected; (c) The top driving system is simplified as a spring, the quality and stiffness coefficient of which is m_1 and k_1 respectively; (d) The vibration is derived from the interaction between drill bit and rock, by which the motion of bit can be regarded as a kind of resonance vibration observing sine or cosine changing alone with time.

Based on the assumptions above, the longitudinal vibration equation of drillstring can be deduced by elastic rod theory, the partial differential equation of drillstring longitudinal vibration can be given as follow:

$$\frac{\partial^2 u}{\partial t^2} = \eta^2 \frac{\partial^2 u}{\partial x^2} \tag{5.44}$$

Wherein, x is the length from the sampled differential element to the top end of drillstring, u is the displacement of sampled differential element along the borehole axis direction, t and a respectively refer to time and wave velocity of longitudinal vibration, here $\eta^2 = E/\rho$, in which E and ρ respectively refers to the elastic modulus and the density of steel. Eq. 1 can be solved by the method of separation of variables, here we set:

$$u = T(t) \cdot X(x) \tag{5.45}$$

Wherein $T(t)$ refers to the variable function of phase position, which is related to the variable of time t and location-independent; $X(x)$ refers to the variable function of vibration amplitude, which is related to the variable of location x and location-independent. With Eq. 5–45, Eq. 5–44 can be rewritten as:

$$X(x)\frac{d^2T(t)}{dt^2} = \eta^2 \frac{d^2X(x)}{dx^2} \cdot T(t) \tag{5.46}$$

Here we introduce p as the natural frequency of the drilling system and set:

$$\frac{d^2T(t)}{dt^2}\Big/T(t) = \eta^2\frac{d^2X(x)}{dx}\Big/X(x) = -p^2 \tag{5.47}$$

Then the solution of the partial differential equation Eq. 1 can be transformed to the solution of the ordinary differential equation as follow:

$$\begin{cases} T(t) = A\sin(pt + \eta) \\ X(x) = C\cos\dfrac{px}{\eta} + D\sin\dfrac{px}{\eta} \end{cases} \tag{5.48}$$

The general solution of Eq. 5.48 is:

$$u = A\sin(pt + \alpha)\left(C\cos\frac{px}{\eta} + D\sin\frac{px}{\eta}\right) \tag{5.49}$$

Wherein A, α, C, and D are integration constant. Here we bring the lower corner mark 1 and 2 referring to the drillstring and the drill collar respectively, the displacement of longitudinal vibration of drillstring and drill collar can be given as:

$$\begin{cases} u_1 = A_1\sin(pt + \alpha_1)\left(C_1\cos\dfrac{px}{\eta} + D_1\sin\dfrac{px}{\eta}\right) \\ u_2 = A_2\sin(pt + \alpha_2)\left(C_2\cos\dfrac{px}{\eta} + D_2\sin\dfrac{px}{\eta}\right) \end{cases} \tag{5.50}$$

Where in, μ_1 and μ_2 respectively represents the displacement of longitudinal vibration of drillstring and drill collar; A_1, C_1, D_1 and A_2, C_2, D_2 are different undetermined integration constant for drillstring and drill collar. Taking the partial derivation of Eq. 5.49 with x and t, the equation group can be obtained as follow:

$$\begin{cases} \dfrac{\partial u}{\partial x} = A\dfrac{p}{\eta}\sin(pt + \alpha)\left(-C\sin\dfrac{px}{\eta} + D\cos\dfrac{px}{\eta}\right) \\ \dfrac{\partial u}{\partial t} = Ap\cos(pt + \alpha)\left(C\cos\dfrac{px}{\eta} + D\sin\dfrac{px}{\eta}\right) \\ \dfrac{\partial^2 u}{\partial x^2} = -A\dfrac{p^2}{\eta^2}\sin(pt + \alpha)\left(C\cos\dfrac{px}{\eta} + D\sin\dfrac{px}{\eta}\right) \\ \dfrac{\partial^2 u}{\partial t^2} = -Ap^2\sin(pt + \alpha)\left(C\cos\dfrac{px}{\eta} + D\sin\dfrac{px}{\eta}\right) \end{cases} \tag{5.51}$$

To solve the equations, some boundary equations must be given according to boundary conditions. On the top end of drillstring, when $x_1 = 0$, on the basis of mechanical equilibrium theory:

$$\mathrm{m}_1 \frac{\partial^2 u}{\partial t^2} = -k_1 u_1 + EF_1 \frac{\partial u_1}{\partial x_1} \tag{5.52}$$

Wherein F_1 is the area of the cross-section of drillstring. On the interface between drillstring and drill collar, when $x_1 = l_1, x_2 = 0$, according to the conditions of equal tension and equal velocity:

$$EF_1 \frac{\partial u_1}{\partial x_1} = EF_2 \frac{\partial u_2}{\partial x_2} \tag{5.53}$$

$$\frac{\partial u_1}{\partial t} = \frac{\partial u_2}{\partial t} \tag{5.54}$$

Wherein F_2 is the area of the cross section of the drill collar. On the bottom of drill collar, when $x_2 = l$, the mechanical equilibrium equation can be given as:

$$\mathrm{m}_2 \frac{\partial^2 u_2}{\partial t^2} = k_2 u_2 + EF_2 \frac{\partial^2 u_2}{\partial x^2} \tag{5.55}$$

By using Eq. 5.51 plug into the boundary condition equations from Eq. 5.52 to Eq. 5.53, the frequency equation of Model 1 can be deduced as:

$$\frac{1}{F_1} \frac{pEF_1 + \eta(k_1 - \mathrm{m}_1 p^2) \tan \frac{pl_1}{\eta}}{-pEF_1 \tan \frac{pl_1}{\eta} + \eta(k_1 - M_1 p^2)} = \frac{1}{F_2} \frac{-pEF_2 - \eta(k_2 - \mathrm{m}_2 p^2) \tan \frac{pl_2}{\eta}}{-pEF_2 \tan \frac{pl_2}{\eta} + \eta(k_2 - M_2 p^2)} \tag{5.56}$$

Then the natural frequency of the longitudinal vibration of drillstring system can be solved, which is directly influenced by the stiffness coefficient of the damper. Because Eq. 5.56 is a nonlinear equation, in order to acquire an effective solution, the finite element method will be selected.

5.3 Simulation Results of Torsional Dynamics

5.3.1 *Drillstring Torsional Vibration without the Effect of* $\mathrm{T_f}$

The specification of the drillstring can be found in Table 5.2. Based on the model developed, simulation is carried out for deterministic case.

Table 5.2 Drillstring parameter

	Drillstring	Drillcollar+VDS
Length	2000 m	200 m
Outer diameter	0.127 m	0.1778 m
Inner diameter	0.1086 m	0.0572 m
Drillstring density	$\rho = 7850.0$ kg/m^3	
Elastic modulus	210×10^9 N/m^2	
Shear modulus	7.96×10^{10} N/m^2	

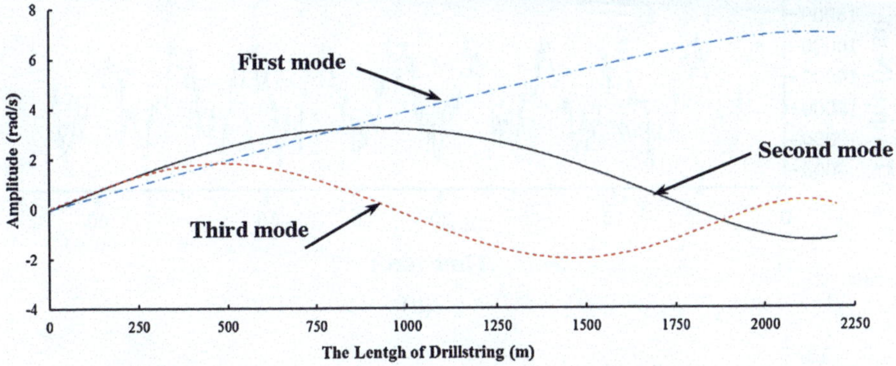

Fig. 5.11 Three modes of mode shape of drillstring free torsional vibration

Table 5.3 The modal vibration energy in the drillstring accounts for the percentage of the total energy of the torsional vibration

Modal order	First order	Second-order	Third-order	Fourth-order	Fifth-order	The other
Percentage (%)	85.402	8.4299	2.5882	1.2054	0.6896	1.6849

The simulation results in the case when the desired table speed is 60 rpm and ratio is 0.6, assume that before the top of the drill pipe begins to rotate, the distributed friction torque and bottom bit torque have been slowly loaded, then the three modes of mode shape of drillstring free torsional vibration are shown in Fig. 5.11.

As depicted in Fig. 5.11, the magnitude larger place of first-order vibration mode concentrates on the high moments section of the drillstring inertia. However, when the second and third-order modes have large amplitudes, they correspond to small moments of inertia of drillstring. As a result, the energy of the first-order modal vibration of the drillstring is greater than other orders. Table 5.3 lists the vibration energy of various modes:

Compared to single drillstring system, this system includes both drill pipe and drill collar, which is more like inertia centralized torsion pendulum. As shown in Fig. 5.12 (b), it can be seen that the drill bit rotary speed experiences large fluctuations. At some points, the drill bit rotation speed slows to near zero and then increase to as high as 14 rad/s (133.7 rpm), this phenomenon is called stick-slip vibration.

In Fig. 5.12, the simulation curve shows that this is two-stage system with drill pipe and drill collar is closer to the concentrated inertia twisting pendulum model, sinusoidal waveform exhibiting free torsional vibration. This also shows that the two-stage drillstring system is more suitable than the single drillstring for simplifying the inertial torsional pendulum. At the same time, comparing the two graphs (a) and (b) in Fig. 5.12, it is observed that the wellhead torque vibration waveform is not as regular as the bottom hole speed vibration waveform. This is mainly due to the

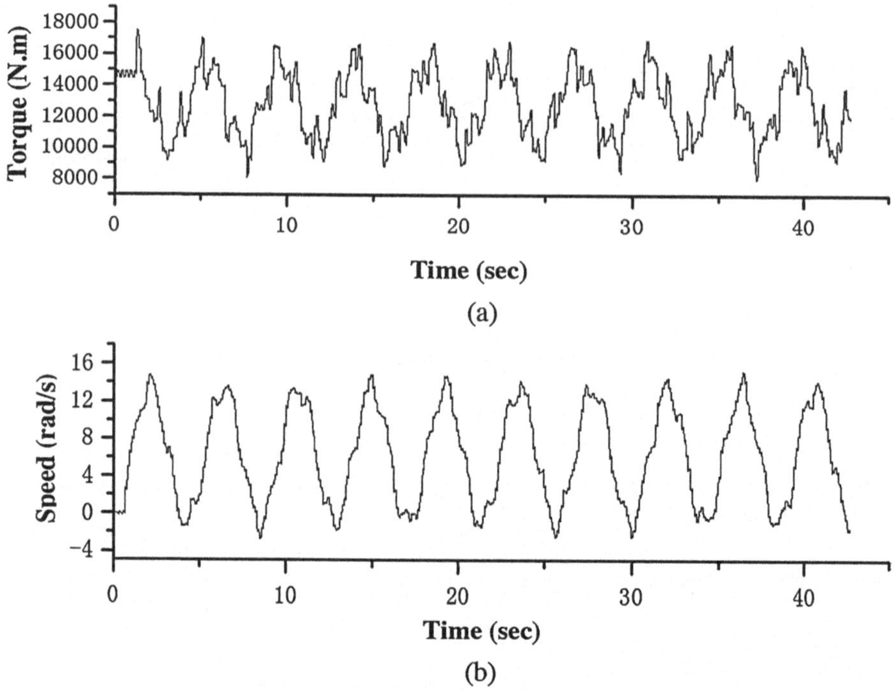

Fig. 5.12 Diagram of wellhead torque and bottom hole speed over time

fact that the torque in the drillstring at the wellhead is greatly affected by higher-order modal vibrations (except for the first-order modal vibrations).

As previously defined, $\varphi(t)$ is the drillstring twist angle, the total energy E of the drillstring torsional vibration can be expressed as:

$$E = \frac{1}{2} \cdot J_s \cdot \varphi^2 + \frac{1}{2} \cdot K_e \cdot \varphi^2 \tag{5.57}$$

The derivative with respect to time for E, get the expression of energy change rate:

$$\dot{E} = -c_e \cdot \dot{\varphi}^2 - [T_{or}(\dot{\varphi} + \Omega_0) - T_{or}(\Omega_0)] \cdot \dot{\varphi} \tag{5.58}$$

As shown in Fig. 5.13, when the vibration amplitude is small, the vibration energy is always increasing. When the amplitude is large, the vibration energy will be decreasing. There is an appropriate amplitude value during this period, so that the vibration energy absorbed and dissipated by the drillstring system during one vibration period is equal. At equilibrium, in most cases, this suitable amplitude is the amplitude at which stick-slip phenomena occur. Energy analysis reflects the energy absorption and dissipation of the drillstring during the entire vibration

Fig. 5.13 The relationship between the energy change rate and the vibration speed

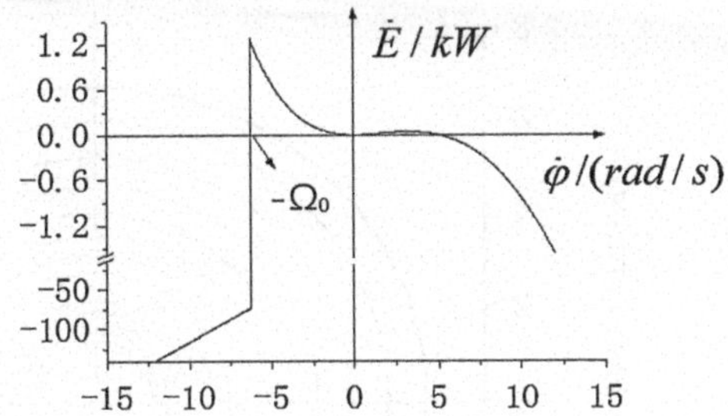

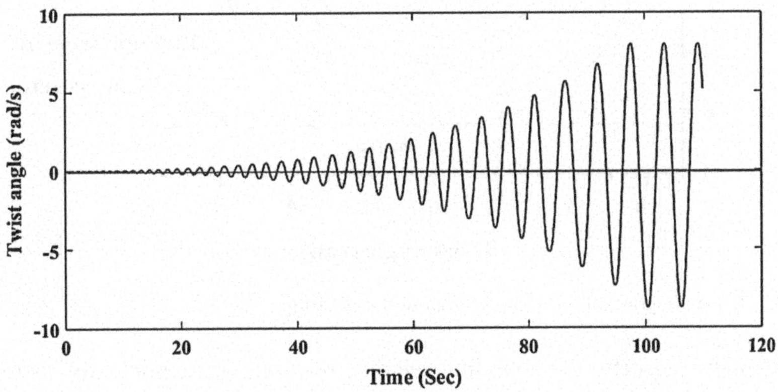

Fig. 5.14 Time-domain waveforms in the formation of stick-slip vibration of drill string

process. Under normal conditions, local instability will lead to stick-slip vibration of the drillstring.

When the drillstring is subjected to small-scale torsional vibration, the rate of change of its vibration energy is always greater than zero, indicating that the energy of the drillstring torsion vibration will increase continuously, and the increase in torsional vibration energy of the drillstring will inevitably lead to the amplitude of the drillstring torsional vibration increase.

When the amplitude increases to a certain extent, the energy dissipated by the drillstring during a vibration cycle is greater than the absorbed energy, then the amplitude of the drillstring decreases. Finally, a proper amplitude value is maintained for torsional vibration.

Through the previous analysis, it can be determined that the steady-state rotational equilibrium state of the drillstring system is unstable. And for any disturbance, it eventually tends to have a constant amplitude for torsional vibration. Figure 5.14 shows the time-domain waveform of the drillstring system under the influence of a small disturbance near the steady-state rotational equilibrium state. It shows how the drillstring develops from stick-slip-vibration to a near-uniform rotational state. Figure 5.15 is a phase diagram of the stick-slip vibration formation process described above. It can be seen how the drillstring starts from near the steady-state equilibrium

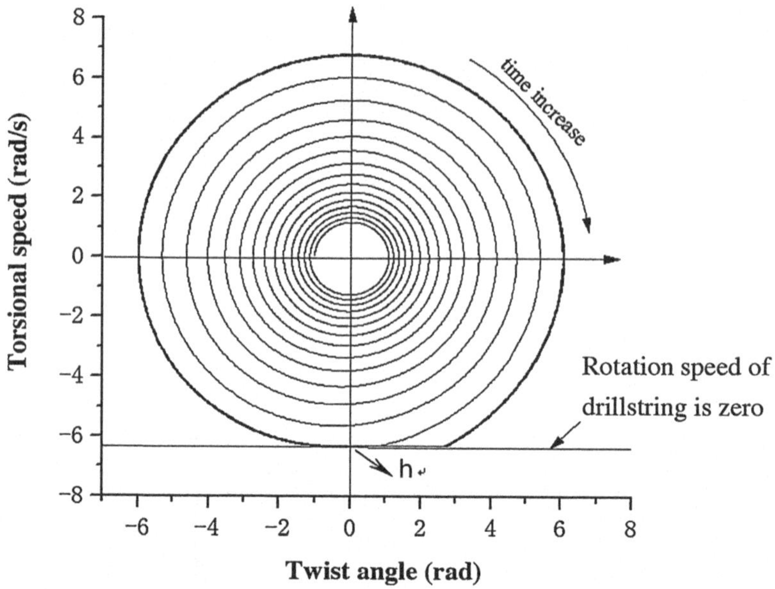

Fig. 5.15 Phase diagram of stick-slip vibration in drill string

and gradually absorbs the vibration energy, resulting in a gradually increasing amplitude, and finally a process of stable stick-slip vibration. The minimum vibration energy of the drillstring system corresponds to point *h* in Fig. 5.15.

5.3.2 *Drillstring Torsional Vibration With the Effect of* T_f

Drillstring torsional vibrations often occurs during the drilling. However, the torsional vibration of the VDS in the Push-the-Bit mode is more severe than that of the conventional drilling system. Statistics show that in most cases, the bottom hole dill bit exhibits stick-slip vibration. In this chapter, we reveal that the main difference between the VDS of Push-the-Bit and the ordinary drilling system is the friction torque (T_f) that generated in the process of pads pushed against the borehole wall. Previous work [2, 40] show that the drill bit torsional vibration phenomena perceptible increased when the friction torque exist. This chapter will reveal the impact mechanism of drillstring free torsional vibration with the effect of T_f.

Figure 5.16 is shown the timing diagram of pushing forces of the three pads. The direction of force follows the rotation of the drillstring. The Fig. 5.14 shows the projection of force F_{ti} on the *x*-axis. Each pad is pushed 180 degrees apart, equivalent to the previous definition:$\Delta\theta = \theta_E - \theta_S = \pi$, then we can get the total friction torque $T_f = \sum_{i=1,2,3} F_{ti} \cdot \mu(v)$, where the friction model is used Dahl model as shown above. When the top drive rotary speed is $\dot{\psi}(t) = \pi/2\ rad$.

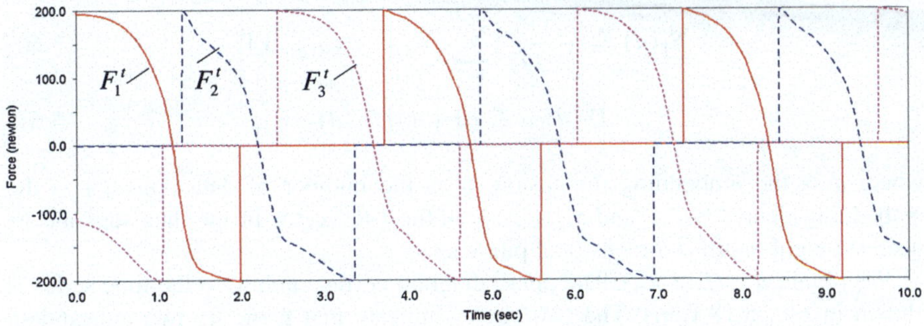

Fig. 5.16 Timing diagram of pushing forces of the three pads

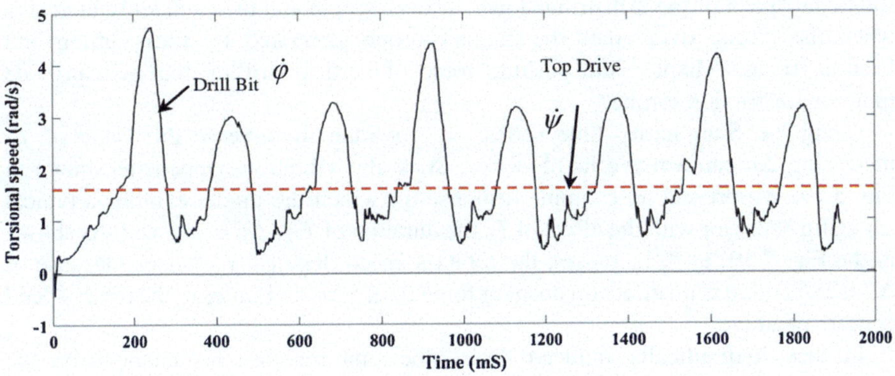

Fig. 5.17 The time series of the angular velocities at the bottom drill bit, torsional oscillations occurring in the simulation. For the parameter values: $J = 0.299\ kg \cdot m^2$, $c = 0.010 N \cdot m \cdot \sec/rad$, $k = 5\,N \cdot m/rad$, $\omega = 4.8\ rad/s$, $W_b = 1kN$

Figure 5.17 presents an example response for which the model exhibits a typical torsional vibration behavior. For time-series data generated in the simulation as shown in Fig. 5.17, we perform phase space reconstruction and probability density estimation, the dynamics of the time series $x_0, x_1, \cdots, x_{n-1}$ are fully captured or embedded in the m-dimensional phase space, $m \geq d$ where d is the dimension of the original attractor.

A vector $\vec{x}_i$ in the reconstructed phase space [46] is constructed from the time series as follows:

$$\vec{x}_i = \left[x_i, x_{i-\tau}, \cdots, x_{i-(m-1)\tau}\right] \tag{5.59}$$

where τ is the delay time. Cao's method [47] computes E_1 and E_2 for the data set of dimension 1 up to a dimension of D, which is the largest embedding dimension, used in the calculation. E_1 and E_2 are defined as follows:

$$E_1(d) = \frac{1}{N - d\tau}\left|\sum_{i=1}^{N-d\tau}\left|x_{i+d\tau} - x_{n(\mathrm{i,d})+\mathrm{d}\tau}\right|\right| \tag{5.60}$$

$$E_2(d) = E_1(d+1)/E_1(d) \tag{5.61}$$

where d is the embedding dimension, N is the number of data points, τ is the embedding delay, $x_{i\ +\ d\tau}$ and $x_{n(i,d)\ +\ d\tau}$ is the *i-th* vector in the data sets and its nearest neighbors of d-dimensional phase space.

We obtain $d = 2$, $\tau = 5$. The phase diagram corresponding to the time series is shown in Fig. 5.18 (left). The two circles indicate that there are two outstanding periods. Figure 5.18 (right) also confirm that two spikes of probability density estimation indicate the existence of two cycles at the same time. For torsional vibration, it often contains stick-slip phenomenon. Stick-slip occurs when the rotational speed of the drillstring slows (or even stops) and then suddenly increases when the torque overcomes the counter-torque generated by rock cutting and friction. In this chapter, the pushing force of vertical drilling tool will provide more major friction torque.

Using the Stationarity time series, we continue to increase the value of T_f, increasing Δp can achieve its objective. Stick-slip vibration happens as shown in Fig. 5.19. We present an example response for which the model exhibits a typical stick-slip behavior with the effect of T_f, the duration of T_f is $\Delta\theta = \pi/3$ *rad*. As shown in the Fig. 5.19, as T_f increases, the rotation speed drastically reduces, because of $\Delta\theta < 2\pi/3$, there is no effect on pushing force for a period of time, so the rotary speed appears negative.

In fact, hydraulically induced thrust does not increase instantaneously, the dynamics of a hydraulic actuator of VDS is more aptly captured by a nonlinear spring. Oscillation frequency can be expressed as, $f_F = \frac{1}{2\pi}M_H K_s$, wherein K_s is the equivalent stiffness. This hydraulic characteristic will make the stick-slip vibration

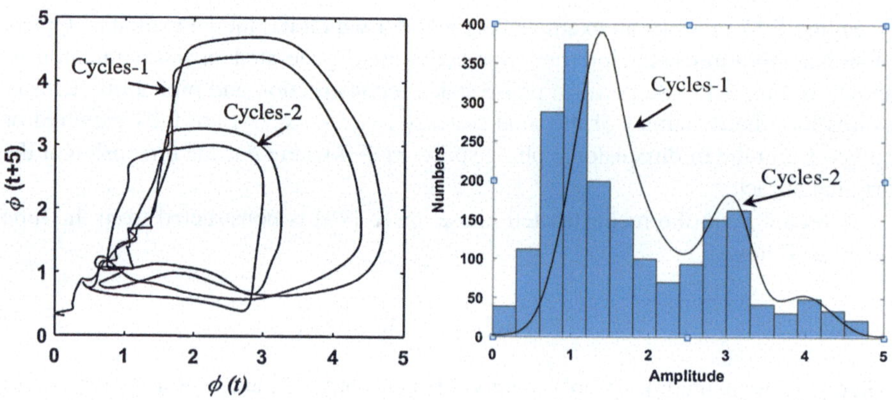

Fig. 5.18 Phase diagram (left) and probability density estimation (right) corresponding to the time histories shown in Fig. 5.17

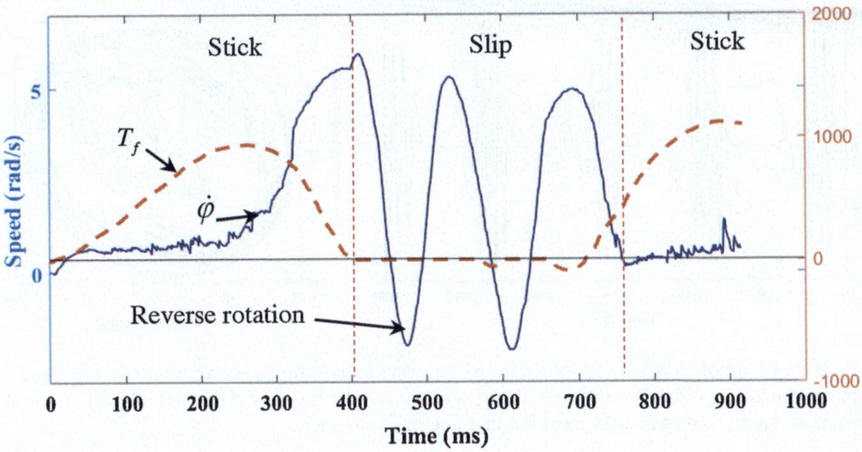

Fig. 5.19 An example of stick-slip oscillations occurring in the simulation with the effect of T_f

show different characteristics. Fig. 5.19 shows another example of stick-slip oscillations occurring in the simulation. When $\Delta\theta = \pi/3$, top drive rotary speed $\dot{\psi} = 100\, r/\min$, since the rotary speed is faster, hydraulic actuator is like a nonlinear spring which does not always act timely, so the stick phenomena always appear after a few revolutions. Through the time histories of the angular position, we can see this phenomenon clearly.

In the time window 700–1200, in Fig. 5.19, the drill bit angular speed ω is positive, and hence the system operates under the slip mode. At 450–700, ω decrease to zero, and thus the system switches to the stick phase. The bottom drillstring is stationary during the stick mode, which can be verified in the angular speed as shown in Fig. 5.19. As the angular displacement ψ of the rotary table increases, the reaction torque T_r shows a tendency to grow due to the increasing elastic energy stored in the flexible shaft. The stick regime finishes when the break-away torque value $T_r > T_b$, as long as the angular velocity of the bit is positive, the system changes the sliding mode of operation and remains there.

As shown in Fig. 5.20 (right), the phase diagram from angular position and angular velocities shows chaotic characteristics of dynamics. Bifurcation and chaos mark the sudden change of the motion in mechanical systems. It is observed that the system response is very sensitive to the changes in parameters. Closely related to bifurcations, chaos is a special kind of motion which is unique to nonlinear oscillation systems. It marks the behavior of a system that is inherently unpredictable. It is observed here that many changes in the parameters lead to chaotic responses. As seen in Fig. 5.20, the clearances between the bit trajectories definitely exceeds the numerical error and therefore it is considered to be chaotic. Bifurcation and chaos will inevitably bring complexity to drillstring system and make down-hole dynamic responses difficult to predict. This requires further in-depth study, especially in the vertical drilling system under the action of pushing force.

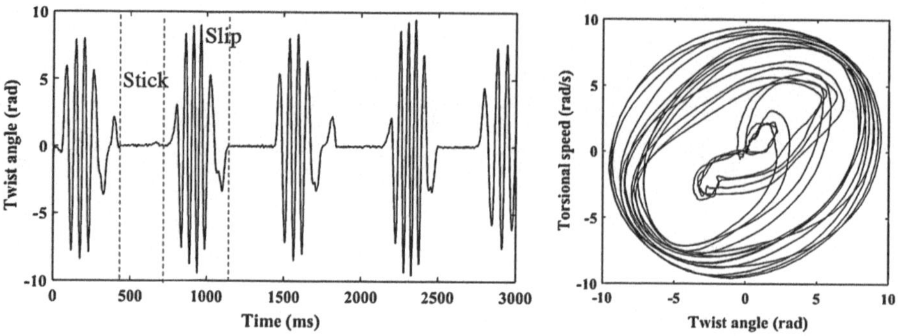

Fig. 5.20 An example of stick-slip oscillations occurring in the simulation when use dynamics of a hydraulic actuator model, left is the time histories of the angular position at the bottom drill bit, right is the phase diagram from angular position and angular velocities

5.4 Vibration Measurement Systems

5.4.1 The Basic Method of Data Acquisition & Processing

We developed a Rotary Steerable Drilling System (RSS), that is a new technology used in directional drilling, the whole drillstring is rotated from the surface by a hydraulically driven top drive, as show in Fig. 5.21a. That mainly included two parts: strap-down measurement while drilling (MWD) surveying system (Fig. 5.21b) and oriented actuator (Fig. 5.5b).

We develop a strap-down measurement while drilling (MWD) surveying system [48] that incorporates three-axis magnetometers and three-axis accelerometers arranged in three mutually orthogonal directions. The sensors are installed inside nonmagnetic drill collar that can avoid the external magnetic interferences. Performance characteristics of the accelerometers and magnetometers are summarized in Table 5.4.

Figure 5.21b shows the installing structure of downhole measurement system. E1~E9 in Fig. 5.21a indicate the field experiments in China between the year of 2011~2013. a_x, a_y, a_z are defined as survey signals of triaxial accelerometers on the xyz axis respectively. m_x, m_y, m_z are defined as survey signals of triaxial magnetometers on the xyz axis respectively, the sampling frequency f_s is 100 Hz. Assume that the Earth's magnetic field strength as M. Obviously, $M = \sqrt{m_x^2 + m_y^2 + m_z^2}$. Under certain sample frequency, measuring signal is time series and can be expressed as time function. We can use $a_h = \pm\sqrt{a_x^2 + a_y^2}$ define the lateral vibration of the BHA and use a_z express the longitudinal vibration.

Triaxial magnetometers are installed 90° phase difference between m_x and m_y, assume m_h is the horizontal projection of the Earth's magnetic field, after Δt, the drillstring rotates an angle of α_n.

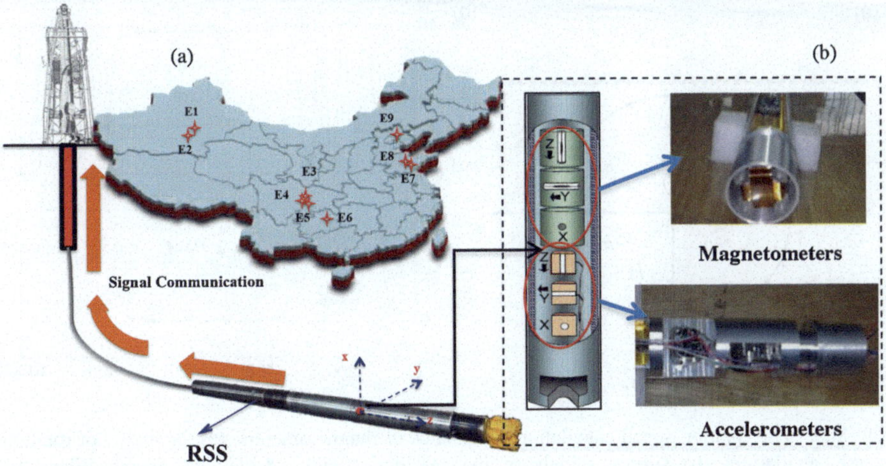

Fig. 5.21 Data acquisition system (**a**), rotary steerable drilling system, E1~E9 indicate the field experiments using the RSS in China, the map is created by the Microsoft PowerPoint; (**b**), construction of the downhole measurement system, three-axis magnetometers and three-axis accelerometers arranged in three mutually orthogonal directions

Table 5.4 Characteristics of Sensors

Parameter	Accelerometers (MS9010)	Magnetometers (CTM-DT06)
Range	±10 g	±100000nT
Scale factor	200 mV/g	5 V/G ± 5%
Nonlinearity	0.8% of SF	—
Bias calibration	<50 mg	±0.005 G
Noise	0.140 mg/√Hz	≤0.1nT
Bandwidth	1000 Hz	350 Hz

$$tg(\alpha_n) = \frac{m_h \cdot \sin(\alpha_n)}{m_h \cdot \cos(\alpha_n)} = \frac{m_x}{m_y} \tag{5.62}$$

Thus, $\alpha_n = arctg\left(\frac{m_x}{m_y}\right)$, as show in Fig. 5.22a and Fig. 5.22b, drillstring rotary from α_n to α_{n+1} though time Δt, it defines the rotary angle as a time series $\alpha(\alpha_1, \alpha_2, \cdots, \alpha_{n-1}, \alpha_n)$, drillstring rotational speed (RPM) is then defined as follows:

$$RPM = \frac{60}{2\pi}\omega = \frac{60}{2\pi} \cdot \frac{(|\alpha_n| - |\alpha_{n+1}|)}{\Delta t} \ (\mathrm{r/\min}) \tag{5.63}$$

There is a more practical approach to quantitative analysis the bottom drilling tool motion characteristics, through the measurement of the bottom drilling bit rotation speed, as well as some other downhole measurement parameters.

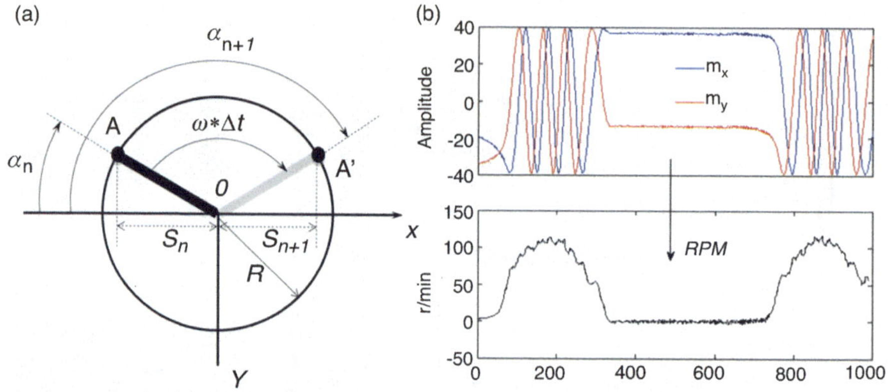

Fig. 5.22 The rotation speed calculation schematics of magnetometers, (**a**) the model of rotation speed calculation; (**b**) survey signals of magnetometers and calculational speed. When the drillstring rotates from α_n to α_{n+1} though time Δt, drillstring rotational speed (RPM) will be obtained using the survey signals of triaxial magnetometers on the *xy* axis

Field observations based on downhole and surface vibration measurements have indicated that drillstrings exhibit severe vibrations. These vibrations are observed to become more severe at the bottom hole assembly (BHA). As shown in the Fig. 5.23b, the value of the angular speed fluctuates is between 0~120 r/min, which indicates that the system is in the state of stick-slip. The survey signals of triaxial accelerometers on the xyz axes as show in the Fig. 5.23a. These measured signals are used to analyses the drilling dynamics.

Moreover, as shown in Fig. 5.23a, when the drillstring rotates, we define the displacement of point A at the x axis as a time series $S(s_1, s_2, \cdots, s_{n-1}, s_n)$, with the circle radius of R. Obviously, $s_i = R \sin(\omega t)$. Then the velocity time series $V(v_1, v_2, \cdots, v_{n-1}, v_n)$, the acceleration time series $A(a_1, a_2, \cdots, a_{n-1}, a_n)$ can be obtained.

5.4.2 Quantification of the Risk Level of Drillstring Vibration

Schlumberger developed an independent quantitative vibration risk technique based on seismic engineering theory [49]. This technique is used to establish the relationship between the measured acceleration value and the failure parameters.

1. Drillstring axial and lateral vibration level

Vibration intensity is an index used to objectively characterize the strength of vibration energy. In the application of drillstring vibration, the parameter

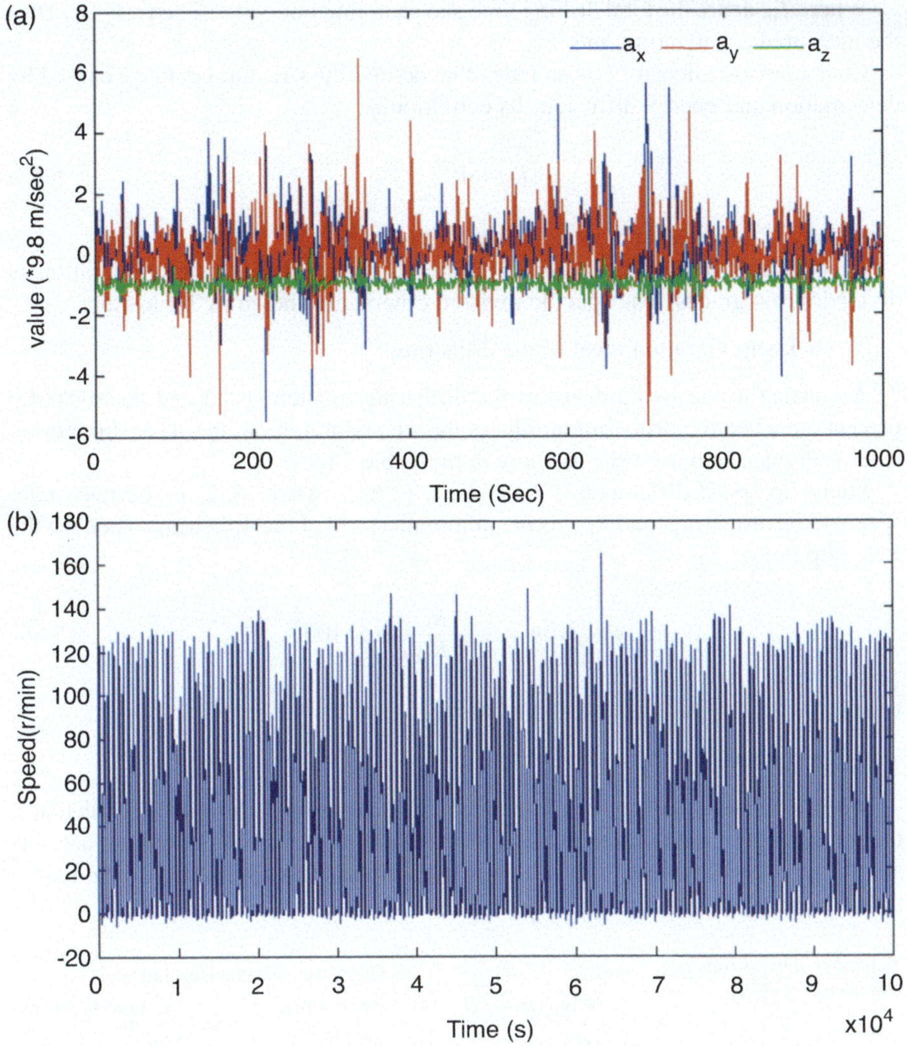

Fig. 5.23 The time series of measurement (**a**), survey signals of triaxial accelerometers on the *xyz* axes; (**b**), the time series of the drillstring rotary speed calculated by Eq. (2), the unstable angular speed indicates that the system is in the stick-slip state

used to describe the vibration intensity is recorded as a_{rms}, and the calculation formula is:

$$a_{rms} = \sqrt{\frac{1}{T_d}\int_0^{T_d} a(t)^2 dt} \tag{5.64}$$

Where T_d, dt are the total drilling time and sampling interval respectively, s; $a(t)$ is the measured acceleration, m/s^2.

Characteristic intensity I_c is an index that defines the structural damage caused by deformation and energy diffusion. Its definition is:

$$I_c = a_{rms}^{1.5} \cdot t_d^{0.5} \tag{5.65}$$

wherein $t_{\rm d}$ is the measurement duration, unit is s.

According to the characteristic intensity I_c value, the drillstring vibration is divided into 4 grades. The specific division criteria are shown in Table 5.5.

2. Stick-slip vibration level of the drillstring

According to the two indexes of the drillstring rotation speed and the stick-slip percentage when the drillstring produces the stick-slip motion, the stick-slip motion level is divided into 4 levels, as show in the Table 5.6.

Stick-slip speed difference is $R_s = R_{\max} - R_{\min}$, where $R_{\max}$ is the maximum speed of the drillstring, and $R_{\min}$ is the minimum speed of the drillstring. Then get the stick-slip ratio:

$$stick/slip = \frac{R_s}{2 \times R_{surface}} \times 100\% \tag{5.66}$$

wherein $R_{\rm surface}$ is the rotary speed of surface.

3. Impact level

When evaluating the degree of impact of the drill string, Schlumberger calculates the impact level based on the peak of the acceleration peak (CPS) with an acceleration peak exceeding 490 m/s^2, and divides the impact risk level into 4 according to the CPS value, as show in the Table 5.7.

Table 5.5 Drillstring vibration level [49]

	Characteristic intensity range	
Vibration level	Axial vibration	Lateral vibration
Low	<0.5	<0.5
Intermediate	0.5–1.5	0.5–2.5
High	1.5–3.5	2.5–5.5
Severe	>3.5	>5.5

Table 5.6 Stick-slip vibration level of the drillstring [49]

Stick-slip ratio	Stick-slip vibration level	Recommend
<0.5	Low	Normal drilling
0.5–1.0	Intermediate	For more than 25 h, the risk of failure is intermediate
1.0–1.5	High	For more than 12 h, the risk of failure is high
>1.5	Severe	For more than 0.5 h, the risk of failure is severe

Table 5.7 Impact level of the drillstring [49]

Impact ratio	Impact level	CPS
0	Low	<1
1	Intermediate	1–5
2	High	5–10
3	Severe	>10

5.4.3 *Mature Products of Vibration Measurement Systems*

1. Measurement system developed by Baker Hughes

The measuring tools developed by Baker Hughes include ground measurement systems and near-bit measurement systems, which can independently measure the vibration signal of the drillstring. The ground measurement system [50] is mounted above the kelly joint and can be used to measure the axial force and torque, axial and torsional acceleration and speed, and transmitted to the field processing device by microwave remote sensing, the data transmission rate is high, real-time and storage, digital filtering from three frequency bands by means of signal processing device, extraction and spectral analysis can realize time domain analysis, spectrum analysis and time-frequency analysis. The ground measurement system has a very significant analysis effect on torsion and stick-slip motion, and can also accurately determine the axial vibration condition, and at the same time, the resonance condition and the corresponding excitation frequency is analyzed and judged. However, the shortcoming of ground measurement technology is that it cannot accurately determine the lateral vibration of the well. This is because the drillstring has ultra-long slender characteristics and frequent contact with the wellbore causes lateral vibration has a large attenuation in the process of transmission to the wellhead [51].

Baker Hughes has also developed a downhole near-bit measurement tool [52] consisting of an axial accelerometer, two orthogonal bending strain gauges, two orthogonal magnetometers, and axial and torsional strain gauges. Axial acceleration, transverse bending moment, torsional angular velocity and axial force and torque are measured, so that vibration modes such as axial vibration, lateral vibration, torsional vibration and stick-slip motion can be analyzed. The system uses mud pulse technology to transmit to the ground device in real time. Downhole vibration signal, the data transmission rate is low. However, due to its close distance from the drill bit, the measured signal is highly realistic.

2. Drillstring Dynamics Sensor (DDS) developed by Halliburton

In order to accurately understand the motion state of BHA, Halliburton developed a drillstring vibration measuring tool DDS [53] equipped with a three-axis accelerometer in 1998, which consists of three mutually orthogonal accelerometers. DDS For the storage type vibration sensor, three kinds of vibration acceleration signals of acceleration mean value, peak value and instantaneous value can be recorded simultaneously for post-drill analysis.

By measuring the lateral, axial, tangential and radial accelerations to determine the lateral vibration, skipping, stick-slip motion and whirl, etc. Based on this information, the technician can pass the drill type, BHA structure, well structure and optimized selection of drilling parameters to reduce harmful vibrations of the drillstring and improve drilling efficiency [54].

The advantage of DDS is that it can deeply reveal the excitation mechanism and inherent characteristics of the vibration of the drill string through the analysis of the instantaneous characteristics of the acceleration obtained by high-frequency acquisition. The disadvantage is that the real-time vibration data cannot be obtained, so the feedback cannot be timely. Downhole vibration, but with the continuous improvement of mud pulse transmission technology, the problem of high-speed transmission of high-frequency data to the ground has been initially solved.

3. MVC (multi-axis vibration chassis) measurement system developed by Schlumberger

Schlumberger has developed a four-axis measuring tool MVC [55–57] mounted on the axis of the drillstring, which uses mud pulse technology to transmit downhole vibration data to the ground in real time, so that the vibration status of the drillstring can be monitored in time to improve drilling efficiency and guarantees the drilling safety.

The MVC tool is mounted on the shaft of the downhole BHA and consists of four axes: the first axis contains the strain gauge and the sensor for measuring torque, and the other three axes contain three accelerometers (measured as a_x, a_y, a_z) and a vibration acquisition board, And the three axes are orthogonal to each other. The a_x value reflects the axial vibration, and a_y, a_z reflects the lateral vibration in two orthogonal directions. The signal transmitted by the MWD (measurement while drilling) to the ground in real time is the root mean square value RMS (root mean square) and the torque signals. In addition, a separate sensor is installed on the MWD to establish a risk level CPS (counts per second) based on the number of impacts (maximum acceleration over $50*9.8\ m/s^2$). When the impact of more than $50*9.8\ m/s^2$ in more than 10 times in 1 s (CPS value is greater than 10), the risk of tool failure must be considered. The shortcoming of MVC real-time uploading method is that it can't analyze the spectrum of its signal, so it can't further analyze the frequency component of the downhole vibration signal. This is because the sampling frequency is low due to the small amount of transmitted data.

5.4.4 Signal Processing Technology of Drillstring Vibration

In the drillstring vibration measurement technology, the vibration signal processing technology is very important, which is the key to understand the vibration information composition and the type of excitation source. The drillstring vibration signal is a typical non-stationary random signal. The processing methods are mainly time domain analysis, frequency domain analysis, time-frequency analysis and wavelet analysis. The time domain analysis reflects the variation of the vibration amplitude of the drillstring with time, which is mainly reflected in the variation of the energy of the drillstring with time. The frequency domain analysis includes spectrum analysis, wavelet analysis, etc. The theoretical basis is Fourier transform. The analysis can determine the distribution of the vibration signal of the drill string in the frequency band and the energy intensity of each frequency component.

5.4.4.1 Time Domain Analysis Method

The vibration signal is an acceleration time domain signal obtained according to a certain sampling frequency. The acceleration peak indicates the impact magnitude of the vibration, the acceleration average indicates the overall trend of the drill string vibration, and the acceleration rms value indicates the strength of the drill string vibration signal.

The acceleration peak, mean and root mean square expression are:

$$a_m = \max |a(t)| \tag{5.67}$$

$$\overline{a} = \frac{\sum_{1}^{N} |a(t)|}{N} \tag{5.68}$$

$$a_{rms} = \sqrt{\frac{\sum_{1}^{N} a^2(t)}{N}} \tag{5.69}$$

5.4.4.2 Frequency Domain Analysis Method

The spectrum is an important feature of the signal in the frequency domain. It reflects the frequency component and distribution of the signal. The theoretical basis is the Fourier transform, which decomposes the vibration signal into a series of superpositions of sine waves of different frequencies.

The mathematical expression of the Fourier transform is as follows:

$$X(\omega) = \int_{-\infty}^{+\infty} x(t) e^{-i2\pi f} t dt \tag{5.70}$$

Where *x*(*t*) is the time domain signal, *X*(*ω*) is the transformed frequency domain amplitude signal, and *f* is the frequency.

Fourier transform has the disadvantages of large computational complexity and high storage space requirements. For this reason, Cooley et al. [58] proposed fast Fourier transformation in 1965. FFT algorithm is an improvement on the algorithm of discrete Fourier transform. It can greatly reduce the number of multiplications required for the computer to calculate the discrete Fourier transform, and the operation speed is significantly improved.

Although FFT spectrum analysis is widely used and the algorithm is simple, its shortcomings are obvious. It is mainly manifested in two aspects: single frequency resolution and lack of time-frequency analysis capability. Morlet [59], who worked on petroleum signal processing in France, first proposed wavelet analysis in 1974. Wavelet analysis is the decomposition of signals into a series of wavelet functions. These wavelet functions are derived from a mother wavelet function through translation and scale expansion. The localization of the wavelet transform is variable, with high time resolution and low frequency resolution at high frequencies; At low frequencies, the time resolution is low and the frequency resolution is high, so the wavelet analysis method has the property of adaptive window.

The definition of wavelet transform is:

$$WT(b,c) = \frac{1}{\sqrt{b}} \int_{-\infty}^{+\infty} x(t)\overline{\varphi}\left(\frac{t-c}{b}\right) dt \tag{5.71}$$

Where $WT(b, c)$ is the transformed frequency domain amplitude signal; $x(t)$ is the time domain signal; b, c are the scale parameter and the displacement parameter, respectively. Using the wavelet transform analysis method to process the spectrum of the drillstring vibration signal, an important breakthrough has been achieved.

5.4.4.3 Time-Frequency Analysis Method

The time-frequency analysis method is a signal analysis method that can reflect both time domain features and frequency domain features. The short-time Fourier transform can analyze non-stationary dynamic signals and analyze the time-frequency characteristics of the drillstring signals. The frequency components of the drillstring at different times are obtained. Commonly used time-frequency analysis methods include short-time Fourier transform, empirical mode decomposition method, etc.

Short time Fourier transform definition:

$$TFT_x(\tau,f) = \int_{-\infty}^{+\infty} x(t)\left[h(t-\tau)e^{j2\pi f}t\right] dt \tag{5.72}$$

Where, $TFT_x(\tau,f)$ is the transformed time-frequency domain amplitude signal, $x(t)$ is the time domain signal, and $h(t-\tau)e^{j2\pi f}$ is the synthesis window.

In summary, the time domain analysis can timely grasp the vibration state and motion form of the drill string, so as to adjust the drilling parameters to avoid serious downhole vibration [60]. In addition, further analysis and analysis of the vibration signal of the drillstring can be performed. In-depth understanding of the excitation characteristics of the induced downhole vibration and the action mechanism of the drillstring and the rock formation, optimize the drilling tool structure, and fundamentally weaken the impact of the vibration of the drillstring body.

5.4.4.4 Wavelet Noise Reduction of Vibration Signals

In the orthogonal wavelet transform, the low frequency portion is decomposed each time, so the frequency resolution is low in the high frequency portion of the signal, and the time resolution is low in the low frequency portion of the signal. Some practical problems may require more detailed analysis of certain time-frequency windows, and wavelet packet analysis provides a solution.

Assumed wavelet packet $\psi^i_{j,k}(t)$ is a function of amplitude modulation i, range parameter is j, The translation parameter is k, which can be expressed as follows:

$$\psi^i_{j,k}(t) = 2^{j/2}\psi^i(2^j t - k) \qquad i = 1, 2, 3, \cdots \tag{5.73}$$

Base wavelet function ψ^t will be obtained:

$$\psi^{2t}(t) = 2^{1/2} \sum_{k=-\infty}^{\infty} h(k)\psi^t(2t - k) \tag{5.74}$$

$$\psi^{2t+1}(t) = 2^{1/2} \sum_{k=-\infty}^{\infty} g(k)\psi^t(2t - k) \tag{5.75}$$

wherein $h(k)$ and $g(k)$is filter coefficient. At each decomposition level N, have 2^N wavelet packet. This allows the input signal to be analyzed at different frequency bandwidths. Wavelet packet analysis provides an effective analytical method for sensor measurement signals that are affected by drillstring vibration. The analysis of the 3-stage decomposition wavelet packet for the input signal is shown in Fig. 5.24. If I have a good understanding of the vibration law of the drillstring, we can use the

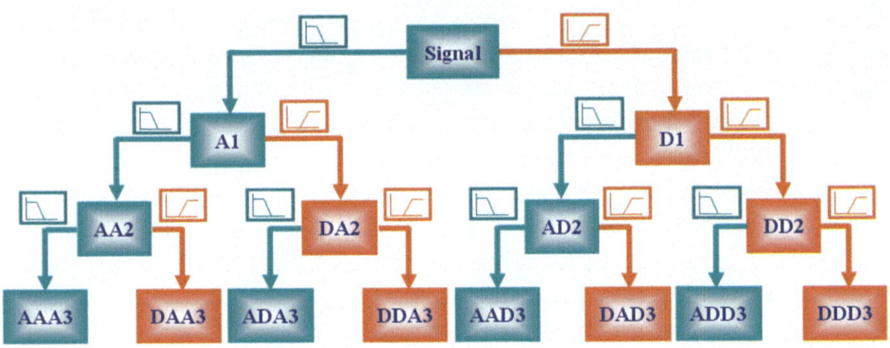

Fig. 5.24 Wavelet packet transform of decomposition of the input signal

wavelet packet analysis to separate the vibration shock signal from the acceleration signal used for attitude measurement.

The vibration signal is decomposed by 6-layer wavelet using the wavelet analysis toolbox in Matlab software. The base wavelet used is Debauches wavelet, and the analysis results are shown in Fig. 5.25.

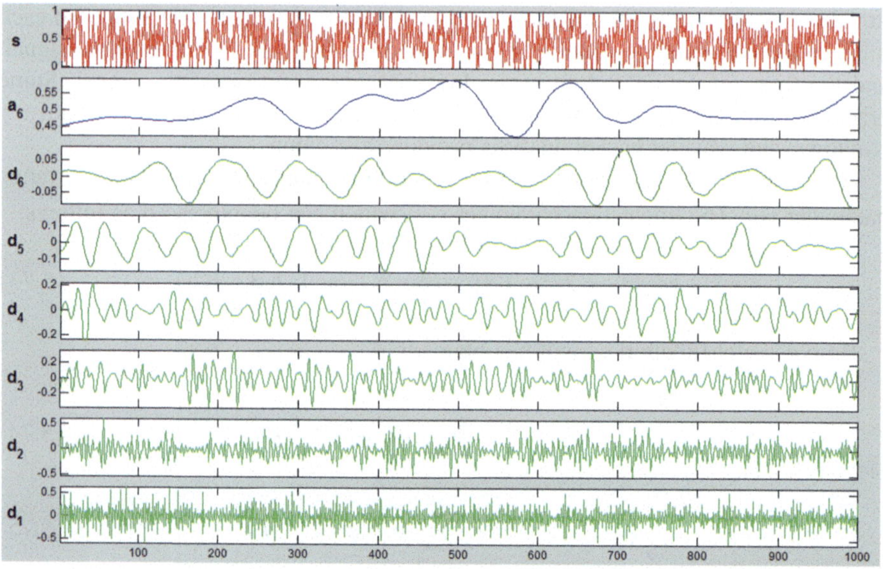

(a) No Stick Slip

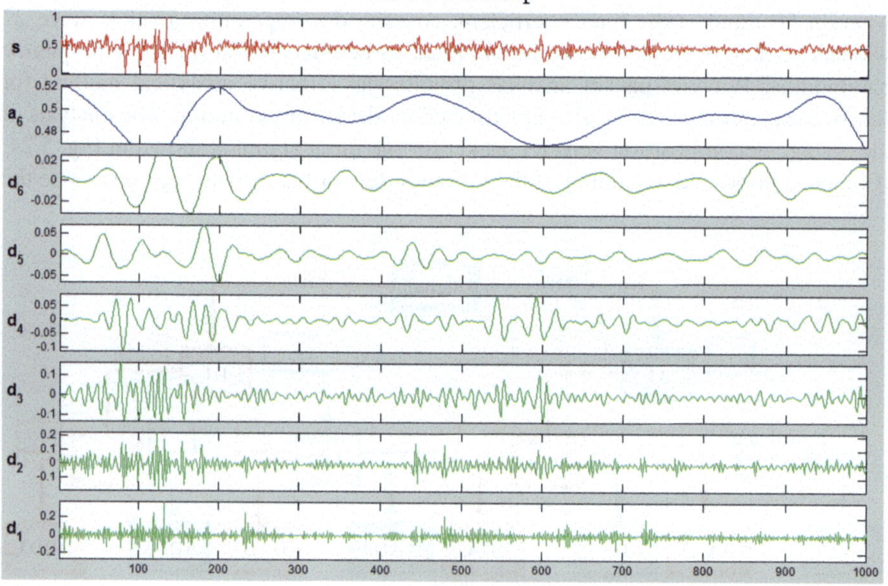

(b) Stick Slip

Fig. 5.25 Wavelet decomposition of accelerometer signal

As can be seen from Fig. 5.25, the d6 signal is closest to the change law of the fluxgate, and the fluxgate signal and the wavelet decomposed d6 signal are put together for comparison. That the decomposed d6 signal is identical to the main frequency of the fluxgate signal except for the phase difference. However, when the drillstring is stick slip, the d6 signal completely shows the opposite trend of the fluxgate signal, which indicates that the wavelet decomposition is still powerless for the stick-slip phenomenon of the drillstring.

From the signal decomposed in Fig. 5.25, d1 represents the high-frequency component of the signal, reflecting the dynamic characteristics of the drill string at this time, we can use the d1 signal at this time as a measure of the degree of vibration of the drill string. Define as following:

$$\|d1\| = \left[\sum_{k=1}^{n_d} D1^2(k)\right]^{1/2} \tag{5.76}$$

where n_d is the coefficient of wavelet packet d1, d1 can well reflect the severity of the vibration or impact in the drilling, and can be used as a parameter in subsequent signal processing.

In the orthogonal wavelet, the orthogonal basis is more dominant than the traditional signal processing method, and the noise signal can be more easily discerned and separated by the wavelet transform. If a signal $f(n)$ is assumed to be changed $s(n)$ after adding noise interference, it can be expressed as the following model:

$$s(n) = f(n) + \sigma e(n) \tag{5.77}$$

where $e(n)$ is the noise, σ is noise intensity. Assume $e(n)$ is Gaussian white noise, and $\sigma = 1$. The purpose of wavelet transform noise reduction is to suppress $e(n)$ as much as possible to recover $f(n)$. From a statistical point of view, this model is a regression model over time. This decomposition method can also be regarded as a non-parametric estimation of the function f on an ortho basis.

Represent the signal as L_ϕ on the wavelet domain, is the decomposition coefficient of the original noise-containing signal under wavelet transform. Threshold operator f_δ retained coefficient terms with large modulus values after the wavelet effect, other coefficients with small modulus values are set to zero. Then:

$$(F_\delta c)_{m,n} = \begin{cases} c_{m,n}, & |c_{m,n}| > \delta \\ 0, & other \end{cases} \tag{5.78}$$

The mask operator is M, Its function is to retain the specific coefficients needed and set the other coefficients to zero. Then,

$$(M_c)_{m,n} = \begin{cases} c_{m,n}, (m,n) \in Q \\ 0, other \end{cases} \tag{5.79}$$

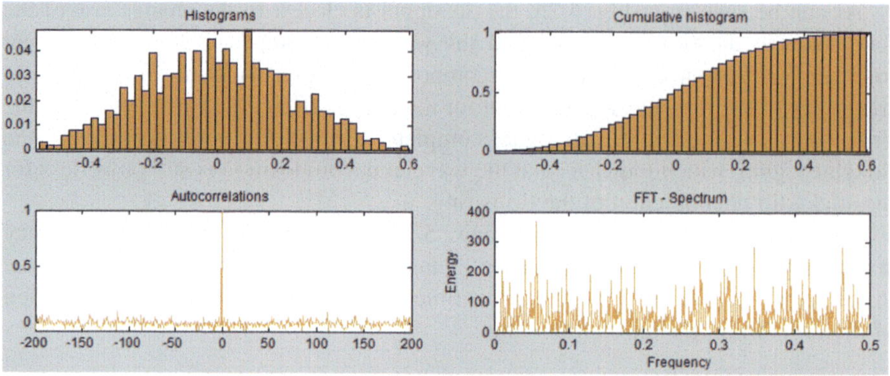

(a) No Stick Slip

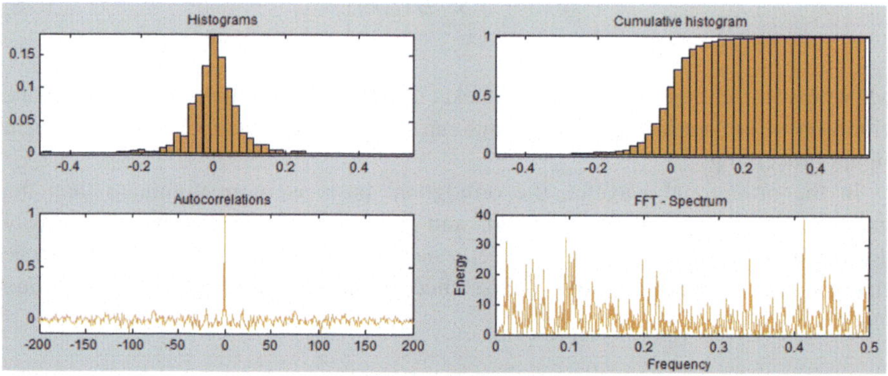

(b) Stick Slip

Fig. 5.26 Statistical analysis of wavelet decomposition of accelerometer signal d6

It can be seen that the generalized threshold operator can obtain the mask operator, and the threshold operator is implemented by introducing the mask operator of the coefficient modulus.

The db6 wavelet packet analysis is used for signal denoising processing, as shown in Fig. 5.26, the statistical analysis results of the original input accelerometer signal and the spectrogram.

From the frequency domain, the stick-slip state and the non-stick slip state are not significantly different. However, on the statistical histogram, there are still significant differences between the two. When the drillstring is stick slip, the velocity amplitude is more concentrated in the zero-speed region. Using Matlab software, the wavelet denoising effect is simulated and analyzed. The noise reduction result is shown in Fig. 5.27. The red line is the filtered signal. It can be seen that the signal amplitude is significantly reduced.

The degree of signal fluctuation can be measured due to the standard deviation σ, we use the standard deviation as a measure of the signal noise reduction effect. After the wavelet denoising process is performed on the three-axis signals of the

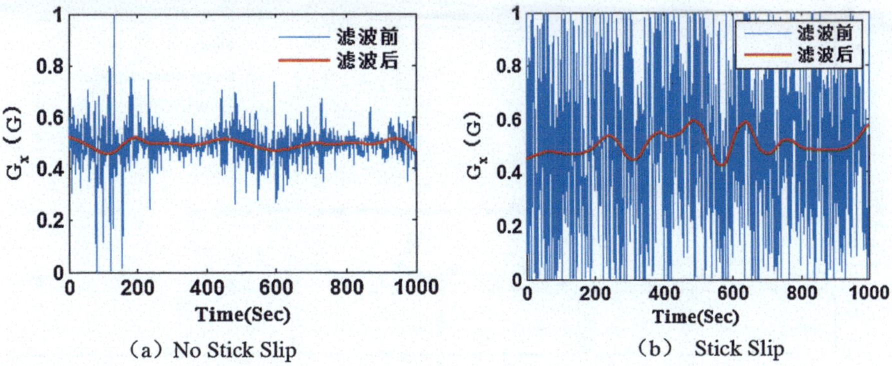

Fig. 5.27 The results of wavelet noise reduction

Table 5.8 Signal output uncertainty of raw and filtered measurements

	G_x	G_y	G_z
Original signal σ	7.09	7.02	1.36
Filtered signal σ	1.13	1.86	0.16
Degree of improvement	84%	73%	88%

accelerometer, the standard deviation of the filtered signal is calculated, and the pair of original signals are shown in Table 5.8.

It can be seen that the amplitude of the signal after noise reduction is above 70%. However, for the gravitational acceleration signal, it has the same sinusoidal variation as the fluxgate signal. The filtered signal does not clearly indicate this characteristic. Especially in the stick-slip state, the signal fluctuates constantly is lost after filtering.

5.5 Vibration Characteristics Based on Dynamics Measurement

5.5.1 Data Analysis of Torsional Vibration

5.5.1.1 Surface Data of Drilling Process

The (Weight on the bit) WOB, surface rotary table torque and speed, pump pressure, as well as the underlying lithology will combine effect of the drilling torsional vibration. We used the drilling field test (Anshun well in China) data to do a detailed analysis. At the period of using rotary steerable system, the well logging data show in Fig. 5.28. It can be seen the surface data has been little changed. Therefore, it can be inferred that the drill bit complex torsional vibration is caused by the underground environment.

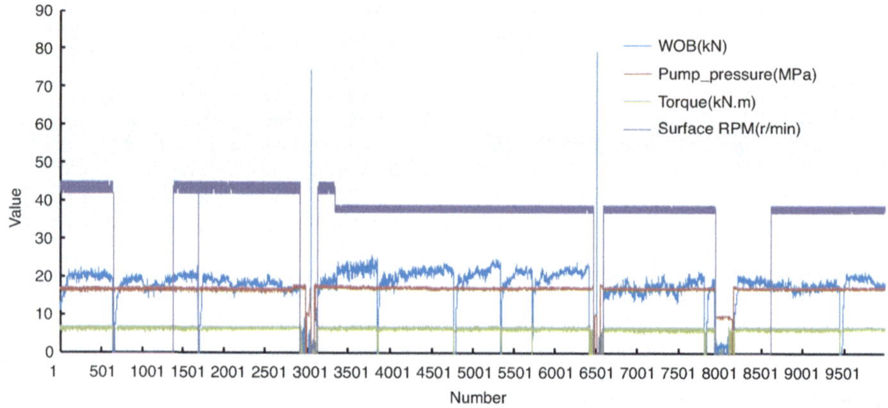

Fig. 5.28 Surface data of drilling process

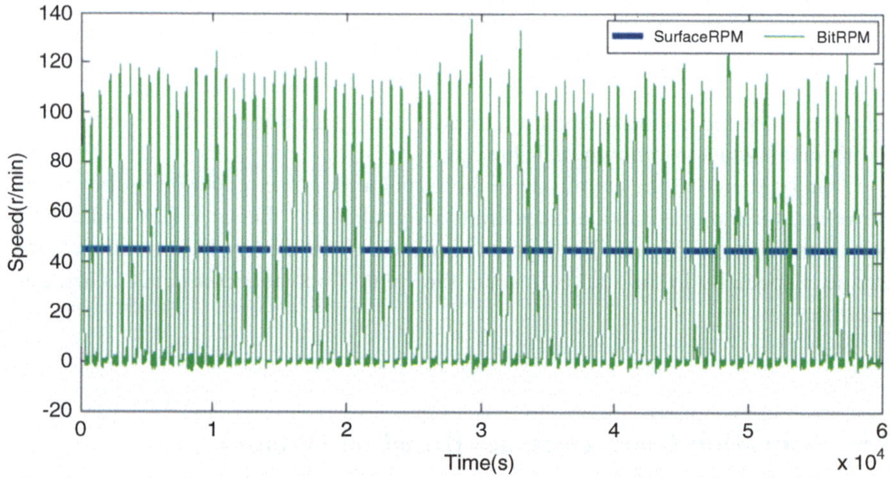

Fig. 5.29 SurfaceRPM and BitRPM of field test (The data is replayed from the underground storage system)

5.5.1.2 Torsional Vibration Patterns

Extracted a period of underground measurement data, as shown in Fig. 5.29, the rotary table speed maintained at 45 r/min, the drill bit speed has fluctuated between 0–140 r/min, stick-slip phenomenon is very serious. Simulation by torsional vibration model as described above, as shown in Fig. 12, the torsional vibration phenomenon significantly increased when T_Y exists. Simulation results show that the torsional vibration phenomenon of push-the-bit rotary steerable drilling system is

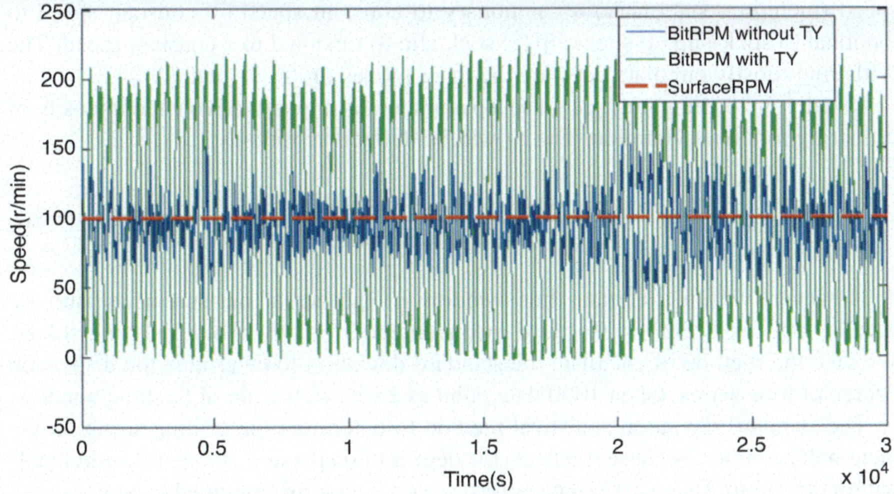

Fig. 5.30 SurfaceRPM and BitRPM of simulation with torsional vibration model

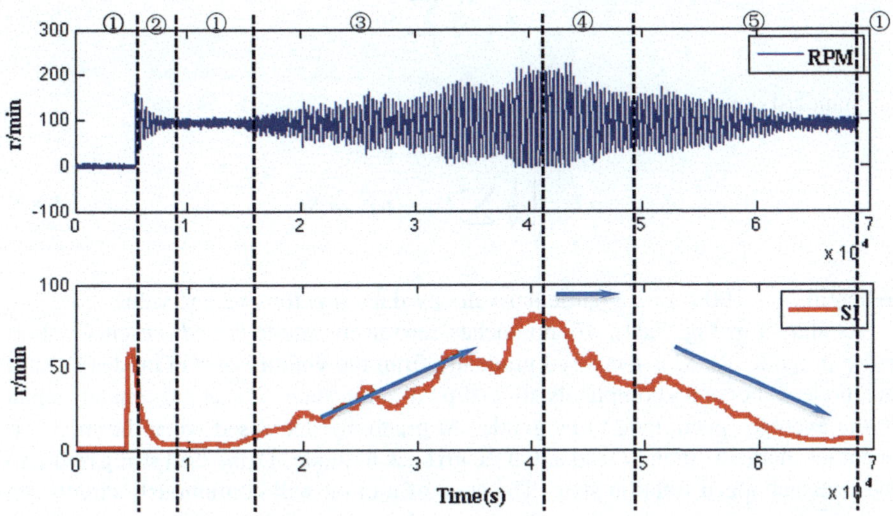

Fig. 5.31 The movement patterns of drill bit torsional vibration (The data is replayed from the underground storage system)

indeed more serious than the conventional of. It can be concluded that the model established in this chapter very good response of the actual drilling process by way of compare the BitRPM with T_Y in Fig. 5.30 and the BitRPM in Fig. 5.29.

Through statistical analysis of speed time series, the bottom of the drill torsional vibration phenomena is divided into the following modes (Fig. 5.31): ① constant speed (including stationary) ② stationary to constant speed ③ constant speed to torsional to stick-slip ④ stick-slip ⑤ stick-slip to torsional to a constant speed. The drillstring mostly circulate between the three modes of ③, ④ and ⑤.

Usually the drilling engineers use a parameter *SI* to quantitatively the intensity of drillstring torsional vibration. The *SI* will be defined as follows:

$$SI = \frac{\max(RPM) - \min(RPM)}{2 \cdot avg(RPM)} \tag{5.80}$$

However, within a certain time window of the actual measurement data, we usually met $avg(RPM) = 0$, cause inconvenience for downhole real-time solver. We take the method of calculate the standard deviation to determine the dispersion degree of time series, taken 1000 data point as each calculation of the time window.

The standard deviation statistical method to determine the drilling string movement will be better, because it reflects the degree of dispersion among the individuals within the group. Using 1000 data points as a time window, assumed to be $x_1, x_2, \ldots, x_{999}, x_{1000}$, the average value defines as u.

$$u = \frac{1}{N}\sum_{i=1}^{N} x_i \tag{5.81}$$

Standard deviation σ obtained:

$$\sigma = \sqrt{\frac{1}{N}\sum_{i=1}^{N} (x_i - u)^2} = SI \tag{5.82}$$

Wherein $N = 1000$, x_i is a particular velocity data, u is the average value.

As shown in Fig. 5.31,, *SI* approaches zero in the mode ①; *SI* presents a short pulse in mode ②; *SI* is increased gradually from the vicinity of 0 in mode ③, until the mode ④ becomes completely stick-slip vibration state, *SI* is approximately equal to the average speed, tend to be gentle., *SI* gradually decreased when the mode ④ becomes mode ③, then *SI* tend to 0 until go back to mode ①, the drill string return to the constant speed rotation state. The state of motion will continuously circulating between these modes when the drillstring work in the underground, we need to focus on these three modes of ③, ④ and ⑤, because the status of these movements is the main reason for the downhole drillstring fatigue damage.

Assume the time series of bit rotational speed finally resolved as $\omega_0, \omega_1, \cdots, \omega_n$, then the time series of rotation angle is $\theta_0, \theta_1, \cdots, \theta_n$, sampling frequency is $f = 1/_{\Delta t} = 100Hz$. Then rotation angle and rotation speed satisfy will be expressed as following:

$$\theta_{j+1} = 2\pi\omega_j t + \theta_j, \theta_0 = 0 \tag{5.83}$$

Using the FFT algorithm, Fourier transforms for the time series of rotational speed is:

$$F_k = \Delta t \sum_{j=0}^{n-1} \omega_j e^{-i\frac{2\pi k}{n}j} \tag{5.84}$$

Then its power spectral density will be obtained:

$$S_k = S\left(\frac{2\pi k}{n\Delta t}\right) = \frac{1}{n\Delta t}|F_k|^2 \tag{5.85}$$

As shown in Fig. 5.32, the power density spectrum has been obtained using periodogram method with Matlab. It is clear that the time series follow a power-law behavior where some characteristic frequencies also exist and are indicated *f*1 (60dB/rad), *f*2 (55dB/rad) and *f*3 (47dB/rad) in the Fig. 14. The power-law behavior is indicative of long-range correlations and self-similar behavior. There is a big difference between model ① and the other model in the lower frequency segments. This is because the drillstring is not rotated at this time, so the sensor signal should be constant. The high frequency component of the spectrum analysis mainly comes from the random noise. The mode④ occurs characteristic frequency at f1, showing the stick-slip vibration period at this time. Mode②, Mode③ and Mode⑤ are the transition stage between the stationary and stick-slip vibration. Their frequency spectrums show more similarity. The high frequency noise always exists in the measurement process, which is why each mode showed the same characteristics in the high frequency.

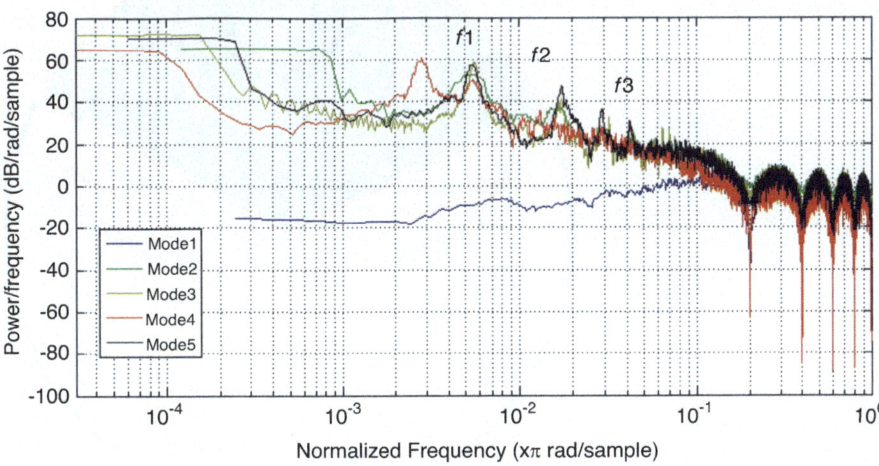

Fig. 5.32 Time-frequency analysis for different movement mode

5.5.1.3 Uneven Distributed Data of the Rotary Speed

The simulation results in the Fig. 5.30 shows that the frictional torque generated by the pads in the RSS will cause the drill bit torsional vibration more severely. The frictional torque T_Y is not constant. It will appear when one of three pads pushed the borehole wall. Hence, the speed of drill bit will decrease when the pads pushed out, this will cause the drill bit torsional vibration. From our measurement data, the drill bit torsional vibration appears all over the drilling process. If the frictional torque generated by the pads is the main reason that leads the drill bit torsional vibration, we can deduce that the rotary speed will slow down when the drill bit rotates to a certain angle.

As show in Fig. 5.33, one of pads pushed out when it rotated to the point A and then keep push through the angle α arrived to the point B. This process will make the speed of drill bit slow down. The drillstring continuing clockwise rotate when the pad will take back at the point B. The speed of the drill bit will increase because of there is no frictional torque generated by the pads. Thus, if we consider the statistic of downhole measurement data in the position of a circle, there must have uneven distribution. In Fig. 5.33, at the arc of AB (red line), the speed will be lower than the other part in the circle. We can find the evidence in the distribution of the downhole measurement data.

The pads of the rotary steerable system were periodically pushed in a fixed position against the borehole wall in the process of drillstring rotating. This is the main difference between RSS and conventional drilling. Frictional force between the pads and borehole wall will make the drill bit instantaneous rotational speed reduces. Assuming the radius of the borehole is R, when the drillstring rotate to a circle, at a

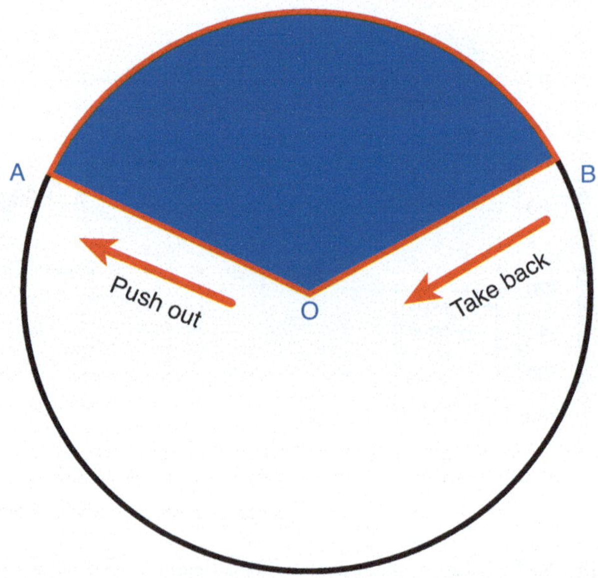

Fig. 5.33 Position of the pads pushing when the drillstring rotating

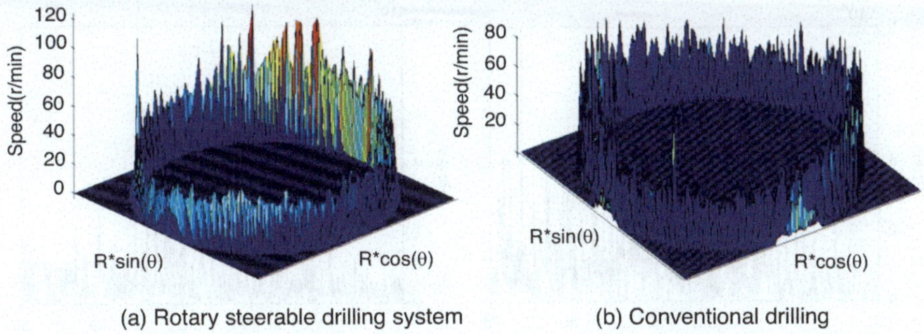

Fig. 5.34 Distribution map of the rotation speed in the rotational position

certain angle θ there will be has a speed value at the position of ($R*sin\theta$, $R*cos\theta$) in the orthogonal coordinates. We draw all the speed data in a three-dimensional figure as shown in Fig. 5.34. We can obtain the drill bit speed distribution map of different angular position, part a is obtained from the RSS data which showing uneven distribution between high and low-speed, part b is obtained from the conventional drilling data which showing evenly distribution. Due to the Fig. 5.34 is generated from all the data of drilling test, there will be have zero speed at each point of the circle. Additionally, in the drilling process of RSS, there will must be having the time that all the pads not pushing out as well as ordinary drilling system.

It can be concluded that the generation mechanism of torsional vibration is completely different between push-the-bit RSS and the ordinary drilling system. Although the conventional drilling also produces torsional vibration (Rotational speed range of 0–80 r/min in Fig. 5.34b), the drilling bit rotational speed at different rotation angles will not be a difference. Because the torsional vibration produced by the drill bit breaking rock is completely disorder and stochastic.

5.5.2 Data Analysis of Lateral Vibration

In the measurement signals, a_x,a_y,a_z are defined as survey signals of triaxial accelerometers on the xyz axis respectively. Under certain sample frequency, measuring signal is time series and can be expressed as time function. We can use $a_h = \pm\sqrt{a_x^2 + a_y^2}$ define the lateral vibration of the BHA and use a_z express the longitudinal vibration.

As shown in the Fig. 5.35, the frequency spectrum of the measurement time series will be obtained. There is no characteristic frequencies in the a_x and a_y surveying signals.

We used Welch's method [61] approach to spectral density estimation. The method is based on the concept of using periodogram spectrum estimates, which

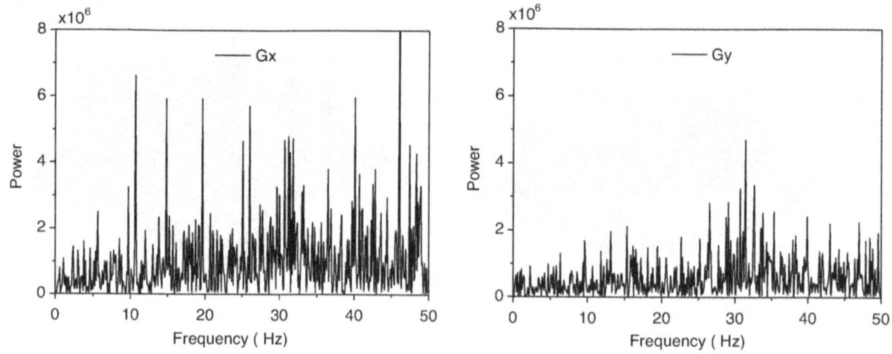

Fig. 5.35 The spectrogram of x and y signals of accelerometer

are the result of converting a signal from the time domain to the frequency domain. Welch's method reduces noise in the estimated power spectra in exchange for reducing the frequency resolution. Due to the noise caused by imperfect and finite data, the noise reduction from Welch's method is often desired.

The time series of lateral vibration as shown in the Fig. 5.36a, frequency spectrum as show in the Fig. 5.36b. Lateral vibration violently amplitude between 0–10G. Since the sampling frequency is 100 Hz, we can obtain frequency spectrum domain at 0~50 Hz, the characteristic frequency of the signal is not significant. It can be inferred that lateral vibration signals are not periodic, essentially a random vibration. The Fig. 5.36c indicate density of amplitude distributing, most of the vibration amplitude concentrated between -5G~ + 5G. Fig. 5.36d show the spectral density estimation using Welch's method. The curve of power spectral density did not show significant spikes, indicating that lateral vibration can be regards as random vibration signals at this time.

Our treatment of frequency analysis would not be complete without discussing the most up-to-date methods involving dynamic spectral analysis, analyzing how the spectrum of a signal evolves in time. This analysis is carried out with Gabor transforms, which are close relatives of wavelet transforms. Gabor transforms are particularly useful in analyzing drillstring vibration signals. It is used to determine the sinusoidal frequency and phase content of local sections of a signal as it changes over time. The function to be transformed is first multiplied by a Gaussian function, which can be regarded as a window function, and the resulting function is then transformed with a Fourier transform to derive the time-frequency analysis.

Using the FFT algorithm for the time series of vibration is:

$$F_k = \Delta t \sum_{j=0}^{n-1} \omega_j e^{-i\frac{2\pi k}{n} j} \tag{5.86}$$

Then its power spectral density will be obtained:

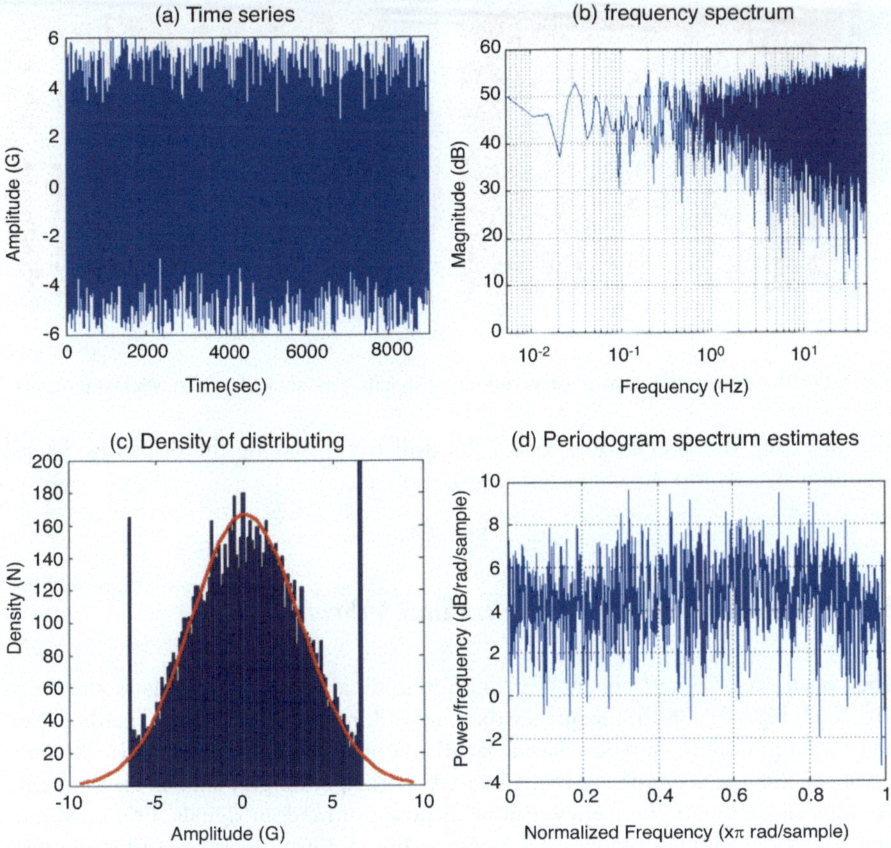

Fig. 5.36 Time-frequency analysis of lateral vibration

$$S_k = S\left(\frac{2\pi k}{n\Delta t}\right) = \frac{1}{n\Delta t}|F_k|^2 \tag{5.87}$$

We using time-frequency box to express the energy distribution of Gabor function. The method of Gabor transform can help us discriminate the rotational movement pattern. Then Fourier transform of the $G_{u,\ \xi}(t)$ will be obtained:

$$\widehat{G}_{u,\xi} = \widehat{G}(\omega - \xi)e^{iu(\omega-\xi)} \tag{5.88}$$

As show in Fig. 5.37a, when the drillstring not exhibit stick-slip, the spectrum not particularly has energy concentration point, the signals can be considered as random. However, when the drillstring exhibit as stick-slip, the spectrum particularly occur energy concentration stripe, as shown in the Fig. 5.37b. The results show that the stick-slip of the drillstring not only reflecting the change of the rotating speed, but

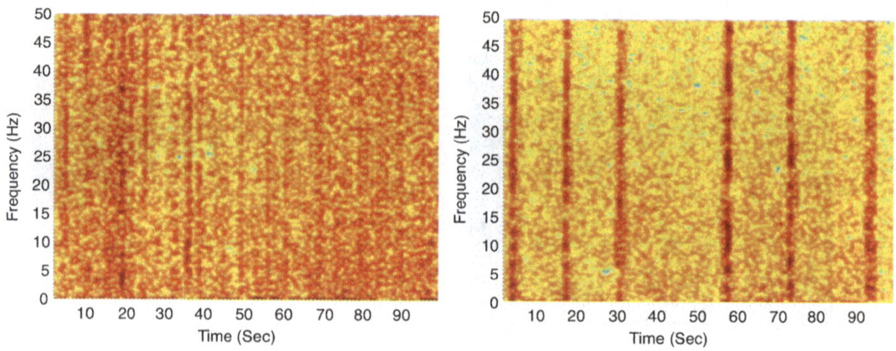

Fig. 5.37 The energy spectrum of lateral vibration signals, non-stick-slip (left), stick-slip (right)

also affecting the lateral vibration of the drillstring, making the drillstring lateral vibration signals lost the characteristics of random.

5.5.3 Data Analysis of Longitudinal Vibration

The spectrum of time-frequency of the longitudinal vibration signals is shown in Fig. 5.38. Figure 5.38(a) is sequence diagram of longitudinal vibration, getting from charts, longitudinal vibration has a smaller amplitude than horizontal vibration. Figure 5.38(b) is a frequency range chart as correspondingly, signals also have no obvious characteristic frequency, can be disposed of random signals. From distribution density of vibration ranges, as shown in Fig. 5.38(c), the range of longitudinal vibration mostly occurred between 0~2G, smaller than horizontal vibration. As shown in Fig. 5.38(d), the power spectrum density worked by period plan estimation, extreme values occurred on three frequencies, proving time sequence of longitudinal vibration intensified at some frequencies, it may be related to formation lithology. The acceleration signals at z-axis have apparent noise spectrum characters, proving under the rotation state, vibration of drilling tools have more regularly influence to the signals of accelerated signals at axial direction.

Comparing with horizontal vibrations, longitudinal vibrations are bare, the acceleration sensors at z-axis are influenced by vibration of drill columns hardly, under the condition of the rotation. Acceleration sensors are put on the center axis of drilling supporter, not influenced by rotation speed of drilling tools theoretically, so under a short time period, the time sequence of accelerated speed measurement signals at z-axis should be a line paralleled to the x-axis. Based on the analysis above, the filtering of signals on the accelerate sensor at *z*-axis is much easier to realize than x-axis and *y*-axis. As the analysis of horizontal vibration, use instant Fourier transform to make vibration time-frequency energy contour under two conditions of stick-slip or not.

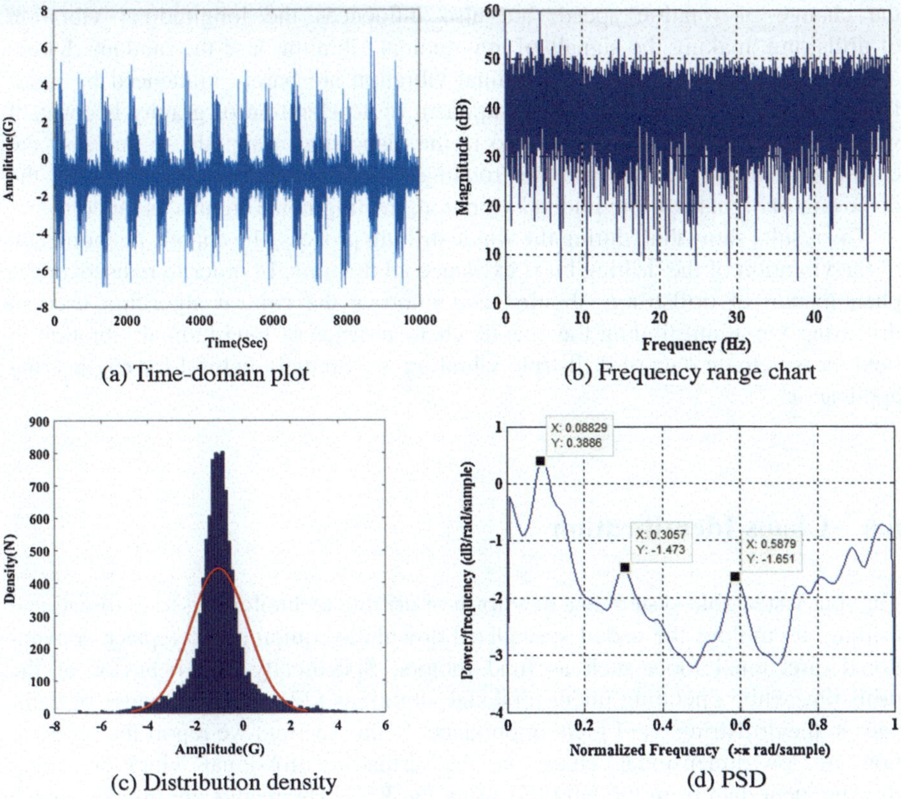

Fig. 5.38 Time-frequency analysis of the longitudinal vibration

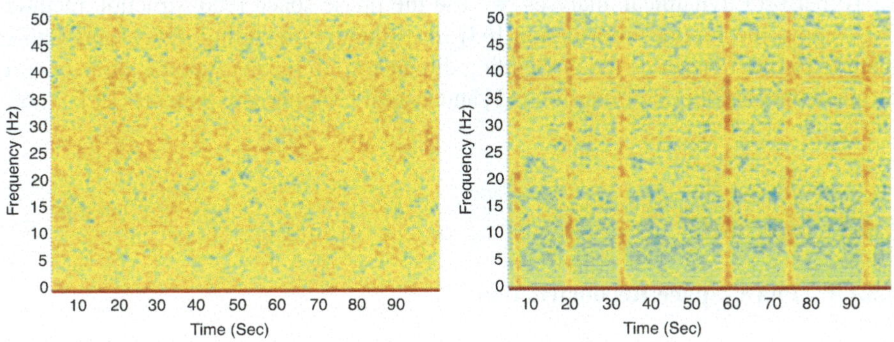

Fig. 5.39 The energy spectrum of the longitudinal vibration signal, non-stick-slip (left), stick-slip (right)

As shown in Fig. 5.39, when there is no stick-slip vibration of drillstring, the time-frequency has no point of concentration, disposing of the random signals like the same before. But when it does exist stick-slip phenomenon, there are stripes occurred on time-frequency contour, illustrating stick-slip vibration not only reflects the change of rotation speed, but also influences the longitudinal vibration of drillstring, making the signals of longitudinal vibration lose the random characters. Because the signals of longitudinal vibration are barely influenced by noise interference, and the longitudinal component of acceleration of gravity is constant whether drillstring rotating or not, so in the disposing of signals, to decrease the calculated amount, the reverse and rotating vibration can be ignored during the disposing of signals of longitudinal vibration, treating noise signals as random.

The results show that, during the whole drilling process, the chaotic phenomenon of the vibration of the drilling bit is existence all the time. In order to rediscover the phenomenon of drillstring vibration and improve the control algorithm used in drillstring vibration, finding the chaotic characteristics of longitudinal vibration in field measurement data of drillstring vibration is extremely helpful for engineering application.

5.6 Chaos Identification

The rotary steerable system is a new form of drilling technology used in directional drilling. It employs the use of specialized downhole equipment to replace conventional directional tools such as mud motors. Specifically, the behavior of the drillstring while operating under torsional vibrations [22], the main cause of damages to the drillstring, is of great importance. In this section, we report the observation of low-dimensional chaos in the drillstring torsional vibration using measurement data from the field test using the RSS. The results are in good match with mathematical models. To our knowledge, this is the first report of the finding of order-chaos transitions in a real drillstring system.

To perform dynamical analyses, we use the phase-space reconstruction method [62] (apply the mutual information [63] to estimate embedding delay, and Cao's method [64] to determine the embedding dimension) to reconstruct the attractor and estimate the correlation dimension [65] and largest Lyapunov exponents (LLE) [66].

5.6.1 Methods

5.6.1.1 Phase-Space Reconstruction

The dynamics of the time series $x_0, x_1, \cdots, x_{n-1}$ are fully captured or embedded in the m-dimensional phase space, $m \geq d$ where d is the dimension of the original attractor. A vector $\vec{x}_i$ in the reconstructed phase space [62] is constructed from the time series as follows:

$$\vec{x}_i = \left[x_i, x_{i-\tau}, \cdots, x_{i-(m-1)\tau}\right] \tag{5.89}$$

where τ is the delay time.

Cao's method [64] computes E_1 and E_2 for the data set of dimension 1 up to a dimension of D, which is the largest embedding dimension, used for calculate. E_1 and E_2 defined as follows:

$$E_1(d) = \frac{1}{N - d\tau} \left| \sum_{i=1}^{N-d\tau} \left| x_{i+d\tau} - x_{n(\mathrm{i,d})+\mathrm{d}\tau} \right| \right| \tag{5.90}$$

$$E_2(d) = E_1(d+1)/E_1(d) \tag{5.91}$$

wherein d is the embedding dimension, N is the number of data points, τ is the embedding delay, $x_{i\,+\,d\tau}$ and $x_{n(i,\,d)\,+\,d\tau}$ is the *i-th* vector in the data sets and its nearest neighbors of d-dimensional phase space.

5.6.1.2 Largest Lyapunov Exponent (LLE)

The basic characteristics of chaotic motion are that the movement is extremely sensitive to initial conditions, two very close initial values resulting in orbit over time by separating exponentially, Lyapunov exponent [66, 67] that describes the amount of this phenomenon.

We use the algorithm of Rosenstein et al. [67] to calculate the LLE. The results were carried out with Tisean package [68], version 3.01. Consider the representation of the time series data as a trajectory in the embedding space, and assume that observe a very close return $s_{n'}$ to a previously visited point s_n. Then consider the distance $\Delta_0 = s_n - s_{n'}$ as a small perturbation, $\Delta l = s_{n+l} - s_{n'+l}$. If one finds that $|\Delta_l| \approx \Delta_0 e^{\lambda l}$ then λ is the largest Lyapunov exponent.

Assuming $S(\varepsilon, m, t)$ exhibits a linear increase with identical slope for all m larger than some m_0 and for a reasonable range of ε, and then this slope can be taken as an estimate of the largest exponent.

$$S(\varepsilon, m, t) = \left\{ \ln \left(\frac{1}{u_n} \sum_{s_{n'} \in u_n} |s_{n+t} - s_{n'+t}| \right) \right\}_n \tag{5.92}$$

5.6.1.3 Correlation Dimension

The correlation dimension method is used for detecting the presence possibility of chaos. An algorithm proposed by Grassberger and Procaccia [65] is the most commonly applied method. According to this method the correlation sum, C(r), is expressed as:

$$C(r) = \frac{2}{N(N-1)} \sum_{i=1}^{N} \sum_{j=i+1}^{N} H(r - \|x_i - x_j\|) \tag{5.93}$$

Where H is Heaviside step function defined as:

$$H(x) = \begin{cases} 0 & \text{when} \quad x \le 0 \\ 1 & \text{when} \quad x \ge 0 \end{cases} \tag{5.94}$$

N is the number of points in time series; r is the radius of a sphere with its center at either of current points. Then the correlation dimension is:

$$\mathrm{D}_2 = \lim_{\substack{D \to \infty \\ r \to 0}} \frac{d \ln (\mathrm{C(r)})}{d \ln \mathrm{r}} \tag{5.95}$$

When the system is chaotic, the slope of *logC(r)* vs *logr* converges to D_2 over an appropriate interval as m increase. The results were carried out with Tisean package [68], version 3.01.

5.6.2 Phase-Space Reconstruction of Vibration Signals

We obtained the first minimum of the mutual information calculated at $\tau = 10$ and the embedding dimension is 6. We then represent the system by a phase-space trajectory $\vec{X}(t) = (A(t), V(t), S(t))$, as shown in Fig. 5.40. We compare the phase-

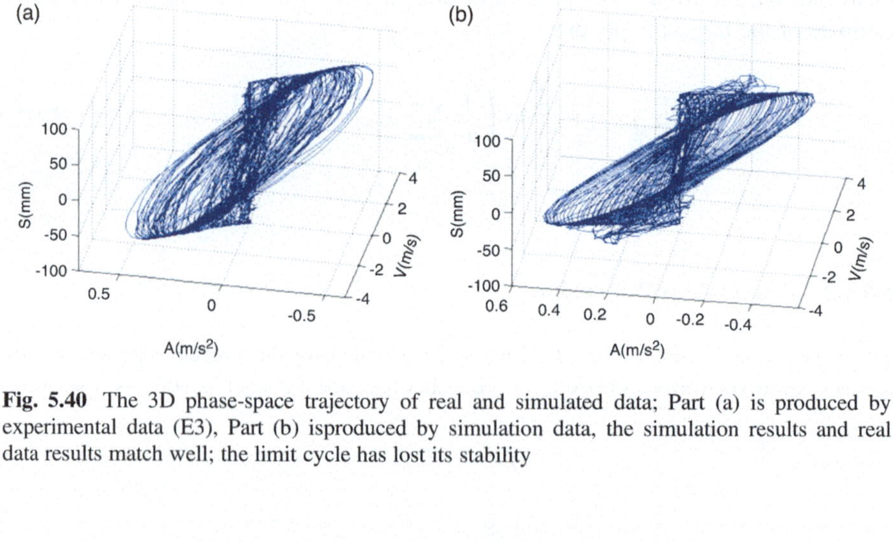

Fig. 5.40 The 3D phase-space trajectory of real and simulated data; Part (a) is produced by experimental data (E3), Part (b) isproduced by simulation data, the simulation results and real data results match well; the limit cycle has lost its stability

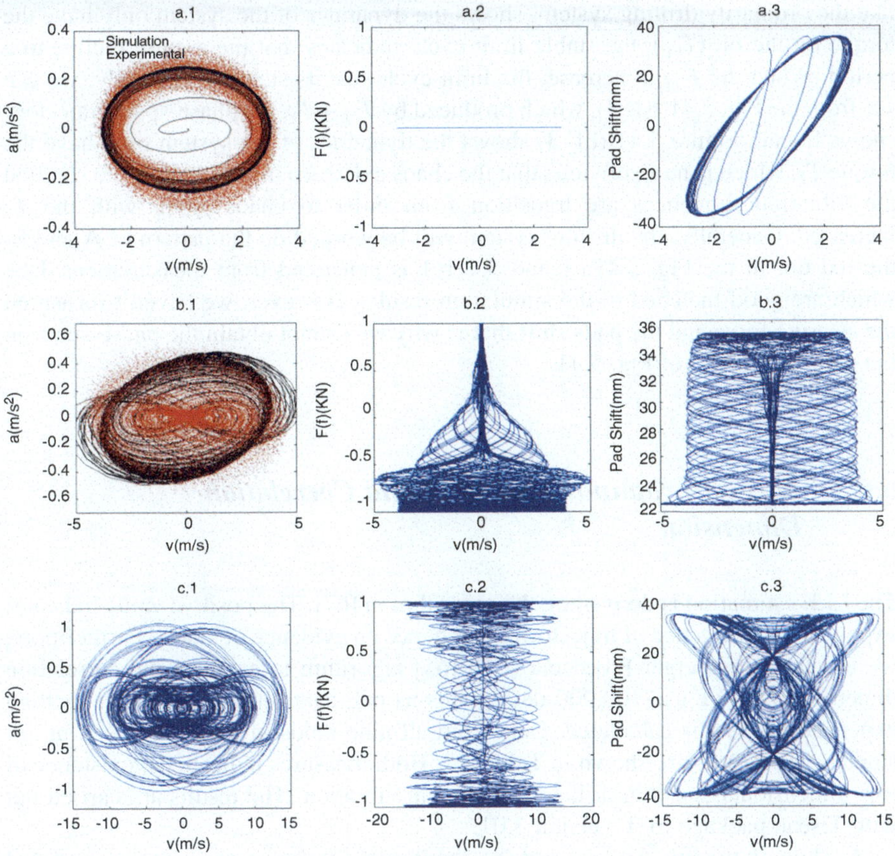

Fig. 5.41 Phase-space of experimental data and simulation data. The first row (a.1-a.3) is produced by the system only has the torque on the bit, the red dot of the A.1 is the experimental data come from the ordinarily drilling system, we can see the phase-space of our system has a limit cycle. The second row (b.1-b.3) is produced when the RSS is worked, Pads pushed to the bit generated the torque T_Y lead to the system to chaotic. The third row (c.1-c.3) is produced by the $T_{bit} = 0$, just the simulation results which will not present to the real drilling process. In addition, the first column is the phase-space of torsional velocity and acceleration, the second column is the phase-space of torsional velocity and steering force, and the third column is the phase-space of torsional velocity and the one of pads shift. Along with the T_Y increased, we first found the sequence of order to chaos transitions in the drill-string system

space trajectory $\overrightarrow{X}(t)$ using the experimental data of E3 and simulated data.The phase-space of our system is at least three dimensional, and the oscillatory tends to give rise to a strange attractor in both real and simulated data.

The main difference between the Push-the-Bit RSS and ordinary drilling system is the friction torque (T_Y) that generated in the process of pads pushed against the borehole wall. The simulation results show that the drilling bit torsional vibration phenomena perceptible increased when the friction torque exist.

As show in the Fig. 5.41, we found that the T_Y will change the kinetic properties of the system. Figure 5.41(a1–3) is the simulation results when we set as $F(f_F) = 0$, like the ordinarily drilling system, shows the dynamics of the system only have the torque on the bit (T_{bit}), the stable limit cycle indicates that the system incline to a period. When the $F(f_F)$ increase, the limit cycle lost its stability gradually, we can see from the Fig. 5.41(b1–3), which produced by $T_{bit} + T_Y$, the phase-space trajectory shows fractal. Figure 5.41 (c1–3) shows the dynamics of the system only have the torque T_Y, which is more obvious that the chaos existence in the system. We can find the vibration sequences are transition from order to chaos along with the T_Y increased. Generally, the drilling system will be worked on the pattern of A and B, the red dot in the Fig. 5.41a.1 and 5.41b.1 is generated from measurement data, which are good matched to the simulation results. However, we haven't measured the steering force and the pads shift that is why we cannot obtain the phase-space in the column 2 and 3 of Fig. 5.41.

5.6.3 Largest Lyapunov Exponent and Correlation Dimension

The LLE estimation is derived on the algorithm in [67]. The positive value indicates exponential divergence of trajectories and hence an evidence of chaos. Furthermore, we use the Grassberger-Procaccia (GP) [65] algorithm to estimate the correlation dimension D_2. We use 100,000 data points as one data set and the whole drilling experimental data is calculated. Data from all nine fields in China are used in our studies. The results are shown in Table 5.9. Both measures indicate the existence of low-dimensional chaos in drillstring torsional vibration. The results are carried out with Tisean package [68], version 3.01.

As show in the Fig. 5.42 (a) and (b), the largest Lyapunov exponent is estimated through least-squares line fit for the time series and is found to be 0.011 of the experimental data of E3 field test and 0.013 of simulation data. This positive value indicates exponential divergence of trajectories and hence an evidence of chaos.

Furthermore, as show in Fig. 5.42 (c) and (d), the slope of $\log(C_D(r))$ gives us an estimation of the correlation dimension D_2. We present the estimated slope as a function of $\log r$ for embedding dimensions m = 1 up to 10. In all cases we obtain a

Table 5.9 Chaos identification of test data

	LLE	D_2
E1	0.013 ± 0.003	1.8 ± 0.2
E2	0.015 ± 0.004	2.1 ± 0.15
E3	0.011 ± 0.002	1.6 ± 0.2
E4	0.011 ± 0.0014	1.8 ± 0.2
E5	0.018 ± 0.003	2.3 ± 0.3
E6	0.013 ± 0.0018	1.8 ± 0.2
E7	0.021 ± 0.0024	2.2 ± 0.3
E8	0.012 ± 0.004	1.7 ± 0.24
E9	0.022 ± 0.005	2.3 ± 0.4

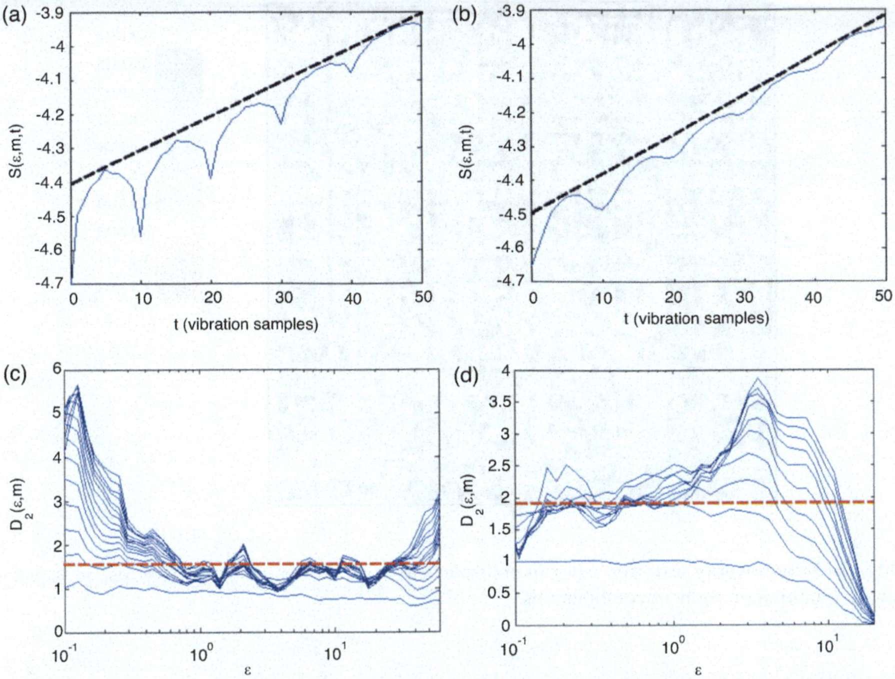

Fig. 5.42 Chaos identification, Part (**a**) and (**b**) estimating the maximal Lyapunov exponent of torsional vibration time series. The part (**a**) shows the result for the experimental data, the straight line indicates $\lambda = 0.011$. For comparison, part (**b**) shows the result for the simulation data, the straight line indicates $\lambda = 0.013$. Part (**c**) and Part (**d**) give us an estimation of the correlation dimension D_2. In all two cases the GP algorithm converges, creating a plateau on the slope of the correlation integral. The red dashed curves give estimates of the correlation dimension in all two cases: experimental time series $D_2 = 1.6$ and simulation time series $D_2 = 1.8$. The all results in these figures were carried out with Tisean package [68], version 3.01

convergence toward a correlation dimension of $D_2 \approx 1.6$ with the experimental data and a correlation dimension of $D_2 \approx 1.8$ with the simulation data.

5.6.4 Stationarity and Determinism Tests

Since linear statistics, such as the mean or standard data deviation, usually do not possess enough discrimination power when analyzing irregular signals, nonlinear statistics have to be applied. We apply the stationarity test [69] program stationarity. exe provided by Matjaˇz Perc [70, 71]. They split the time series into several short non-overlapping segments and then use a particular data segment to make

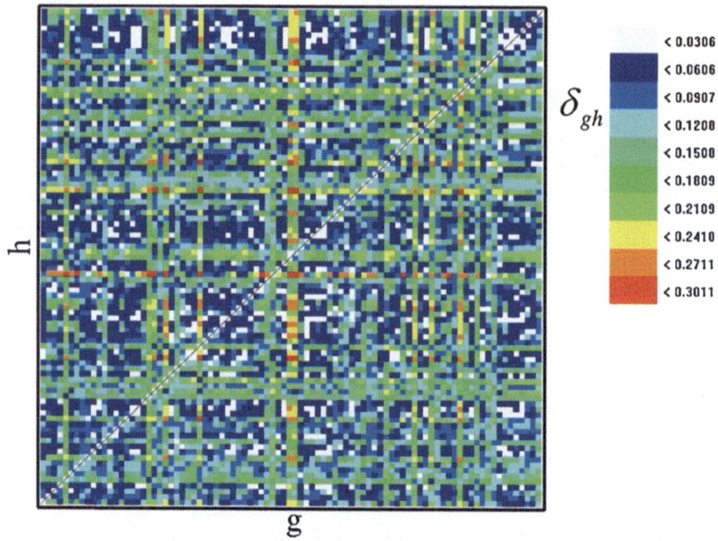

Fig. 5.43 Stationarity test. The color map displays average cross-prediction errors δ_{gh} in dependence on different segment combinations

predictions in another data segment. By calculating the cross-prediction error (δ_{gh}) when considering points in segment g to make predictions in segment h, the cross-prediction error as a function of g and h then revels which segments differ in their dynamics. They obtained a very sensitive statistic capable of detecting changes in dynamics and thus a very powerful probe for stationarity.

The average cross-prediction errors for all possible combinations of g and h are presented in Fig. 5.43. The whole time series was partitioned into 56 non-overlapping segments each occupying 1000 data points. The average value of all δ_{gh} is 0.17, while the minimum and maximum values are 0.03 and 0.3, respectively. Since the maximal cross-prediction error is not one time larger than the average, we can determine that the studied time series is stationarity. We just consider only 1000s of the torsional vibration time series. Otherwise, longer data sets of the vibration time series almost yield non-stationary.

Additionally, the results of the surrogate data analysis could be further confirmed by applying the determinism test [72], which enables us to verify if the time series we have obtained originates from a deterministic process. In this chapter, we use the method developed by Kaplan and Glass [73] in order to examine the possible deterministic nature of the underlying process, if the system is deterministic, the average length of all directional vectors k will be 1, while for a completely random system $k \approx 0$.

As shown in Fig. 5.44, the value of the determinism factor k is in the range 0.5–0.9, indicating the possible stochastic nature of the underlying process.

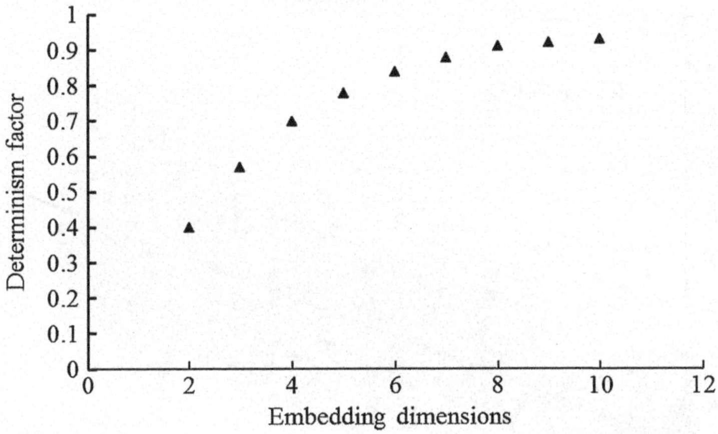

Fig. 5.44 Deterministic test for measurement time series. The values of determinism factor k are given for embedding dimensions in the range m = 2–10. It is evident that k $\leq$ 0.95, indicating the possible random (stochastic) behavior

5.7 Quantification of Drillstring Integrity Failure Risk

The drillstrings are mechanical systems which undergo complex dynamical phenomena, often involving non-desired oscillations. Three main types of oscillations are distinguished: torsional, axial and lateral vibration [2, 74~76]. These oscillations are a source of failures that reduce penetration rates and increase drilling operation costs. Stick–slip phenomenon appearing at the bottom-hole assembly (BHA) is particularly harmful for the drillstring and it is a major cause of drill pipes and bit failures, in addition to well bore instability problems [76].

The risk of a vibration-related integrity failure of drillstring can be predicted by using vibration measurement to calculate some severity index [77], defined as $I_c = a_{rms}^{1.5} t_d^{0.5}$. It relates to the average root mean square (RMS) acceleration (a_{rms}) and duration of the run time (t_d), where the RMS acceleration is defined as Eq. (5.64).

RMS increases with the presence of increasing vibration level. The number of vibration samples used to calculate each feature sample is set to be $N = 50{,}000$, which means that the time interval between two successive feature samples is $\Delta t = N/f_s = 500$ second. The authors and field engineers have found that this time frame is acceptable for the requirement of time accuracy. Additionally, Field data indicates that higher rotary speeds generally lead to increased levels of lateral vibration, as shown in the Fig. 5.45.

This failure detection is a very important issue in drilling. It is difficult to extract the weak signals in such a drilling system. The measured data from such a system contain detailed dynamic characteristics of the measured structure, but the amount of data is often huge. How to efficiently and accurately extract concise, clear, and unique system dynamic characteristics and health information from such a large set of data is very difficult. Moreover, dynamics-based system identification is a very

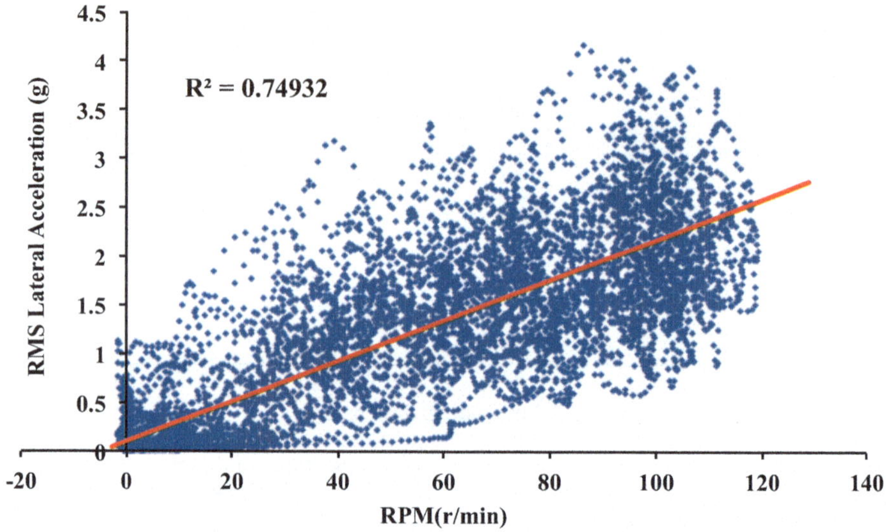

Fig. 5.45 Statistical analyses indicated there is direct relationship between RPM and lateral vibration levels (RMS Lateral Acceleration)

challenging reverse engineering task. It is even more important to have a signal processing and data mining method that can extract from each set of experimental data as many system parameters as possible. A damaged structure shows nonlinearities and intermittent transient response in its time traces of measured points. The acquisition and analysis of this data provided new insights of the dynamic behavior of drillstring.

Nonlinear dynamical systems sometimes exhibit chaotic behavior, and the Lyapunov exponent is a useful tool to distinguish and measure the extent of chaos. Previous studies on chaos and on the Lyapunov exponents have found applications to several fields such as turbulence, communication, heartbeats, and so on. However, little research has been done on the relationship between the behavior of Lyapunov exponents and fault detection. In the history of the field of fault detection, many methods have been proposed to extract and analyze experimental data in order to detect faults and assist in diagnosis [78–80]. Difficult to detect faults early on due to certain weakly developing faults usually covered by background noise and other chaotic elements. If the signal to noise ratio (SNR) is low, weak fault signals may not be extracted from background noise using only the above-mentioned methods. In recent years, as chaos theory has developed, some new technologies (especially phase space reconstruction) have begun to be applied to extract information hidden beneath experimental data [81, 82]. The largest Lyapunov exponent is usually used to distinguish and to measure chaos of dynamical systems. The exponent or its change can have some relationships with system faults. Usually, the equations of the dynamical systems are difficult to obtain directly; only time series data sets are observable. There are varieties of faults which are hard to detect directly from the data itself. The LLE is the indicator of divergence or convergence of two trajectories with nearby initial conditions, and it could be sensitive to small changes of the systems. However, the relationship between the LLE and the system damage level is not clear and rarely studied up to now. Based on the LLE calculation method and considering the conditions of a real system and the measured data, the changes of the LLE are applied to fault detection of the RSS.

The severity index of vibration measurement is lack of sensitivity. Assume that L_n is the LLE of nth data set; we define the drillstring integrity failure risk parameter F_r as,

$$F_r = \sum_{n \to N} \left(\frac{L_n - L_{n+1}}{L_n} + 1 \right) \cdot a_{rms}^{15} \cdot t_d^{05} \tag{5.96}$$

We got the results of the risk parameter F_r and compare it with the I_c as shown in Fig. 5.46.

We chose 6 field test data sets as shown in Fig. 5.21. (a). Three of them are from drillstring failure and three of them are in good condition. We calculate both indices at the beginning and the end of the drilling process. By comparing the variation of coefficients, we observe that the proposed measure provides a more reliable solution to support risk-management processes with regard to drillstring failures. The chaos detected is related to the inner dynamic characteristics of drillstring.

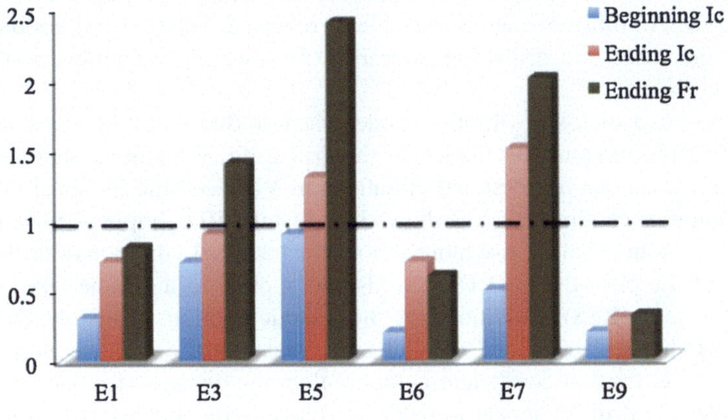

Fig. 5.46 Vibration index charts.We use 1e6 data points as one data set to calculate the vibration index. In the six field tests as shown in Fig. 1.a, E3, E5 and E7 are failure cases. Both indices I_c and F_r are less than one at the beginning of the drilling process, and the indices increase at the end of the drilling process. This is different from the normal cases E1, E6 and E9, in which the drillstrings are in good condition for the entire drilling process. It is noted that I_c cannot identify the failure in sample 2 while the ruptured dynamical approach can detect the weak failure signal

5.8 Discussion

As a booming technology catering the twenty-first century, automatic vertical drilling system features extended reach capacity, well trajectory control accuracy and flexibility, etc. which can notably increase drilling efficiency and safety. The one of the technical difficulties in the VDS is that how to dynamic measurement of spatial attitude accurately at the bottom rotating drillstring. However, the dynamics of bottom drillstring will greatly affect the accuracy of the solution. Detailed analysis of the dynamic characteristics of VDS has great benefits for improving measurement accuracy while drilling, improving the reliability of closed-loop control, and improving the life of drilling tools.

In this chapter, the dynamic model of the drillstring with Automatic Vertical Drilling System (VDS) is introduced. The friction between the pads and the borehole wall will reduce the instantaneous speed of the drill bit. The pads of the actuator in the VDS is continuously pushed to the borehole wall during the work process, and a cyclically acting nonlinear damping force is formed on the drill bit at the bottom of the well, resulting in chaotic movement of the bottom drillstring. The investigation is concerned with the modelling the oil drilling process, to better understand the effect of friction upon the propagation of torsional waves. One type of frictional model, the continuous frictional model, is considered using the Dahl model. During the drillstring rotation, the drill bit breaking rock torque and friction torque between the drillstring and borehole wall decreases with the increase in rotating speed. Which will introduce negative damping to the drillstring system. When the negative damping is over-assisted, the drillstring rotates at a constant speed and the equilibrium state is unstable or reversed. When the vibration energy gradually increases, the drillstring appears to be torsional vibration, resulting in stick-slip vibration.

Furthermore, a stick-slip vibration model was introduced as a new mechanism to explain the large torsional oscillation of the drillstring. We aim to improve understanding of the causes for torsional vibrations in VDS and the torsional vibrations with or without stick-slip case. It is shown here that the VDS implementing agencies pushing the borehole wall cause more serious the torsional vibration of drill bit. The dynamics of the push-the-bit VDS can also be applied to the same type of rotary steerable system (RSS). Additionally, the chaotic motion is mainly caused by elasticity of the drillstring and changing frictional forces at the bottom tool of VDS. The static friction coefficient is higher than the dynamic friction coefficient, which makes the drill bit appear to work alternately with sticking and sliding., and presents complex dynamic behavior, which makes down-hole dynamic responses difficult to predict. Under the guidance of the above dynamic characteristics, our measurement algorithms and control programs need to be further improved from a nonlinear perspective.

References

1. Ritto TG, Escalante MR, Sampaio R, Rosales MB. Drillstring horizontal dynamics with uncertainty on the frictional force. J Sound Vib. 2013;332(1):145–53.
2. Qilong X, Leung H, Ruihe W, Baolin L, Leilei H, Shenglai G. The chaotic dynamics of drilling. Nonlinear Dyn. 2016;83(3):2003–18.
3. Belokobyl'skii SV, Prokopov VK. Friction induced self-excited vibration of drill rig with exponential drag law [J]. Sov Appl Mech. 1982;18(12):1134–8.
4. Zamudio CA, Tlusty JL, Dareing DW. Self-excited vibrations in drillstrings [C]. SPE 16661, 1987. SPE annual technical conference and exhibition, 27–30 September, Dallas, Texas.
5. Brett JF. The genesis of bit-induced torsional Drillstring vibrations [J]. SPE Drill Eng. 1991;7 (3):168–74.
6. Dowson R, Lin YQ, Spanos PD. Drill string stick-slip oscillations [C]. Spring conference of the society for experimental mechanics, Houston, Texas, June 14–19, 1987.
7. Lin YQ, Wang YH. Stick-slip vibration of drill strings [J]. Trans ASME J Eng Ind. 1991;113 (1):38–43.
8. Bailey JJ. An analytical study of drill-string vibration[J]. ASME J Eng Ind [J]. 1960;82 (2):122–8.
9. Elsayed MA, Phung CC. Modeling of drillstrings [C]. Proceedings of the 24th ASME international conference on offshore mechanics and arctic engineering. Halkidiki, 2005.
10. Lubinski A. Dynamic loading of Drillpipe during tripping[J]. J Pet Technol. 1988;40 (8):975–83.
11. Aadnoy BS, Fazaelizadeh M, Hareland G. A 3D analytical model for wellbore friction [J]. J Can Pet Technol. 2010;49(10):25–36.
12. Yigit AS, Christoforou AP. Coupled axial and transverse vibrations of oil well drillstrings [J]. J Sound Vib. 1996;195(4):617–27.
13. Yigit AS, Christoforou AP. Coupled torsional and bending vibrations of drillstrings subject to impact with friction [J]. J Sound Vib. 1998;215(1):167–81.
14. Khulief YA, Al-Naser H. Finite element dynamic analysis of drillstrings [J]. Finite Elem Anal Des. 2005;41(13):1270–88.
15. Sampaio R, Piovan MT, VeneroLozano G. Coupled axial/torsional vibrations of drill-strings by means of non-linear model [J]. Mech Res Commun. 2007;34(5):497–502.
16. Kapitaniak M, Hamaneh VV, Chávez JP, et al. Unveiling complexity of drill-string vibrations: experiments and modelling[J]. Int J Mech Sci, 2015, s101–102:324–337.
17. Gupta SK, Wahi P. Global axial-torsional dynamics during rotary drilling [J]. J Sound Vib. 2016;375(4):332–52.
18. Huang W, Gao D, Wei S, Li X. A generalized quasi-static model of drill string system [J]. J Nat Gas Sci Eng. 2015;23(3):208–20.
19. Lian Z, Zhang Q, Lin T, Wang F. Experimental and numerical study of drill string dynamics in gas drilling of horizontal wells [J]. J Nat Gas Sci Eng. 2015;27(3):1412–20.
20. Jones S, Feddema C, Castro J. Fully mechanical vertical drilling system delivers RSS performance in vertical drilling applications while providing an economical alternative conventional rotary steerable systems set-Up for vertical hold mode [C]. SPE-178788, 2016, IADC/SPE Drilling Conference and Exhibition, 1–3 March, Fort Worth, Texas.
21. Warren T. Rotary steerable technology conclusion: implementation issues concern operators [J]. Oil Gas J. 1998;96(12):23–4.
22. Wang R, Xue Q, et al. Torsional vibration analysis of push-the-bit rotary steerable drilling system [J]. Meccanica. 2014;49(7):1601–15.
23. Hoffmann OJ, Jain JR, Spencer RW, Makkar N. Drilling dynamics measurements at the drill bit to address Today's challenges. IEEE international instrumentation and measurement technology conference: smart measurements for a sustainable environment, Graz, p. 1772–1777, 13–16, May 2012.

24. Nayfeh AH, Balakumar B. Applied nonlinear dynamics: analytical, computational, and experimental methods. Weinheim: Wiley-VCH; 1995.
25. Ertas D, Bailey JR, Wang L, et al. Drill string mechanics model for surveillance, root cause analysis, and mitigation of torsional and axial vibrations. Louisiana: Society of Petroleum Engineers; 2013.
26. Jansen JD. Nonlinear rotor dynamics as applied to oil well drillstring vibrations. J Sound Vib. 1991;147(1):115–35.
27. Navarro-Lopez EM, Cortes D. Avoiding harmful oscillations in a drillstring through dynamical analysis. J Sound Vib. 2007;307:152–71.
28. Millheim K, Jordan S, Ritter CJ. Bottom-hole assembly analysis using the finite-element method. SPE J. 1978;30(2):265–74.
29. Costa FS, Rebeiro PR. Finite element modeling of the mechanical behavior of unbalanced drill collars. Rio de Janeiro: Society of Petroleum Engineers; 1997.
30. Ledgerwood LW, Jain JR, OJ H®m, et al. Downhole measurement and monitoring lead to an enhanced understanding of drilling vibrations and polycrystalline diamond compact bit damage. Louisiana: Society of Petroleum Engineers; 2013.
31. Raap C, Craig AD, Graham RB. Drill pipe dynamic measurements provide valuable insight into drill string dysfunctions. Ohio: Society of Petroleum Engineers; 2011.
32. Oueslati H, Jain JR, Reckmann H, et al. New insights into drilling dynamics through high-frequency vibration measurement and modeling. Louisiana: Society of Petroleum Engineers; 2013.
33. Chien ML, Nicholas V, Hamad K, et al. Parametric studies on drill-string motions. Int J Mech Sci. 2012;54(1):260–8.
34. Besaisow AA, Payne ML. A study of excitation mechanisms and resonances inducing BHA vibrations. New Orleans: Society of Petroleum Engineers; 1986.
35. Schen AE, Snell AD, Stanes BH. Optimization of bit drilling performance using a new small vibration logging tool. Amsterdam: Society of Petroleum Engineers; 2005.
36. Field DJ, Swarbrick AJ, Haduch GA. Techniques for successful application of dynamic analysis in the prevention of field-induced vibration damage in MWD tools. Amsterdam: Society of Petroleum Engineers; 1993.
37. Wolf SF, Zacksenhouse M, Arian A. Field measurements of downhole drillstring vibrations. Las Vegas: Society of Petroleum Engineers; 1985.
38. Lutz J, Raynaud H, Gstalder S, et al. Dynamics theory of drilling and instantaneous logging. Los Angeles: Society of Petroleum Engineers; 1971.
39. Stephen WL, Mitch W, Aaron E, et al. Stick-slip detection and friction factor testing using surface-based torque and tension measurements. Netherlands: Society of Petroleum Engineers; 2014.
40. Xue Q, Leung H, Huang L, Zhang R, Liu B, Wang J, Li L. Modeling of torsional oscillation of drillstring dynamics. Nonlinear Dyn. 2019;96(1):267–83.
41. Jansen JD. Nonlinear dynamics of oil well Drillstrings. PhD thesis, Delft University, Netherlands, 1993. p. 38–55.
42. Spanos PD, Sengupta AK, Cunningham RA, Paslay PR. Modeling of roller cone bit lift-off dynamics in rotary drilling. J Energy Resour Technol. 1995;117(3):197–207.
43. Lanczos C. The variational principles of mechanics. Dovers Publications, Inc. 1970, p. 90–106.
44. Zhao G, Gong W. Fundamentals of drilling mechanics. China: Petroleum Industry Press; 1988.
45. Qilong X, Ruihe W, Feng S. Study on lateral vibration of rotary steerable drilling system. J Vibroeng. 2014;16(6):2702–11.
46. Kennel MB, Brown R, Abarbanel HDI. Determining embedding dimension for phase-space reconstruction using a geometrical construction. Phys Rev A. 1992;45(6):3403–11.
47. Cao L. Practical method for determining the minimum embedding dimension of a scalar time series [J]. Physica D. 1997;110(1–2):43–50.
48. Russel MK, Russel AW. Surveying of boreholes. United States Patent US4163324. 1979.

49. Yezid IA, Yonnellybeth M, Andre N. Vibration risk index offers tool for preventing drillstring failure. World Trends and Technology for Offshore Oil and Gas Operations, 2011.
50. Macpherson JD, Mason JS, Kingman JEE. Surface measurement and analysis of drillstring vibrations while drilling. Amsterdam: Society of Petroleum Engineers; 1993.
51. Dubinsky VSH, Henneuse HP, Kirkman MA. Surface monitoring of downhole vibrations: Russian, European, and American approaches. Cannes: Society of Petroleum Engineers; 1992.
52. Ledgerwood LW, Ho®mann OJ, Jain JR, et al. Downhole vibration measurement, monitoring, and modeling reveal stick/slip as a primary cause of PDC-bit damage in today. Florence: Society of Petroleum Engineers; 2010.
53. Zannoni SA, Cheatham CA, Chen CKD, et al. Development and field testing of a new downhole MWD drillstring dynamics sensor. Houston: Society of Petroleum Engineers; 1993.
54. Chen DCK, Smith M, LaPierre S. Integrated drilling dynamics system closes the model measure optimize loop in real time. Amsterdam: Society of Petroleum Engineers; 2003.
55. Akinniranye G, Megat A, Elsweisy H, et al. Implementation of a shock and vibration mitigation process: achieving real-time solutions and savings. SPE J. 2009;24(2):1–10.
56. Ashley DK, McNary XM, Tomlinson JC. Extending BHA life with multi-axis vibration measurements. Society of Petroleum Engineers, Amsterdam, 2001.
57. Adam B, Lojini L, Junichi S, et al. Continuous high-frequency measurements of the drilling process provide new insights into drilling system response and transitions between vibration modes. Amsterdam: Society of Petroleum Engineers; 2014.
58. Cooley JW, Tukey JW. An algorithm for the machine calculation of complex Fourier series. Math Comput. 1965;19:297–301.
59. Morlet J. Continuous wavelet transform and continuous multiscale analysis. J Math Anal Appl. 2003;169(1):179–96.
60. Craig AD, Hanley C, McFarland B, et al. A proven approach to mitigating drilling vibration problems in offshore Western Australia. Doha: International Petroleum Technology Conference; 2009.
61. Welch PD. The use of fast Fourier transforms for the estimation of power spectra: a method based on time averaging over short, modified periodograms. IEEE Trans Audio Electroacoust. 1967;15(62):70–3.
62. Kennel MB, Brown R, Abarbanel HDI. Determining embedding dimension for phase-space reconstruction using a geometrical construction. Phys Rev A. 1992;45(6):3403–11.
63. Fraser AM, Swinney HL. Independent coordinates for strange attractors from mutual information. Phys Rev A. 1986;33(2):34–40.
64. Cao L. Practical method for determining the minimum embedding dimension of a scalar time series. Physica D. 1997;110(1–2):43–50.
65. Grassberger P, Procaccia I. Measuring the strangeness of strange attractors. Physica D. 1983;9 (1–2):189–208.
66. Wolf A, Swift JB, Swinney HL, Vastano JA. Determining Lyapunov exponents from a time series. Physica D. 1985;16(3):285–317.
67. Rosenstein MT, Collins JJ, De Luca CJ. Apractical method for calculating largest Lyapunov exponents from small data sets. Physica D. 1993;65(1–2):117–34.
68. Rainer H, Kantz H, Schreiber T. Practical implementation of nonlinear time series methods: the TISEAN package. Chaos. 1999;9(2):413–35.
69. Schreiber T. Detecting and analyzing nonstationarity in a time series with nonlinear cross-predictions. Phys Rev Lett. 1997;78(5):843–6.
70. Perc M. Nonlinear time series analysis of the human electrocardiogram. Eur J Phys. 2005;26 (5):757–68.
71. Kodba S, Perc M, Marhl M. Detecting chaos from a time series. Eur J Phys. 2005;26(1):205–15.
72. Kostić S, et al. Stochastic nature of earthquake ground motion. Physica A. 2013;392:4134–45.
73. Kaplan DT, Glass L. Direct test for determinism in a time series. Phys Rev Lett. 1992;68 (4):427–30.

74. Chen SL, Blackwood K, Lamine E. Field investigation of the effects of stick–slip, lateral and whirl vibrations on roller-cone bit performance. SPE Drill Complet. 2002;17(1):15–20.
75. Christoforou AP, Yigit AS. Fully coupled vibrations of actively controlled drillstrings. J Sound Vib. 2003;267(5):1029–45.
76. L.W. Ledgerwood III, et al. Downhole vibration measurement, monitoring and Modeling reveal stick-slip as a primary cause of PDC bit damage in Today's applications. SPE annual technical conference and exhibition, Florence. SPE-134488-MS, 19–22 2010.
77. Yezid A, Ashley F. Quantification of Drillstring integrity failure risk using real-time vibration measurements. SPE J. 2012;27(2):216–22.
78. Altmann J, Mathew J. Multiple band-pass autoregressive demodulation for rolling element bearing fault diagnosis. Mech Syst Signal Process. 2001;15(5):963–77.
79. Junsheng C, Dejie Y, Yu Y. The application of energy operator demodulation approach based on EMD in machinery fault diagnosis. Mech Syst Signal Process. 2007;21(2):668–77.
80. Ocak H, Loparo KA, Discenzo FM. Online tracking of bearing wear using wavelet packet decomposition and probabilistic modeling: a method for bearing prognostics. J Sound Vib. 2007;302(4–5):951–61.
81. Müller PC, Bajkowski J, Söffker D. Chaotic motions and fault detection in a cracked rotor. Nonlinear Dyn. 1994;5(2):233–54.
82. Zhao Z, Wang F-L, Jia M-X, Wang S. Intermittent-chaos-and-cepstrum-analysis-based early fault detection on shuttle valve of hydraulic tube tester. IEEE Trans Ind Electron. 2009;56 (7):2764–70.

Chapter 6
Data Processing and Mining in Seismic While Drilling

Jin Wang

Abstract Seismic While Drilling (SWD) is a well seismic method developed in recent years which is based on reverse vertical seismic logging. It is a newly-developed well seismic technology which combines the seismic exploration technology with petroleum drilling engineering technology. Compared to conventional VSP, SWD has its own characteristics and unique advantages in that it uses bit vibration in the process of drilling as the source for seismic measurement, without interfering with the drilling or occupying drilling time, and without any risk to the hole, especially the bit can be predicted in real-time structure details of the formation in front of the bit through the field seismic imaging processing, with the main purpose of reducing drilling risks. The key of this technology is how to collect and recover the weak bit reflection signal under strong disturbance noise and make it the equivalent formation impulse response.

Keywords Seismic While Drilling (SWD) · Data processing · Formation · Drilling risks

6.1 Introduction

Oil drilling is an underground construction with high investment and high risk. The fuzziness and uncertainty of underground conditions bring great risks to drilling construction. Therefore, it is very important to master various parameters of formation, predict formation pressure and monitor downhole conditions. The sustainable use of the global oil and gas resources is becoming more and more prominent. The oil and gas exploration and development are developing towards new exploration area, deep wells and ultra-deep wells, which puts forward higher technical requirements on the drilling efficiency, cost, safety, and reservoir protection. Thus, positive and effective exploration should be carried out in scientifically drilling engineering design and construction so as to prevent and reduce drilling accidents.

Contributions by Fangtao Li, Jin Wang and Qilong Xue.

Q. Xue, *Data Analytics for Drilling Engineering*, Information Fusion and Data Science, https://doi.org/10.1007/978-3-030-34035-3_6

Vertical Seismic Profiling (VSP) [1] is a seismic exploration technology for conventional wells. When VSP is collecting data, it is necessary to interrupt the drilling, put a detector in the well, and measure through the surface energy source excitation. Considering the risk of operating in open hole wells, VSP measurements are usually carried out after well cementing. The hysteresis of such measurements makes it impossible to timely grasp the relative position of the drill bit and the target layer in the drilling process. If the bit deviates from the target layer at this time, we have lost the best time to adjust the bit's drilling trajectory. In addition, during VSP observation, the idled drilling equipment, geophone running and long drillstring running (well pressure cannot be controlled) will bring high cost and greater risk to further drilling operations. In fact, more often than not, the required borehole seismic data may not be available due to high costs or surface conditions that do not allow for the placement of surface sources.

Reverse Vertical Seismic Profiling (RVSP) [2] is a seismic exploration technology which is completely opposite to the way of VSP observation both geometrically and physically, it is also the further development of VSP technology. Therefore, seismic data collected by RVSP has the same advantages as those collected by VSP. RVSP technology adopted the downhole excitation ground receiving mode of operation, reducing the exploration cost and improving the working efficiency. And there is no risks and inconvenience caused during the geophone drilling. RVSP can effectively overcome the effect of the noise caused by the bad coupling of downhole instruments and the formation, and there is no uncertainty in the direction of wave field received by multi-component downhole instruments. The recording wave field is further purified, and the separation of wave field is relatively simple and easy to carry out. However, RVSP requires that the downhole seismic source must have enough energy to reach the surface through the formation without destroying the borehole, which cannot be met at present. Moreover, it is difficult for underground cables to maintain normal working conditions under high temperature and high pressure. In addition, the RVSP measurement still requires the interruption of drilling operations and cannot be continuously measured. These factors limit the wide application of RVSP technology.

Seismic While Drilling (SWD) [3–6] is a well seismic method developed in recent years which is based on reverse vertical seismic logging. It is a newly-developed well seismic technology which combines the seismic exploration technology with petroleum drilling engineering technology. Different from surface seismic and VSP (Vertical Seismic Profile), it uses the continuous random vibration signal generated from the drill bit breaks as a downhole seismic source for the seismic exploration. If the geophone or geophone arrangement is placed on or near the surface of the earth, it is RVSP in essence, which is called an VSP while drilling or an RVSP while drilling. SWD method overcomes the limitation of the conventional VSP development and the drawbacks of VSP observation, is a combination of the advantages of surface seismic and vertical seismic VSP. It features in the non-interference with the drilling construction, continuous measurement in the depth, high exploration efficiency (especially for 3D observation and offset and directional observation) etc. Through real-time seismic measurement while drilling, it can predict the structural details of the formation around the wellbore and in front of the drill bit, can timely acquire pore pressure, target layer and downhole status monitoring of the formation to be drilled in front of the bit, and can timely find abnormal high pressure and faults, thus providing real-time downhole working parameters for the well team, timely adjusting the drilling plan for decision makers, and providing valuable information for safe drilling, improving drilling efficiency and optimizing casing design. Therefore, SWD is a high and new technology with strategic significance, which has become a new drilling technology in the field of drilling engineering.

Drill-bit seismic technology was a topic of rather intense research and development in the 1980s and 1990s, and the application should not be forgotten. The principle of geologic imaging with a rotary-cone drill bit is illustrated in Fig. 6.1. The key to the imaging procedure is to position a reference sensor at the top of the drill string, near the swivel. This reference sensor records each impact of each tooth of a rotary-cone bit as rock strata are being drilled. As shown in Fig. 6.1, the drill-tooth impulses propagate along direct paths to sensors deployed on the Earth surface (or on the seafloor if the well is offshore) at stations that allow specific target geology to be imaged. An imaging capability is created by the drill-bit wavefields that propagate downward and reflect upward from rock interfaces below the drill bit as depicted by the ray path diagram.

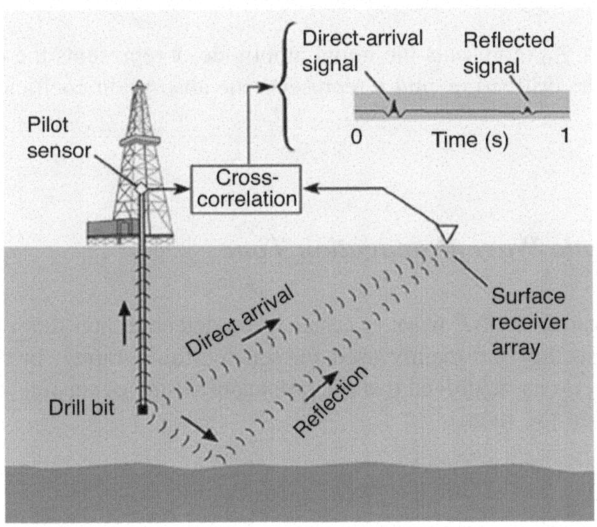

Fig. 6.1 In drill-bit seismic technology, the mechanical energy produced by an active rotary-cone drill bit propagates up the drill string to a reference sensor positioned on the swivel and radiates into the Earth as a seismic wavefield. Seismic events are created that travel direct and reflected ray paths from the drill bit to surface-positioned sensors. A conventional seismic trace is created by continuously correlating the response of the reference sensor at the top of the drill string with the responses of the surface sensors

6.2 Law of Seismic Wave Propagation While Drilling

SWD is different from the surface seismic, and also from the conventional VSP exploration. Since SWD uses drilling bit vibration in the drilling process as the seismic source for exploration, it has its special seismic wave field and the spatial propagation characteristics [7] (including the effective wave and interference wave). Systematic analysis of the wave field and its spatial propagation characteristics is of great significance for effective separation of the effective wave field and interference wave field.

6.2.1 Propagation Speed of the Drill String

During the propagation of the bit signal along the drill string, the drill string absorbs the bit signal, so the energy of the bit signal will be lost in the form of heat energy. The absorption of bit signals is related to the physical characteristics of drill string. When the characteristics of drill string are invariable, the longer the drill string is, the more obvious the absorption effect will be, and the more energy will be lost. Amplitude attenuation caused by drill string absorption can be expressed as:

$$A = A_0 e^{-\eta x} \tag{6.1}$$

In the equation, A_0 represents the initial amplitude, x represents the distance propagated along the drill string, and n represents the absorption coefficient of the drill string.

6.2.2 Seismic Wave Propagation Time

The propagation time of P wave is easier to be detected than that of S wave in a seismic map, so they are mainly used for seismic tomography. In the asymptotic approximation, it can be proved that the propagation time of seismic waves satisfies Eikonal equation [8], then:

$$\nabla^2 \mathrm{t}(\mathrm{x}) = \frac{1}{c(x)} \tag{6.2}$$

Here, only longitudinal waves are considered, and $c(x)$ represents the longitudinal wave velocity at a given position x of the underground medium. If only vertical propagation waves are considered, then the sum of the depth z and time t is:

$$\begin{cases} z = \int_0^t c(t')dt' \\ t = \int_0^z \frac{1}{c(z')} dt' \end{cases} \tag{6.3}$$

For the velocity model, the one-way travel time associated with each depth can be calculated by using the above equation.

Using the bit's seismic wave field to locate the bit requires searching for multi-channel data in the xz plane to find possible coherent direct bit waves. In the traditional offset scheme, the buried source location can be offset based on the diffraction sum, which is the sum of the amplitudes of the source points along the diffraction curve in the xz plane [9]. In the xz plane, when the correct coherent hyperbola is matched, the sum produces high amplitudes in the offset domain. In uniform medium, the wavefront of the underground sound source is circular in space, and its diffraction shape (or emission shape) can be defined by the travel time trajectory as:

$$t = \sqrt{t_0^2 + \frac{(x - x_0)^2}{v^2}} \tag{6.4}$$

Where t_0 is the one-way zero offset time from the underground source to the receiver at position x_0; x_0 is the position perpendicular to the source, and $x - x_0$ is the offset time of the receiver at position x_0. The summation in the standard offset scheme can be replaced by coherence to overcome weak signal detection or velocity macro model defects.

6.2.3 3D Space-Time Relationship of Main Seismic Wave Fields

The so-called 3D space-time relationship refers to the change of seismic wave propagation time with 3D spatial coordinates. We consider the space-time relations of various wave fields in 3D uniform half space from the viewpoint of 3D surface seismic reflection. Assume that multiple geophone arrays are arranged on the ground for receipt, and the array orientation is parallel to the x or y axis (equivalent to 3D surface seismic observation). R_n is the position of the detector at n channel in any array, and the coordinate is $(x, y, 0)$. In lead-straight drilling, S is the position of downhole bit, the coordinate is $(0, 0, z)$, and z is the bit depth.

The one-way time t_d of direct wave propagating from the position of the bit through the formation to the point R_n is:

$$t_d = \frac{1}{v_e}\left|\vec{r}_n - \vec{r}_s\right| = \frac{\sqrt{x^2 + y^2 + z^2}}{v_e} = \frac{\sqrt{X^2 + z^2}}{v_e} \tag{6.5}$$

Where $X = \left|\vec{r}_n\right| = \sqrt{x^2 + y^2}$, v_e is the propagation speed of the formation above the drill bit. Similarly, the travel time of the reflected wave (reflected from the position of the drill bit through the interface in front of the drill bit and propagated to the point R_n through the formation) is:

$$t_r = \frac{1}{v_e}\left(\left|\vec{r}_n - \vec{r}_s - \vec{r}_{sr'}\right| + \left|\vec{r}_{s'r'}\right|\right) = \frac{\sqrt{X^2 + (z + 2h)^2}}{v_e} \tag{6.6}$$

In normal drilling, if the velocity of seismic wave propagating along the drill string is greater than the near-well formation velocity, and the radial displacement of the drill bit can transfer sound energy into the well fluid, which is then transmitted to the formation to form the drill string refraction wave (first wave), then the space-time relationship of the drill string first wave is:

$$t_h = \frac{z - Xtg\theta_c}{v_p} + \frac{X}{v_e \cos\theta_c} = \frac{X\cos\theta_c}{v_e} + \frac{z}{v_p} \tag{6.7}$$

In the equation $\theta_c = \sin^{-1}\frac{v_e}{v_p}$, v_p is the propagation speed along the drill string.

Since wellsite generators, mud pumps, and drilling operations produce a variety of strong noise interference, such as surface waves and surface refraction waves, for 3D seismic observation, these interference sources can all be considered to be from the ground wellhead location. Assume the propagation speed of noise on the surface be v_n, and the time to reach the geophone point R_n be t_n, then:

$$t_n = \frac{X}{v_n} + t_{0n} \tag{6.8}$$

Where, for surface waves, $t_{0n} = 0$, v_n is group velocity of surface waves (frequency dispersion variation of surface waves is not considered); For the refraction wave of the surface layer, $t_{0n} \neq 0$, it is the high-speed layer velocity that produces the refraction wave.

6.3 SWD Data Processing

The key of SWD reference signal processing is to extract weak bit source signals by eliminating interference from many noise components. To do this, the vibration signals from the drill bit must first be isolated from the acceleration sensor records at the top of the drill string as a source function correlated with the ground records.

Because of the unique source of SWD – drill bit vibration, and a variety of transmission paths from the source, SWD earthquake signals are characterized with complexity. The analysis method based on periodicity and mixed signal characteristics is very suitable for fast and effective analysis of SWD signal, a continuous mixed signal with strong periodicity.

Seismic interferometry is based on interference theory proposed by Claerbout [10], developed by Schuster [11], which includes correlation and offset, also known as virtual shoting method or acoustic solar imaging method. Seismic interference processing technology can be applied to extract weak effective seismic signal from strong noise. Seismic interference measurement is widely used in the SWD data processing.

During SWD exploration, it is a common method to improve the signal-to-noise ratio and resolution by cross-correlation of the lead signals on the drill string through deconvolution [12] and ground signals or by cross-correlation of a selected certain ground track as a reference signal and other channels. Because of the complexity of the SWD wave field, low signal-to-noise ratio, and susceptible to the influence of the drill string, interferometry can synthesize the data recorded on any two receivers, and the seismic waves propagating between the receivers can be approximately regarded as a source.

6.3.1 Cross-Correlating

In the field of seismic interference measurement, the researchers retrieved surface waves and body waves by cross-correlating unrelated noise sources to extract useful subsurface information.

Cross-correlating is used to describe the correlation degree between two time series, that is, the degree of correlation between the values of signals at any two different times. SWD correlates the reference signal received by the sensor on the drill string with the reference signal recorded by the geophone on the ground to reflect the effective information of underground structure. Cross-correlating compresses the continuous bit signals into pulse signals, where the different pulses correspond to different seismic waves (such as direct wave, emission wave, etc.). Considering the cross-correlation test bit signal extraction effect, noise needs to be attenuated and signal-to-noise ratio improved. The travel time required for seismic waves to reach each receiver along different paths can be measured by the corresponding time of pulses. Theoretically, we hope to cross-correlate the downhole drill bits as reference signals. Since the sensor of the reference signal is on the derrick, there is a time lag between the downhole signal and the reference signal, called delay time of the drill string. To add the delay time of the drill string to the cross-correlating timeline, the delay time of the bit signal in cross-correlating function can become bit signal travel time in the formation.

Cross-correlating is used to measure the similarity of two signals, and its calculation formula is as follows:

$$\phi_{pg}(\tau) = \sum_{k=1}^{N} p(k)g(k+\tau) \tag{6.9}$$

In the equation, $p(k)$ and $g(k)$ represent the cross-correlating two signals respectively, τ represents the amount of time shift of $g(k)$ relative to $p(k)$, N represents the signal length in cross-correlating. Cross-correlating is a function relative to time shift. For a certain relative shift, if the two signals are similar, then the corresponding item is usually a, is the product of cross-correlation values is big;If two signals are not very similar, the product of the corresponding terms is positive, the cross-correlating value is big. If not similar, then the product may be either positive or negative, and the sum of them is small.

6.3.1.1 Reference Signal Auto-Correlation

Since the drill bit rotates in the downhole, the seismic signals produced are continuous. Continuous bit signals can be converted into sharp pulses by auto-correlation of reference signals [13]. If$B(z)$ is used to represent the transformation of the discrete bit signal. $D(z)$ is used to represent the transmission response of the drill string, the reference signal $P(z)$ received by the pilot sensor on the drill string top can be expressed as:

$$P(Z) = B(Z)D(Z) \tag{6.10}$$

Auto-correlation of the reference signal can be expressed as:

$$\psi_{pp}(Z) = P(Z)P\left(\frac{1}{Z}\right) = B(Z)B\left(\frac{1}{Z}\right)D(Z)D\left(\frac{1}{Z}\right) \tag{6.12}$$

The frequency spectrum of the drill bit signal conforms to the characteristics of white noise, then:

$$|B(Z)|^2 = C_1 \tag{6.13}$$

Where C_1 is a constant, the auto-correlation of the reference signal is only the transmission effect of the drill string,

$$\psi_{pp}(Z) = C_1 \times D(Z)D\left(\frac{1}{Z}\right) \tag{6.14}$$

In the Eq. (6.14), for SWD, assuming that the bit signal is white noise signal, the auto-correlation of the reference signal is the auto-correlation of the drill string impulse response.

6.3.1.2 Auto-Correlation of Reference Signal and the Ground

The reference signal is collected by the sensor at the top of the drill string and is not a true bit signal. If *E*(*Z*) is used to represent the *Z* transformation of the bit signal formation pulse response, the signal *G*(*Z*) received by the ground sensor can be expressed as:

$$G(Z) = B(Z)E(Z) \tag{6.15}$$

The cross-correlation between reference signals and ground records is:

$$\psi_{PG}(Z) = P\left(\frac{1}{Z}\right)G(Z) = B(Z)B\left(\frac{1}{Z}\right)E(Z)D\left(\frac{1}{Z}\right) = |B(Z)|^2E(Z)D\left(\frac{1}{Z}\right) \tag{6.16}$$

The reference signals are cross-correlated with the ground to obtain not only an ideal formation impulse response, but also the interference generated by drill string transmission effect is coupled within the impulse response, so deconvolution (anti-filtering) processing is necessary.

6.3.1.3 Theory of Interferometric Method [14, 15]

Assume that G_0 is a non-direct wave, G_s includes all scattered waves; Only acoustic waves are processed in SWD. Background medium and medium disturbance can be arbitrary, non-uniform and anisotropic. The frequency domain wave field $\mu(r_A, s, \omega)$ recorded at r_A are the superposition of undisturbed scattering green function $G_0(r_A, s, \omega)$ and $G_S(r_A, s, \omega)$, the convolution of source function $W(s, \omega)$ related with the excitation at *s*, so:

$$\mu(r_A, s, \omega) = W(s, \omega)[G_0(r_A, s, \omega) + G_S(r_A, s, \omega)] \tag{6.17}$$

In the equation,G_0 and G_s represents undisturbed wave and wave field disturbance respectively. Besides, $W(s, \omega)$ can either be the complicated function of frequency or function of *s*.

The wave fields of r_A and r_B in the frequency domain are cross-correlated as follows:

$$C_{AB} = |W(s)|^2 G(r_A, s)G*(r_B, s) \tag{6.18}$$

In the equation, $*$ represents the complex conjugate. From the Eq. (6.18), cross-correlation C_{AB} depends on the power spectrum of $W(s)$. Integrate the cross-correlation in Eq. (6.18) to the curved surface ∂V containing all sources, then:

$$\int_{\partial V} C_{AB}\mathrm{d}s = \left\langle |W(s)|^2 \right\rangle [G(r_A, r_B) + G*(r_A, r_B)] \tag{6.19}$$

In the equation, $\langle |W(s)|^2 \rangle$ is the source mean value of the power spectrum, $G(r_A, r_B)$ and $G*(r_A, r_B)$ are respectively the causal and inverse causal green functions of excitation r_B and receiver excitation r_A. For Eq. (6.19), G corresponds to the pressure response in the acoustic medium. If it is the particle velocity response, the plus sign on the right of Eq. (6.19) is replaced by the minus sign. Eq. (6.19) is strictly applicable to nondestructive medium.

Given the complexity of $\langle |W(s)|^2 \rangle$, it can be difficult to restore the response between r_A and the local receivers r_B through the above equation. Most researchers recommend deconvolution after the integral of the above equation, which presumes that the power spectrum of the source can be independently estimated. In fact, such an estimate can be obtained only in some applications, not in most of the cases.

Suppose the source function in Eqs. (6.17)–(6.19) is independent of the source position s ($W(s)=W$). Combing (6.17) with (6.18), C_{AB} can be extended as follows:

$$\begin{aligned} C_{AB} &= \mu(r_A, s)\mu*(r_B, s) \\ &= \underbrace{\mu_0(r_A, s)\mu_0*(r_B, s)}_{C_{AB}^1} + \underbrace{\mu_S(r_A, s)\mu_0*(r_B, s)}_{C_{AB}^2} + \underbrace{\mu_0(r_A, s)\mu_S*(r_B, s)}_{C_{AB}^3} \\ &\quad + \underbrace{\mu_S(r_A, s)\mu_S*(r_B, s)}_{C_{AB}^4} \end{aligned} \tag{6.20}$$

In the equation, $\mu_0 = WG_0$ and $\mu_s = WG_s$. Four items $C_{AB}^1 \sim C_{AB}^4$ are inserted into the Eq. (6.20), and get:

$$\begin{aligned} &\int_{\partial V} C_{AB}^1 ds + \int_{\partial V} C_{AB}^2 ds + \int_{\partial V} C_{AB}^3 ds + \int_{\partial V} C_{AB}^4 ds \\ &= |W(s)|^2 [G_0(r_A, r_B) + G_S(r_A, r_B) + G_0*(r_A, r_B) + G_S*(r_A, r_B)] \end{aligned} \tag{6.21}$$

Considering only G_S term (scattered wave) in Eq. (6.21), the first integral is related to the undisturbed term on the right side of Eq. (6.21),

$$\int_{\partial V} \mu_0(r_{a,}s)\mu_0*(r_{a,}s)ds = |W|^2 [G_0(r_A, r_B) + G_0*(r_A, r_B)] \tag{6.22}$$

The interference measurement of the undisturbed wavefield to the left of Eq. (6.22) must yield the causal and anti-causal unperturbed wavefields (at right side of Eq. 6). The main contribution of the causal scattered wavefield between r_A and $r_B$comes from the correlation between the unperturbed wave field at r_B and scattered wave field at r_A,

$$\int_{\partial V_1} \mu_S(r_A, s)\mu_0 * (r_B, s)ds = |W|^2 G_S(r_A, r_B) \tag{6.23}$$

where ∂V_1 is a portion of ∂V that yields stationary-phase contributions to $G_S(r_A, r_B)$;

Eq. (6.23) is an approximate relation, ignoring the influence of volume integral. However, in exploration experiments, it is impossible to surround the underground with seismic source. Therefore, only a partial source integral can be yielded rather than the closed surface integral required for the Eq. (6.19).

Truncation of the surface integral can introduce spurious events in the final interferometric gathers. This holds for general 3D models as well, and it can be verified.

$$\int_{\partial V_1} C_{AB} ds + \int_{\partial V_2} C_{AB} ds = |W|^2 [G_S(r_A, r_B) + G_S * (r_A, r_B)] \tag{6.24}$$

In the equation, ∂V_1 and ∂V_2 are curved section, $\partial V_1 \cup \partial V_2 = \partial V_0$

6.3.1.4 Interferometric Migration

1. Single seismic source

Interferometric offset [16, 17] converts relevant seismic data into reflectivity or source distribution. Figure 6.2 illustrates interferometric imaging with unknown source location and its wavelet. It shows only direct waves (S_A, S_B) and the ray path of the first-order multiples $S_A r_B$.

In uniform nondestructive medium, the receivers at position A and B can be modeled as:

$$d_A(w) = s(\omega)e^{-i\omega t_{sA}} \tag{6.25}$$

$$d_B(w) = s(\omega)e^{-i\omega tsA} + s(\omega)\, Re^{-i\omega\left(t_{sA+}t_{Ar} + t_{rB}\right)} \tag{6.26}$$

In the equation, $s(\omega)$ is the source function at position s, ω is the angular frequency, t_{sA} and t_{sB} represent the travel time from s to A and B, t_{Ar} and t_{rB} are

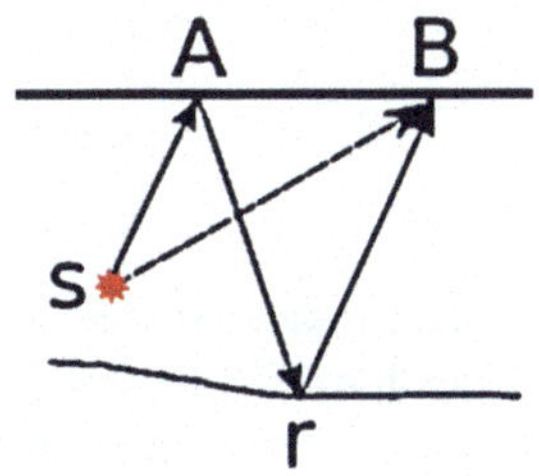

Fig. 6.2 Source location and interferometric offset of reflectivity imaging

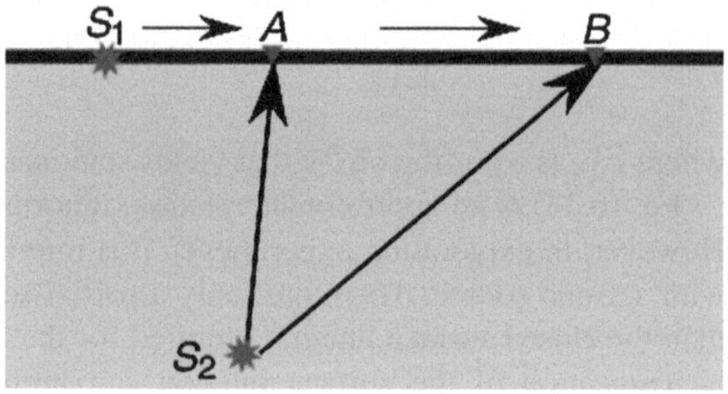

Fig. 6.3 Direct wave cross-correlation diagram of receiver A and B with two vibration sources (S1 and S2)

travel time of the first order multiples from A to r and from r to B. Taking $dA(\omega)$ as a reference, the correlation between A and B is:

$$\Phi(\omega) = d_A * d_B = \underbrace{|s(\omega)|^2 e^{-i\omega(tsB - tsA)}}_{\text{direct wave time delay}} + \underbrace{|s(\omega)|^2 \, Re^{-i\omega\left(t_{Ar} + t_{rB}\right)}}_{\text{reflectivity}} \tag{6.27}$$

Where $*$ represents complex conjugate.

2. Two seismic sources

For spatially separated two-point source model (S_1 and S_2, as shown in Fig. 6.3), in uniform nondestructive medium, the direct arrival of receiving points A and B can be modeled in the frequency domain as follows:

$$d_A(w) = s_1(\omega)e^{-i\omega t_{s1A}} + s_2(\omega)e^{-i\omega t_{s2A}} \tag{6.28}$$

$$d_B(w) = s_1(\omega)e^{-i\omega t_{s1B}} + s_2(\omega)e^{-i\omega t_{s2B}} \tag{6.29}$$

In the equation, s_i represents the source function, ω represents the angular frequency, and t_{siA} represents the propagation time of the direct wave from the i source position to the receiver at A. Our goal is to determine the time delay of the receiver array relative to the reference channel. Taking $dA(\omega)$ as a reference, complex conjugate of dA is multiplied by db to get:

$$\begin{aligned}\Phi &= d_A * d_B \\ &= \underbrace{s_1 s_1 * e^{-i\omega(t_{s1B} - t_{s1A})}}_{\text{rig crossterm}} + \underbrace{s_2 s_2 * e^{-i\omega\left(t_{s2B} + t_{s2A}\right)}}_{\text{bit crossterm}} \\ &\quad + \underbrace{s_1 s_2 * e^{-i\omega(t_{s1B} - t_{s1A})} + s_1 s_2 * e^{-i\omega\left(t_{s2B} + t_{s2A}\right)}}_{\text{crosstalk}}\end{aligned} \tag{6.30}$$

The image of the underground seismic source location with direct wave time delay term in the relevant domain is obtained by summation of the time delay response of the receiving array of pilot channel [18], that is:

$$m(x) = \sum_{A,B} \sum_{\omega} \Phi(A, B; \omega) e^{i\omega(t_{xB} - t_x A)} \tag{6.31}$$

In the equation, (*A*, *B*) represents the sum in paired channel used for correlation, *A* represents the reference channel, and *B* is the receiving array with indexes from 1 to *n*. Φ is correlation between channel *A* and channel *B*. The core of the interferometric offset at the unknown source location is $e^{i\omega(t_{xB} - t_x A)}$.

Cross-correlation is usually used to compress passive data records and construct the source receiver set. Through deconvolution interference, reference channels can be deconvolved to expand the cross-correlation spectrum, improve the detectability of narrow-band weak signals in the relevant domain, and improve the time resolution of cross-correlation. Correlation in the frequency domain can be expressed as:

$$\Phi(\omega) = \frac{d_A * (\omega) d_B(\omega)}{|d_A(\omega)|^2} = \frac{d_B(\omega)}{d_A(\omega)} \tag{6.32}$$

In the equation, $*$ is the complex conjugate. In an ideal nondestructive medium, deconvolution by using the reference trace d*A* eliminates the receiving function d*A*, removes the source wavelet, and normalizes the process to expand the relevant spectrum. Therefore, the program improves the time delay (improves the resolution of the time delay). Another benefit of this program is that it can remove contamination from other sources in open scattered medium [19].

Here, we apply coherence measurement in the offset process. The destructive interference in the superposition can be minimized by using the cross-correlation of imaging, which is more suitable for weak correlation signals.Correlation offset can be expressed as,

$$m(x) = coherency\left(\sum_{\omega} D(A, B; \omega; \tau) e^{i\omega(t_{xB} - t_{xA})} \right) \tag{6.33}$$

In the equation, *t* is correlation time window, (-*w*, +*w*), in which *w* represents sampling point.

6.3.1.5 Semblance Algorithm

In uniform medium, the wavefront of the underground acoustic source is circular in space, and its diffraction shape (or emission shape) can be defined through propagation time trajectory as:

$$t = \sqrt{t_0^2 + \frac{(x - x_0)^2}{v^2}} \tag{6.34}$$

Semblance is most commonly used for coherency analysis. Semblance measurement is made along the diffraction travel time curve given by Eq. (6.34) at a finite time window to measure the bit-correlating straight wavefront. Semblance is the normalized output-input energy ratio of signals normalized along the window hyperbola, as shown below:

$$S(x_0, t_0) = \frac{1}{M} \frac{\sum_{\tau=-w}^{w} \left(\sum_x u(x, t(x) + \tau)\right)^2}{\sum_{\tau=-w}^{w} \sum_x u(x, t(x) + \tau)^2} \tag{6.35}$$

where $\boldsymbol{M}$ is the number of traces indexed by $\boldsymbol{x}$, and $\boldsymbol{\tau}$ ranges over a time window $(-w, +w)$. The semblance has value in range $0 < S < 1$. The advantage of using semblance over the simple summation is that it considers of the similarity of the signals in the given time window. The length of the time window controls the trade-off between a reduced resolution in time domain and low *S/N* detection.

6.3.1.6 MUSIC Coherency Measurement

Multiple Signal Classification (MUSIC) has the potential advantage of higher imaging resolution than Semblance resolution.

MUSIC is a classic high-resolution algorithm for direction-of-arrival (DOA) estimation. It was first proposed by Schmidt [20], and is applied to the position estimation of the narrow-band, irrelevant multi-emission points and the signal parameter estimation. The classical MUSIC algorithm uses the covariance matrix of the receiving array as input and is robust to ambient noise. Asgedom [21] integrates the window-controlled versions of classical MUSIC into coherency-migration schemes, so that they can deal with broadband seismic data and coherent sources.

An important prerequisite for using MUSIC is that the number of features to be addressed must be smaller than the number of sources and receivers. When the surface multiple-receiver array is used to calculate the position of drill bit, this premise is easy to be satisfied. Window-steered MUSIC algorithm is described as below. In the case of a receiver array with m receivers, the data can be modeled as:

$$d_m(t) = s(t - \Delta t_m) + n_m(t) \tag{6.36}$$

Where, the modeling data at receiver m consists of delayed source wavefield $\boldsymbol{s}$ and random noise n. Δt_{m} represents the time delay of the receiver m.

In the time domain, the steering time window is adopted for coherency measurement, and the it is obtained through time-shift operation, so there is no wavelet stretch. In this process, adopting interpolation in the time domain is suggested. MUSIC algorithm is closely related to singular value decomposition (SVD).

The W sample of time window data was decomposed into signals and complementary noise subspaces by SVD, as shown below:

$$D = [U_s\ U_n]\begin{bmatrix} \sum_s & 0 \\ 0 & \sum_n \end{bmatrix}\begin{bmatrix} V_s^T \\ V_n^T \end{bmatrix} \tag{6.37}$$

where U, V, and $\sum$ represent eigenvector matrices and the singular matrix, respectively. The $\sum_s$ denotes a diagonal matrix with the R largest singular values, which are associated with the first R eigenvectors of U_s, and V_s (referred to as the signal subspace). The $\sum_n$ denotes the remaining diagonal matrix (corresponding to the noise subspaces).

Consistency matrix d in steered finite-window data has the similarity between row vectors. After SVD, the columns of U can be interpreted as seismic wavelet, and the columns of V shows normalized amplitudes of each recording track wavelet. Therefore, the MUSIC algorithm uses the matrix V to project the steering vector onto the noise subspace of V, so as to determine the DOA and the number of scatterers. The steering vector used for scanning space is a function of angle, which can be obtained by $a = \left[1, e^{-i\omega\Delta t_1}, e^{-i\omega\Delta t_2}, \cdots, e^{-i(M-1)\omega\Delta t_{M-1}}\right]^T$, where Δt is a function of angle θ. The angle is called the direction of the signal reaching the receiver array [20], and ω is the angular frequency.

In the window-steered MUSIC, use the fixed steering vector [21, 22], $a = [1, 1, 1, \cdots, 1]^T$. For coherent flat events with relatively constant amplitude in the data window, the signal space is divided by fixed steering vector A. Because of the orthogonality of signal and noise subspace, the projection of A onto noise subspace is the minimum. Therefore, placing the projection in the denominator result can lead to a rush of such a coherency event. MUSIC coherency in imaging spatial position (x_0, z_0)can be expressed as follows:

$$P_{MU}(x_0, z_0) = \frac{1}{a^T[P_n]a} = \frac{1}{\sum_{i=1}^{M}\sum_{j-1}^{M}[P_n]_{i,j}} \tag{6.38}$$

where the estimated source positions (x_0, z_0) and velocity model (v) are used to estimate the direct arrival of the source wavefield in a time window. Note that $P_n = V_{\mathrm{n}} \cdot V_n^T$ is the projection matrix onto the noise subspace. This defines the coherency value at the imaging position (x_0, z_0), and is referred to as the MUSIC pseudospectrum.

$$P_{MU}(t_0, x_0) = \frac{a^T a}{a^T\left[E_n \cdot E_n^T\right]a} \tag{6.39}$$

where $\boldsymbol{a} = [1,1,1,\cdots,1]$, $\boldsymbol{T}$ is a fixed steering vector. $\boldsymbol{E_n} = [\boldsymbol{e_{W+1}}, \cdots, \boldsymbol{e_M}]$, where W split the signal and noise subspaces of left unitary matrix of singular value decomposed data matrix $\boldsymbol{D_w}$. The $\boldsymbol{D_w}$ is time shifted windowed data $\boldsymbol{D}$.

The above equation can be used for velocity analysis and imaging, and Music can be well integrated into seismic coherency analysis. Semblance is the normalized energy ratio with its value ranging from 0 to 1, while Music algorithm does not have such boundary, and its value changes more randomly.

6.3.1.7 Cross-Correlation Algorithm in the Wavelet Transform Domain

The cross-correlation algorithm in the wavelet transform domain [23] can perform the wavelet transform of the reference channel and other receiving channels respectively under low SNR, and obtain the time delay of the receiving channel relative to the reference channel, then extract the direct wave and reflected wave information. Suppose there are two data $x(t)$, $y(t)$, where $x(t)$ is reference channel, $y(t)$ is the sampling data on other channels. Suppose the common components of $x(t)$ and $y(t)$ are s_1 and s_2. The time delay of the common component s_1 and s_2 in $y(t)$ are τ_1 and τ_2 relative to that in $x(t)$, then $x(t) = s_1(t) + s_2(t)$. The actual seismic record contains noise, its mathematical model is expressed as $y(t) = s_1(t - \tau_1) + s_2(t - \tau_2) + z(t)$, where $z(t)$ is environmental noise. The cross-correlation of the two time domains can be defined [24, 25] as:

$$R_{xy}(\tau) = \int_{-\infty}^{\infty} x(t)y(t-\tau)dt \tag{6.40}$$

The discrete form can be expressed as:

$$R_{xy}(\tau) = \sum_{n=-\infty}^{\infty} x(n)y(n-\tau) \tag{6.41}$$

In actual SWD processing, since the seismic source signal is difficult to be directly obtained, suppose the wavelet transform of the reference channel $x(t)$ is:

$$\begin{aligned} W_{\varphi,x}(a,b) &= \int_{-\infty}^{+\infty} x(t)\varphi * \left(\frac{t-b}{a}\right)dt = \int_{-\infty}^{+\infty} \left[s^1(t) + s^2(t)\right]\varphi * \left(\frac{t-b}{a}\right)dt \\ &= W_{\varphi,s^1}(a,b) + W_{\varphi,s^2}(a,b) \end{aligned} \tag{6.42}$$

The wavelet transform of the other channel is:

$$\begin{aligned} W_{\varphi,y}(a,b) &= \int_{-\infty}^{+\infty} y(t)\varphi * \left(\frac{t-b}{a}\right)dt \\ &= \int_{-\infty}^{+\infty} \left[s^1(t-\tau^1) + s^2(t-\tau^2) + z(t)\right]\varphi * \left(\frac{t-b}{a}\right)dt \\ &= W_{\varphi,s^1}(a,b-\tau^1) + W_{\varphi,s^2}(a,b-\tau^2) + W_{\varphi,z}(a,b) \end{aligned} \tag{6.43}$$

The cross-correlation algorithm in the wavelet transform domain between the reference channel $x(t)$ and the other channel $y(t)$ can be expressed as:

$$C^F_{\mathrm{y,x}}(\tau) = \int_a \int_b W_{\varphi,\mathrm{y}}(a,b) W_{\varphi,x}(a, b-\tau) dadb \tag{6.44}$$

By substituting the above equation into the Eqs. (6.40) and (6.41):

$$\begin{aligned} &\int_a \int_b \left[W_{\varphi,s^1}\left(a,b-\tau^1\right) + W_{\varphi,s^2}\left(a,b-\tau^2\right) + W_{\varphi,z}(a,b)\right] \\ &\quad \left[W_{\varphi,s^1}(a,b-\tau) + W_{\varphi,s^2}(a,b-\tau)\right] dadb \\ &\quad = W_{\mathrm{s^1s^1}}\left(\tau-\tau^1\right) + W_{\mathrm{s^1s^2}}\left(\tau-\tau^1\right) + W_{\mathrm{s^1s^2}}\left(\tau-\tau^2\right) \\ &\quad + W_{\mathrm{s^2s^2}}\left(\tau-\tau^2\right) + W_{\mathrm{zs^1}}(\tau) + W_{\mathrm{zs^2}}(\tau) \end{aligned} \tag{6.45}$$

In the equation, $W_{s^1s^1}(\tau - \tau^1)$ is the wavelet transform domain auto-correlation of the component s^1. $W_{\mathrm{s^1s^2}}(\tau - \tau^1)$ and $W_{\mathrm{s^1s^2}}(\tau - \tau^2)$ is the cross-correlation of the component s^1 and s^2 wavelet transform domain. $W_{\mathrm{s^2s^2}}(\tau - \tau^2)$ is wavelet domain auto-correlation of the component s^2. The latter two terms of Eq. (6.45) are interference elements.

When $\tau = \tau^1$:

$$\begin{aligned} C^F_{y,x}\left(\tau^1\right) &= W_{\mathrm{s^1s^1}}(0) + W_{\mathrm{s^1s^2}}(0) + W_{\mathrm{s^1s^2}}\left(\tau^1-\tau^2\right) + W_{\mathrm{s^2s^2}}\left(\tau^1-\tau^2\right) \\ &\quad + W_{\mathrm{zs^1}}\left(\tau^1\right) + W_{\mathrm{zs^2}}\left(\tau^1\right) \end{aligned} \tag{6.46}$$

In the equation, the first term on the right side is the maximum value of the auto-correlation function in the component s^1 wavelet transform domain, which is far greater than the latter terms, then this component is highlighted. When $\tau = \tau^2$, the composition s^2 is highlighted. When $s^1 = s^2$, the Eq. (6.45) becomes:

$$C^F_{y,x}(\tau) = 2W_{\mathrm{s^1s^1}}\left(\tau-\tau^1\right) + 2W_{\mathrm{s^1s^1}}\left(\tau-\tau^2\right) + 2W_{\mathrm{zs^1}}(\tau) \tag{6.47}$$

At this time, other channels can be interpreted as echoes of seismic signals with different time delays. According to different formation wave impedance, the echos number of source signals received by seismic receiving channels is different, then $C^F_{y,x}(\tau)$ can be expressed as:

$$\begin{aligned} C^F_{y,x}(\tau) &= \int_a \int_b \left[\sum_{i=1}^{M} a_i W_{\varphi,y}\left(a,b-\tau^i\right) + W_{\varphi,z}(a,b)\right] W_{\varphi,x}(a,b-\tau) dadb \\ &= \sum_{i=1}^{M} a_i W_{\mathrm{ss}}\left(\tau-\tau^i\right) + W_{zs}(\tau) \end{aligned} \tag{6.48}$$

In the equation, $W_{zs}(\tau)$ is the time-frequency cross-correlation in the wavelet domain between the source signal and noise, and it is the interference caused by noise; $W_{ss}(\tau)$ is the time-frequency correlation wavelet in the wavelet domain. As the signal-to-noise ratio decreases, the optimal effect of one-time cross-correlation in the wavelet domain can not be achieved, so multiple time-frequency cross-correlation algorithms in the wavelet domain can be further performed. The specific steps are as follows:

(1) Perform the wavelet transform of the reference channel signal and the receiving channel signal respectively; (2) Calculate the wavelet coefficients of each scale to the power of N after the two transformations. Let N be a natural number and select the suitable N according to the analysis; (3) Carry out cross-correlation algorithm after the coefficients of the two wavelet transforms to the power of N. The specific time-frequency cross-correlation function in the wavelet domain is:

$$C^F_{y,x}(\tau)^N = \int_a \int_b \left[W_{\varphi,y}(a,b)W_{\varphi,x}(a,b-\tau)\right]^N dadb \tag{6.49}$$

In the equation, the selection of N depends on the actual signal processing. In the cross-correlation algorithm in wavelet transform domain, only the time domain and frequency domain of noise overlap with the effective signal, thus causing interference. If the noise delay information is the same as the signal delay information, it can be extracted from the signal by the wavelet transform as long as the frequency information is different, and then suppress the noise by cross-correlation algorithm, highlighting the time delay of the useful signal.

The cross-correlation algorithm in the wavelet transform domain processes the data by utilizing the characteristics of time overlapping between the noise and the useful signals, no overlapping of the frequency domain. And its noise suppression ability has obvious advantages over cross-correlation in time domain. Under low signal-to-noise ratio, multiple cross-correlation can also be performed to suppress interference and improve the signal-to-noise ratio.

6.3.2 Deconvolution-Deconvolution Interference

SWD, different from surface seismic and vertical seismic profiling (VSP), uses the vibration of the drill bit in the drilling process as the source, arranges the sensor on the top of the drill string to receive the bit signal propagated by the drill string, and arranges the geophone on the ground to receive the seismic signal propagated by the drill bit through the formation. And perform the cross-correlation of the two signals, then the pulse response of the formation can be obtained.

The reference signal is obtained by convolution of drill bit signal and drill string system function. The signal received by the sensor at the top of the drill pipe is affected by the drill string resonance effect or path propagation effect, which results in spectrum distortion and high-speed multiple wave interference. The purpose of deconvolution is to eliminate the influence of drill string propagation by processing the reference signal into the bit signal. The application of deconvolution interference processing method [12] can eliminate the processing results of seismic source interference. And deconvolution of the cross-correlation output can eliminate the high-speed multiple wave and spectral distortion of the reference signal in the cross-correlation output, and improve the signal-to-noise ratio of SWD data.

Deconvolution interferometry, based on the theory put forward by Claerbout [10], is developed based on cross-correlation. It can extract effective source signal excited by bit from strong environmental seismic data, make use of the random noise to obtain green function between the seismic stations, underground structure and velocity structure imaging. There is no need to make independent estimates of the source function and no need to depend on the source function information, and it is the most suitable for pilot unrecorded or unreliable bit excitation.

6.3.2.1 Deconvolution Principle, Seismic Channel Convolution Model

The ideal seismic record is the time series of reflection coefficient, assuming $\xi(t)$ is the reflection coefficient series, then the ideal seismic record $x(t)$ can be expressed as:

$$x(t) = N_0 \xi(t) \tag{6.50}$$

In the equation, N_0 is the intensity value of source pulse, it is a constant.

In fact, seismic channel $x(t)$ is composed of the superposition of effective wave $s(t)$ and interference wave $n(t)$, $x(t) = s(t) + n(t)$. The effective wave only refers to the first reflection wave, and all the others are interference waves. The first reflection wave of layered medium is usually represented by linear convolution model:

$$s(t) = w(t) * r(t) \tag{6.51}$$

In the equation, $w(t)$ is the system wavelet; $r(t)$ is the function of reflection coefficient.

System wavelet is formed by seismic source wavelet $o(t)$ passing through the formation filter $g(t)$ to form the ground sub-wave $w_1(t)$, then it is received by the ground receiver $d(t)$ and recorded by the instrument $i(t)$ after layer reflection-refraction(penetrating response $\tau(t)$-reflection). It is the result of the comprehensive responses other than the reflection coefficient. Stratigraphic response, penetrating response, receiving response, instrument response are all produced by filtering, the source wavelet filtering is equivalent to continuous convolution in time domain:

$$w(t) = o(t) * g(t) * \tau(t) * d(t) * i(t) = w_1(t) * \tau(t) * d(t) * i(t) \tag{6.52}$$

By substituting the above equation into the Eq. (6.51):

$$s(t) = o(t) * g(t) * \tau(t) * i(t) * r(t) = o(t) * f_\tau(t) * f_d(t) * r(t) \tag{6.53}$$

In the equation, $f_\tau(t) = g(t) * \tau(t)$ is the earth filter; $f_d(t) = d(t) * i(t)$ is the receiving filter;

The convolution of Eq. (6.53) in the frequency domain is shown as follows:

$$S(\omega) = W(\omega)R(\omega) \tag{6.54}$$

In the equation, $S(\omega)$, $W(\omega)$ and $R(\omega)$ are the seismic record spectrum, wavelet spectrum and reflection coefficient spectrum respectively. Apparently,

$$R(\omega) = \frac{1}{w(\omega)} \cdot S(\omega) = A(\omega)S(\omega) \tag{6.55}$$

Where, $A(\omega) = \frac{1}{w(\omega)}$.

By inversely Fourier transform to the time domain, it can be obtained:

$$R(\omega) = a(t) \cdot S(t) = a(t) \cdot w(t) \cdot R(t) \tag{6.56}$$

In the equation, $a(t)$ is the time function of $A(\omega)$, so, $a(t) \cdot w(t) = \delta(t)$, $a(t)$ is the deconvolution factor;

In the case of known seismic wavelet, $a(t)$ can be obtained by using mathematical method. And perform the convolution of the inverse wavelet with the seismic record $S(t)$ by using the Eq. (6.56):

$$R(t) = \sum_\tau a(\tau)S(t - \tau) \tag{6.57}$$

6.3.2.2 Convolution Model of SWD

According to the Yilmaz's [9] definition of the convolution model of ground seismic data before vibroseis cross-correlation in 1987, if B(z) represents the Z transform of discrete bit signal and D(z) represents the drill string impulse response, then the lead signal P(z) received by the sensor at the top of the drill string can be expressed as:

$$P(z) = B(z) \times D(z) \tag{6.58}$$

Discrete signal received by the ground geophone G is:

$$G(z) = B(z) \times E(z) \tag{6.59}$$

In the equation, $E(z)$ represents the impulse response of the formation.

Bit deconvolution is mainly used to eliminate the drill string transmission effect in the reference signal and to suppress long-period drill string multiples and BHA multiples. The deconvolution operator can be obtained by auto-correlation of the reference signal or bit deconvolution before and after cross-correlation.

After perform the bit deconvolution of the reference signal, it can be obtained:

$$P(z) \times OP(z) = B(z)D(z)/D(z) = B(z) \tag{6.60}$$

The cross-correlation of the reference signal and the ground records:

$$\Psi_{PG}(z) = P\left(\frac{1}{z}\right)G(z) = B\left(\frac{1}{z}\right)B(z)E(z) = |B(z)|^2E(z) \tag{6.61}$$

Bit deconvolution is performed after the cross-correlation, then the impulse response of the formation can be:

$$\Psi_{PG}(z) \times OP(z) = P\left(\frac{1}{z}\right)G(z)OP(z) = |B(z)|^2E(z) \tag{6.62}$$

The length of bit deconvolution operator is determined by the pulse response of drill string and the length of the response is determined by the time span of direct wave and main drill pipe multiple waves. The drill string will affect the deconvolution operator.

This is the ideal case without noise. The actual SWD reference signals are recorded under the strong noise interference. Improving the signal-to-noise ratio of SWD signals is still the key to SWD signal processing.

6.3.2.3 Elastic Deconvolution Interferometry

The deconvolution interferometry is developed from acoustic to elastic medium, it can also be used for extracting elastic scattered waves. Then, the application of deconvolution interferometry in dealing with drilling noise passive recording will be described. The application of deconvolution interferometry to elastic waves should meet the following definition:

$$D_{AB}^{(p,q)} = D_{AB,K}^{(p,q)} = \frac{u_{(p,K)}^{(v,\Phi)}(r_A, s, \omega)}{u_{(q,K)}^{(v,\Phi)}(r_B, s, \omega)} = \frac{G_{(p,K)}^{(v,\Phi)}(r_A, s, \omega)}{G_{(q,K)}^{(v,\Phi)}(r_B, s, \omega)} \tag{6.63}$$

In the equation, $u_{(p,K)}^{(v,\Phi)} = \omega_K G_{(p,K)}^{(v,\Phi)}$ is the response measured in the frequency domain, subscript p (or q) represents a specific component of the measured particle velocity (superscript u represents the measured field quantity, particle velocity). All subsequent equations meet the dependence on angular frequency ω.

Similar to acoustic deconvolution interferometry, the denominator of Eq. (6.63) can be expanded into a power series when the scattered wave field is weaker than the direct wave field, then:

$$\left\{G_{(q,K)}^{(v,\Phi)}(r_B,s)\right\}^{-1}=\frac{1}{G_{0(q,K)}^{(v,\Phi)}(r_B,s)}\times\sum_{n=0}^{\infty}(-1)^n\left(\frac{G_{S(q,K)}^{(v,\Phi)}(r_B,s)}{G_{0(q,K)}^{(v,\Phi)}(r_B,s)}\right)^n \tag{6.64}$$

It is called the elastic deconvolution interference series, which has the same form as the acoustic deconvolution interference series. As with deconvolution interferometry in acoustic medium, we put Eq. (6.64) into Eq. (6.63), and then integrate the available sources to obtain:

$$\begin{aligned}\int_{\partial V_1} D_{AB,K}^{(p,q)}ds &= \underbrace{\int_{\partial V_1}\frac{G_{0(p,K)}^{(v,\Phi)}(r_A,s)\left\{G_{0(q,K)}^{(v,\Phi)}(r_B,s)\right\}*}{\left|G_{0(p,K)}^{(v,\Phi)}(r_B,s)\right|^2}ds}_{Term\,1}\\ &+\underbrace{\int_{\partial V_1}\frac{G_{s(p,K)}^{(v,\Phi)}(r_A,s)\left\{G_{0(q,K)}^{(v,\Phi)}(r_B,s)\right\}*}{\left|G_{0(p,K)}^{(v,\Phi)}(r_B,s)\right|^2}ds}_{Term\,2}\\ &-\underbrace{\int_{\partial V_1}\frac{G_{0(p,K)}^{(v,\Phi)}(r_A,s)\left\{G_{S(q,K)}^{(v,\Phi)}(r_B,s)\right\}*}{\left\{G_{0(p,K)}^{(v,\Phi)}(r_B,s)\right\}^2}ds}_{Term\,3}\end{aligned} \tag{6.65}$$

Only the term whose integrand function is linear in the scattered wave G_s is retained, it can be obtained:

$$\int_{\partial V_1}\frac{G_{S(p,K)}^{(v,\Phi)}(r_A,s)\left\{G_{0(p,K)}^{(v,\Phi)}(r_B,s)\right\}*}{\left|G_{0(p,K)}^{(v,\Phi)}(r_B,s)\right|^2}ds\approx\kappa G_{S(p,q)}^{(v,f)}(r_A,r_B) \tag{6.66}$$

where κ is a constant related to the source averaging of spectra $\left|G_{0(p,K)}^{(v,\Phi)}(r_B,s)\right|^2$. In our elastic case, setting $r_A=r_B$ and $p=q$ in Eq. (6.66) results. In $D_{BB,K}^{(q,q)}=1$, this translates to the time-domain condition, then,

$$D_{BB,K}^{(q,q)}(t)=\delta(t) \tag{6.67}$$

Eq. (6.67) imposes a clamped-point boundary condition for the pseudo-source experiment reconstructed by deconvolution interferometry.

6.3.2.4 Deconvolution Calculation of the Pilot Signal

Ideally (without noise interference), the auto-correlation method described above can be used to eliminate the effect of drill string transmission. The reference signal contains the recorded random noise in time. In the time domain, the method for calculating the deconvolution of the pilot signal with random noise will be discussed. The pilot signal model with random noise can be expressed as,

$$p(t) = b(t) * d(t) + n(t) = \sum_{\tau=0}^{n} b(\tau)d(t-\tau) + n(t) \tag{6.68}$$

Where $b(t)$ represents the bit signal, $d(t)$ represents the impulse response of the drill string, and $n(t)$ represents the recorded random noise.

The interference signal is maximum suppressed, and the error sum of squares between the actual output and the desired output of the narrow pulse after the action of the pilot signal $p(t)$ and filter $a(t)$ is the smallest, finally:

$$\sum_{\tau=0}^{m} r_{pp}(\tau - s)a(\tau) = r_{\tau p}(s), (s = 0, 1, \cdots m) \tag{6.69}$$

Where $r_{pp}(\tau - s)$ represents the auto-correlation of the pilot signal with a time delay of $(\tau - s)$, and $r_{\tau p}(s)$ represents the cross-correlation between the pilot signal with a time delay of s and the desired output.

The cross-correlation between the drill string impulse response and the desired output spike pulse is also a spike pulse, so $r_{ld}(j) = \begin{cases} \neq 0, j = 0 \\ = 0, j \neq 0 \end{cases}$, the above matrix form can be expressed as:

$$\begin{bmatrix} r_{pp}(0) & r_{pp}(1) & \cdots & r_{pp}(m) \\ r_{pp}(1) & r_{pp}(0) & \cdots & r_{pp}(m-1) \\ \cdots & \cdots & \cdots & \cdots \\ r_{pp}(m) & r_{pp}(m-1) & \cdots & r_{pp}(0) \end{bmatrix} \begin{bmatrix} a(0) \\ a(1) \\ \cdots \\ a(m) \end{bmatrix} = \begin{bmatrix} 1 \\ 0 \\ \cdots \\ 0 \end{bmatrix} \tag{6.70}$$

Adding an artificial white noise, called pre-whitening processing:

$$\begin{bmatrix} (1+\varepsilon)r_{pp}(0) & r_{pp}(1) & \cdots & r_{pp}(m) \\ r_{pp}(1) & (1+\varepsilon)r_{pp}(0) & \cdots & r_{pp}(m-1) \\ \cdots & \cdots & \cdots & \cdots \\ r_{pp}(m) & r_{pp}(m-1) & \cdots & (1+\varepsilon)r_{pp}(0) \end{bmatrix} \begin{bmatrix} a(0) \\ a(1) \\ \cdots \\ a(m) \end{bmatrix} = \begin{bmatrix} 1 \\ 0 \\ \cdots \\ 0 \end{bmatrix} \tag{6.71}$$

Where ε=0.1%-1%, the pilot signal deconvolution operator can be obtained from the above equation, then the impulse response of drill string can be eliminated through direct convolution of the deconvolution operator with the pilot signal or inverse convolution of the deconvolution operator in the time domain, and then through the convolution of it with the record after cross-correlation.

6.3.3 Independent Component Analysis

Independent Component Analysis (ICA) [26], originated in the 1990s, is a statistical signal processing method that makes use of the different statistical characteristics of various components to separate different components. It is a new blind source separation method, which is characterized by the independent statistics of signal sources and various interference sources without the need to know its characteristics. It can solve the problem of aliasing many kinds of SWD signals and improve the quality of SWD data processing.

An effective independent component analysis method can find its inherent statistical independent factors or components from multivariant or multidimensional statistical data. The general linear model of ICA is:

$$x = As \tag{6.72}$$

Where $s = (s_1, s_2, \cdots, s_m)$ is the signal sent by an independent signal source; A is mixed matrix of order $N \times M$; $x = (x_1, x_2, \cdots, x_m)$ is the mixed signal. x is obtained by observation, while s and A are waiting to be calculated. Since the mixed system s is unknown, the process obtained from x is "blind". The mixed matrix W of the order $N \times M$ needs to be found out.

$$y = Wx \tag{6.73}$$

Where $y = (y_1, y_2, \cdots, y_m)$ requires to get close to s as much as possible.

The central limit theorem indicates that mixed signals of multiple independent random variables tend to be Gaussian distributed. Therefore, in the ICA model, mixed signals composed of multiple independent source signals are closer to Gaussian distribution than any source signal. Therefore, the non-gaussianism of separated signals can be used as the measure of independence of the signals, which provides a criterion for the separation of independent sources. For non-gaussian signals, the independence of signals cannot be determined only by using irrelevance, and high-order cumulants need to be considered, mainly taking extremum of fourth-order cumulants (kurtosis).

The independent component analysis (ICA) algorithm generally consists of two steps: spherification and non-gaussification. After spheroidizing calculation, the input mixed signal is transformed into a set of unrelated signals with the mean of 0 and the equal maximum amplitude. Spheroidizing calculation can make the mixed signal separated at the second order cumulants, and make the expression of fourth-order cumulants simplified, thus spheroidizing calculation is necessary pre-processing steps of ICA. After orthogonal transformation of $z(t)$ obtained from spheroidizing calculation can be expressed as $y = Uz$. Due to the orthogonality of U, then,

$$E\{yy^T\} = E\{UZZ^T U^T\} = UIU^T = I \tag{6.74}$$

$y(t)$ is also orthogonal. U is the Givens rotation matrix. Through rotation search, the extremum of kurtosis of $y(t)$ can be found out.

6.3.4 Compression of SWD Data

Downhole seismic data receivers can obtain more accurate seismic data signals, but due to the limitation of downhole space and limited data storage capacity, it is necessary to compress SWD data. Sensors are installed downhole near the drill bit to better record seismic signals. However, the transmission of downhole data is a difficult problem at present, downhole storage is a practical solution. Although storage technology has improved greatly with large storage capacity in small storage particles, internal storage will be filled up in a few hours at high acquisition rate. It is necessary to compress SWD data.

Giancarlo Bernasconi has proposed a lossy data compression algorithm based on a new downhole data in the angular domain [27], which is suitable for downhole implementation and can be successfully applied to online and offline solutions. Numerical tests based on field data can achieve high compression ratio without causing significant information loss. This results in a significant increase in downhole time acquisition and real-time information transmitted by mud pulse telemetry. Downhole instruments are usually equipped with magnetometers or radial accelerometers to calculate the rotation speed (rad/s or r/min). A magnetometer measures the projection of the earth's magnetic field along two orthogonal radial relative to the vertical borehole shaft.

$$m_x(t) = A_x M \cos[\phi(t)] \tag{6.75}$$

$$m_y(t) = A_y M \sin[\phi(t)] \tag{6.76}$$

Where, $m_x(t)$ and $m_y(t)$ are the recorded signals; A_x and A_y is the gain of the sensor; M is the earth's magnetic field, and the angle law can be expressed as follows:

$$\phi(t) = \tan^{-1}\left[\frac{A_x m_y(t)}{A_y m_x(t)}\right] \tag{6.77}$$

When the time sampling period is much higher than the bit rotation speed, the one-dimensional expansion of rotation angle cannot be a problem. To separate centrifugal acceleration from pure radial acceleration, radial accelerometers are usually installed in pairs along the vertical axis of the well and in opposite positions. In this case, the rotation speed is calculated as follows:

$$\Omega(t) = \sqrt{[a_{x1}(t) + a_{x2}(t)]/(2r)} \tag{6.78}$$

Where t is the distance between the accelerometer and the rotation axis, $a_{x1}(t)$ and $a_{x2}(t)$is the radial acceleration. Relative to the rotation axis, the positive sign is located in the outward direction of the rotation velocity, and the angular position can be obtained through simple integration.

$$\phi(t) = \int_0^t \Omega(\tau) d\tau \tag{6.79}$$

Due to the interaction with rocks, the low torsional stiffness of the drill string and the fluctuation of the motor speed, the bit speed is not constant: the conventional sampling of signals in the time domain becomes irregular sampling in the angular position domain.

The key of the compression algorithm is re-sampling the signal as a function in the angular position domain. As expected, the new representation restores the periodicity implicit in the signal and unrecognized in the time domain. Apparently, the periodicity presents the frequency shifts. In the angular frequency domain, energy is compressed, spectral peaks appear sharper than that in the time-frequency domain: this means that the main events occur at the same angle position. In addition, the information contained in the signal is also compressed. The compression process organizes the data in a matrix as a function of rotation angle. Each row represents a complete bit rotation, and the column is a continuous circulation.

6.4 Noise Processing and Filtering

The cone bit can generate enough energy, so there are few researches on suppressing the noise of drilling rig in the early times. However, for hard rock drilling, it is very important to extract useful bit signals by overcoming the noise of drilling rig because of the weak vibration energy of the diamond bit.

Influenced by the characteristics of drill string and acceleration sensor, the wave propagated by drill string attenuated slightly with faster speed compared with that propagated by earth. Therefore, the frequency spectrum of SWD signal monitored by the sensor on the top of the drill pipe is different from that recorded by the geophone on the ground. The drill string data record contains a wide band and multiple frequency peak energy, which reflects the instrument characteristics of the frequency response of seismic waves in different paths and the noise of the drill rig. Therefore, in the process of data processing, we do not use it as a reference wavelet. The frequency band is mainly concentrated from the 0 Hz to 200 Hz, while the spectrum peaks from 30 Hz to 75 Hz.

6.4.1 Classification of the Noise

1. Coherent noise

The actual bit vibration signal is very weak compared with the noise interference signal, and it is superimposed with the noise interference information continuously. For weak bit, vibration signals are required. The interference caused by various human factors in the drilling process directly affects the direct waves and reflected waves of the drill bit. The typical coherent noise wave field is mainly composed of the following factors:

1. Low-speed waves (usually 200 to 600 m/s) are surface waves generated by drilling vibration and human activities in the well area.
2. Direct wave and refracted wave generated on the drilling platform pass through the topmost layer. In common-source gathers, the apparent velocity of these waves is related to the surface velocity (usually between 2000 and 4500 m/s).
3. The drill string can radiate conical head-waves when the propagation velocity in the surrounding strata is below the extensional velocity in the drill pipe (such as 4750 m/s).
4. In the process of drilling inclined wells, strong coherent noise will be generated due to the increased contact between drill string and wellbore wall (see Figure 2). This noise is not fixed throughout the drilling phase and is difficult to eliminate during processing.
5. Other coherent noise components include the bottomhole assembly (BHA) and drillstring multiples, which will affect the signal and noise.

2. Random noise

In addition to the coherent noises, there are several types of time random-noise:

1. Spatially organized noise filtered by array. Part of this noise, such as the noise of cars and human activity, is not stationary.
2. Pure random noise, such as the noise generated by rain and wind. One way to eliminate the noise is to bury the receiver, which can reduce random noise.
3. Random scattered noise in vertical and inclined directions of well .As mentioned above, all noise has its own characteristics. According to these characteristics, we can effectively eliminate noise components and extract bit signals from various environmental noise fields.

6.4.2 The F-K Filter

The separation of direct and reflected waves from SWD records in the common detection point domain can be carried out in the F-K domain. The direct and reflected waves in the F-K domain are separated and do not overlap with less computation time by using fast Fourier transform. The seismic signal is filtered by F-K filter, and the signal is removed from the remote source according to the difference frequency, wave propagation direction or wave number characteristics of the seismic data. According to the difference between the effective signal generated by the bit excitation and the direction and speed of ground propagation, the F-K filter is used to separate the seismic wave field and study the characteristics of the effective signal. The F-K filter eliminates the influence of noise and provide high quality data or subsequent seismic interference analysis.

If $f(t, x)$ is the set of SWD records of common detection points, then its 2-dimensional Fourier transform is:

$$F\left(\hat{f}, k\right) = \int_{-\infty}^{\infty} \int_{-\infty}^{\infty} f(t, x) \exp\left[2\pi j(ft + kx)\right] dt dx \tag{6.80}$$

According to the difference in the apparent velocity of the direct wave and the reflected wave, a filter $H(f, k)$ with a certain frequency-wave number characteristic is designed to filter the above $F\left(\hat{f}, k\right)$:

$$F(f, k) = F\left(\hat{f}, k\right) H(f, k) \tag{6.81}$$

Then perform two-dimensional Fourier transform:

$$F(\hat{t}, x) = \int_{-\infty}^{\infty} \int_{-\infty}^{\infty} f(t, k) \exp\left[-2\pi j(ft + kx)\right] dt dk \tag{6.82}$$

In other words, the SWD records of common detection points are obtained which eliminate direct or reflected wave fields. The separation of direct and reflected waves in the F-K domain requires the space sampling interval to be dense enough, and the basic principle is the same as that of multi-channel velocity filter in the time-space domain.

6.4.3 Karhunen-Loéve Transform

In the SWD process, KL transform [28] can be used to suppress the coherent noise. The wave field separation in the receiver gathering domain and the Karhunen-Loéve (KL) transform can be used to suppress rig noise and improve SNR of the bit signal. Compared with traditional 2D filters (such as f-k filters), KL transform is a more effective SWD processing method.

The wave field of the bit can be extracted by KL transform in the case of bit movement. KL transform is closely related to singular value decomposition (SVD), which can decompose any complex data set into limited patterns. In seismic processing, kl transform is usually used to suppress the coherent noise in windowed data segment. The seismic signal distortion is small by using KL transform.

The multi-channel seismic data set can be represented as matrix D of $m \times n$, where each column represents the seismic trace. D can be expressed as,

$$D = \begin{bmatrix} D_{11} & D_{12} & \cdots & D_{1n} \\ \vdots & \vdots & \vdots & \vdots \\ D_{m1} & D_{m2} & \cdots & D_{mn} \end{bmatrix} \tag{6.83}$$

There are different ways in derivation of KL transform. Using SVD, matrix D can be decomposed into:

$$D = U \sum V^H \tag{6.84}$$

Where the superscript H is Hermitian transposition, and U is composed of left singular vectors$\vec{u}_i(i = 1, 2, \cdots, m)$, $m \times n$ is a unit matrix. $\sum$ is a $m \times n$ diagonal matrix whose diagonal elements are singular values in descending order $\{\sigma_1 \geq \sigma_2 \geq \dot{c}\sigma_n\}$. V is a $n \times n$ unit matrix with columns of singular vectors $\vec{v}_j(j = 1, 2, \cdots, n)$. Any column vector $d_i = \{D_{ji}\}_{j=1}^{m}$ can be approximately a linear combination of the first r feature vector u:

$$\widehat{d}_i = \sum_{k=1}^{r} (U_k U_k^H) d_i \tag{6.85}$$

Where U_k^H is called the KL transform operator on the vector d_i, because $\sum_{k=1}^{m} U_k U_k^H = I$ is the full rank unit matrix of the unit matrix U. Since r=m, the vector d is completely reconstructed. However, if only the first r<m singular vector is used, then $\widehat{d}_i$ is equal to the lower order approximation of d_i. KL transform of data matrix D is expressed as:

$$\Psi = U^H D \tag{6.86}$$

Where, the primary data $D = U\Psi$, the optimal estimation of matrix D can be obtained by maintaining the first r rows of the matrix Ψ, and $\widehat{D}_r$ is the low-order approximation of matrix D, which is expressed as follows:

$$\widehat{D}_r = U\Psi_r \tag{6.87}$$

In the case of dealing with contaminated coherent noise, take rig noise as an example, the best estimation matrix D represents rig noise, as described in the following section. So we need to set the first r rows of the matrix Ψ to zero. Then, the coherent noise suppression data can be expressed as,

$$\widehat{D}_{m-r} = U\Psi_{m-r} \tag{6.88}$$

SVD can produce the same results. The row vector of Ψ is $\sigma_i^H i$, then, the low-order approximation method is used to reconstruct the data to get:

$$\widehat{D}_r = \sum_{i=1}^{r} \sigma_i u_i v_i^H = U_r \sum_r V_r^H \tag{6.89}$$

Similarly, $\widehat{D}$ can be reconstructed by using the remaining m-r feature vectors. Then,

$$\widehat{D}_{m-r} = \sum_{i=m-r+1}^{m} \sigma_i u_i v_i^H = U_{m-r+1} \sum_{m-r+1} V_{m-r+1}^H \tag{6.90}$$

SVD provides an effective method for calculating KL transform. In the application of suppressing the noise in the coherent drilling field, the first r singular vector is used to reconstruct the matrix, which greatly reduces the dimension of data compression, and the filtering is realized by ignoring the first singular vector or basing on the first few singular vectors of eigenvalue spectrum.

6.4.4 Wellsite Noise Suppression

In the early phase of de-noising the ground geophone signals, the former AGIP company in Italy believed that the noise source of the MWD earthquake was mainly from the derrick vibration, and the convolution model of the ground array track is as follows:

$$G_i = NH_i + AL_i \tag{6.91}$$

Where G_i represents the record received from the ground geophone in the channel of i; N represents the derrick vibration noise; H_i represents the transmission function of the above noise H_i from the noise source to the geophone in the channel of i. A represents the bit signal; L_i represents the transfer function from bit position to channel i geophone.

The interference noise of seismic wave field is very complex, the wellsite noise can be interfered by the derrick vibration and the strong noise such as the mud pump and generator. Considering the interference of various noises, the pilot signal is deconvolved first, and the linear constrained least square method is used to fit the wellsite noise, then eliminate it respectively. The convolution model of the modified ground array track is as follows:

$$G_i = \sum_{i=1}^{n} N_l H_{li} + AL_i = H_i \sum_{i=1}^{n} N_l + AL_i = H_i \widehat{N}_i + AL_i \tag{6.92}$$

N_l represents the strong interference noise generated by the l interference source near the well site (l=1,··, 1, n, assume that there be a total of n strong interference noise sources). As the noise source is relatively close to the gathering site, H_i is approximately the same for all the above noises.

Then, we can place multiple strong interference noise monitoring geophones near each interference source, set there are M geophones, then the average estimation of l interference noises N_l (l=1, ··, m) received by the channel i geophone in the above formula can be obtained by the following formula:

$$\widehat{N}_i = \frac{1}{M} \sum_{j=1}^{M} N_{ij} e^{jwt_j} \tag{6.93}$$

Where N_{ij} represents the interference wave ($j = 1, \cdots, M$) generated by the 1 interference source received by the j interference noise monitoring geophone; t_1 is the propagation time from the lth interference source to the jth monitoring geophone mentioned above. The above formula is equivalent to multiple strong interference monitoring geophone combinations, which is better than single geophone receiving.

Similarly, the unbiased estimation $\widehat{H}_i$ of the transmission function H_i can be obtained:

$$\widehat{H}_i = \frac{E\left[G_i \cdot \widehat{N}_i\right]}{E\left[\left|\widehat{N}_i\right|^2\right]} \tag{6.94}$$

Therefore, the above method is applied to obtain the geophone recording track after eliminating the above M strong interference noises:

$$\widehat{G}_i = G_i - \widehat{N}_i \widehat{H}_i = \widehat{AL}_i \tag{6.95}$$

6.5 Bit Signal Extraction

6.5.1 Cepstrum Analysis of Drill String Vibration

Cepstrum analysis [29] is a technology used in reverberation signal processing. The outstanding advantage lies in that it is insensitive to the influence of signal transmission path and detection point position, so it can find unique applications in transmission path, system identification and machine fault identification.

Bit random excitation source signals are susceptible to drill string and other machine noise due to the wide frequency range and short duration. The propagation of bit signal along the drill string is a typical convolution and reverberation problem. Cepstrum technique can be used for the separation of bit signal. The cepstrum analysis method is a nonlinear filtering technique, which uses window function to filter in inverse time domain, and can eliminate structure reverberation, so as to separate signal and extract source signal. Cepstrum filtering can be used to process SWD reference signal, which can strengthen bit source signal.

An acceleration sensor is installed at the top of the drill string to record signals from the drill bit. For the convenience of research, it is generally assumed that the drill string system is a linear system. Due to the influence of drill string structure reverberation and transmission path, the bit signal received at the top of the drill string has been seriously interfered. In other words, the reference signal $p(n)$ recorded at the top of the drill string is the convolution of the signal $b(n)$ sent by the bit and the drill string transmission path and other factors $h(n)$:

$$p(n) = b(n) * h(n) = \sum_{k=0}^{n-1} h(n-k)b(k) \tag{6.96}$$

Thus, the cepstrum of the above equation can be obtained as:

$$\widehat{p}(q) = F^{-1}[\ln B(f) + \ln H(f)] = \widehat{b}(q) + \widehat{h}(q) \tag{6.97}$$

Where, $B(f)$ and $H(f)$ are the frequency spectrum of the bit signal and the drill string transmission path, $\widehat{p}(q)$, $\widehat{b}(q)$ and $\widehat{h}(q)$ are the complex cepstrums of signal $P(n)$, $b(n)$ and $h(n)$, and q is the inverse frequency. From Eq. (6.97), it can be seen that cepstrum is the spectrum of spectrum, and the reference signal complex cepstrum $\widehat{p}(q)$ is the sum of the signal complex cepstrum $\widehat{b}(q)$ sent by the drill bit and the signal complex cepstrum $\widehat{h}(q)$ of the drill string transmission path.

In this way, the convolution signal in the time domain becomes the addition signal in the cepstrum domain, and the linear filtering method can be used to realize the separation of signal in the time domain and extraction of the periodic components of the signal, and carry out reverberation demodulation.

The purpose of complex cepstrum analysis of the reference signal is to separate the convolution signal. Generally, the nonlinear filtering method, also known as cepstrum filtering analysis method, is used to process the complex cepstrum with a non-negative smoothing window, so as to obtain the desired cepstrum component. Select a "inverse filtering rectangular window" *w(q)* to pick out the required cepstrum component, then

$$\widehat{b}(q) \approx \widehat{p}(n)w(q) \tag{6.98}$$

In the formula, $w(q)$ has the following form: $W(q) = \begin{cases} 1, & 0 \leq q \leq q_0 \\ 0, & other \end{cases}$

Where q_0 is the cut-off point of the source and the transmission path in the self-cepstrum domain, and the value depends on the specific situation. Different window functions should be selected according to the data, such as Hamming window, Gaussian window, etc. to filter and separate the complex cepstrum of the reference signal. Under the Fourier transform, exponential operation, inverse Fourier transform and other mathematical inverse transformation of the separated signal, the separated signal is returned back to the time domain, so as to achieve the source signal extraction.

6.5.2 Sparse Representation Theory (Recovery of Bit Source Signal)

The traditional linear time-frequency representation usually refers to the use of orthogonal basis or non-orthogonal basis of signal space to represent the signal. Using the unique signal form (Fourier transform, Gabor change and wavelet transform, etc.), with the help of a complete set of orthogonal basis, perform decomposition, and the corresponding decomposition coefficient can be obtained, then suppress the coefficient related to noise, finally the purpose of denoising through inverse transformation can thus be achieved. If the specified signal form does not exactly match the basis function, then the decomposition result is not necessarily a sparse representation of the signal, then it is difficult to achieve effective signal-noise separation.

The sparse representation theory is introduced into the processing of SWD reference signals. Based on Hilbert transform, the instantaneous attributes of actual SWD reference signal can be calculated. Using Morlet wavelet [30], build appropriate complete atom dictionary and perform sparse decomposition of the SWD reference signal. The L1 norm spectral projected gradient algorithm is adopted to realize basis pursuit de-noising, which can realize the sparse decomposition and de-noising of the SWD reference signal, improve signal-to-noise ratio, and recovery bit source signal. Morlet wavelet in time domain is expressed as:

$$m(t) = e^{-(t-u)^2 f^2 \ln(2/k)} e^{[2\pi i f(t-u)+\varphi]} \tag{6.99}$$

Where u is the time delay, f is the frequency, k is the scale, and φ is phase. In Eq. (6.99), the phase difference between the real part and the imaginary part is 90°. The form of Morlet wavelet is primarily determined by parameter pairs (f, u, k, φ). The information of instantaneous amplitude, instantaneous frequency and instantaneous phase can be calculated by Hilbert transformation.

The actual seismic signal $f(t)$ can be expressed as the superposition of the linear combination of N Morlet atoms in A and random noise, that is,

$$f(t) = S + Q = \sum_{n=0}^{N-1} a_n m_n(f_n, u_n, k_n, \phi_n, t) + Q \tag{6.100}$$

Where S is the effective signal, Q is the noise, and m_n is the decomposed Morlet atom of S based on dictionary A; a_n is the decomposition coefficient of effective signal in A.

6.5.3 Bit Signal Extraction Algorithm

The method of extracting weak bit signal is another form of filtering. On the basis of the analysis of the original seismic field, the noise is preliminarily evaluated and the Fourier transform is carried out on the original seismic field. The core of it is to summarize the contribution of a large amount of data $\log G(\omega)$ to eliminate the strong coherent noise generated during deviation drilling and strengthen the bit signal. Finally, inverse Fourier transform is performed to obtain time domain seismic data.

The time domain raw seismic traces recorded by the receiver on the surface can be expressed as:

$$g(t) = b(t) * h_B(t) * h_{DS}(t) * h_R(t) * h_o(t) + n(t) \tag{6.101}$$

Where the symbol $*$ represents convolution, $b(t)$ represents bit signal, $h_B(t)$, $h_{DS}(t)$, $h_R(t)$and $h_o(t)$ respectively represent the wave response, drill string response, drill rig response and other coherent noise source responses propagated from the drill bit to the receiver in the formation. $n(t)$ represents pure random noise. Through the random noise attenuation technique, we can remove pure random noise, which can be expressed in the frequency domain as follows:

$$G(\omega) = B(\omega) H_B(\omega) H_{DS}(\omega) H_R(\omega) H_O(\omega) \tag{6.102}$$

Here, the low-cut filter can remove the low-speed surface waves caused by drilling rig vibration and human activities. The direct and refracted waves generated by the drill rig can be removed by applying Wiener filtering, radial predictive filtering, wavelet transform and curvelet transform. In addition to the multiples, some other coherent noise component can be eliminated at the same time. It is not recommended to use de-noising techniques with aliasing effect (such as F-K filtering) to attenuate these coherent noises. Because they can cause serious damage to the required signal components and distort the signal. After eliminating the above noise, the frequency response can be obtained:

$$G(\omega) = B(\omega) H_B(\omega) H_{DS}(\omega) \tag{6.103}$$

In fact, the drill string response $H_{DS}(\omega)$ includes head wave response $H_{hw}(\omega)$, strong coherent noise response $H_{sc}(\omega)$ (produced during the hole deviation), response with BHA and drilling string multiples $H_m(\omega)$. Then (6.103) can be expressed as,

$$G(\omega) = B(\omega)H_B(\omega)H_{hv}(\omega)H_{sc}(\omega)H_m(\omega) \tag{6.104}$$

Therefore, there are still three types of interference waves in the seismic track in frequency domain. The multichannel statistical deconvolution can be used to suppress BHA and drill string multiples. Next, we can eliminate the head wave in the radial trace transform. However, it is difficult to eliminate the strong coherent noise generated by deviated drilling in frequency domain. Then (6.104) becomes:

$$G(\omega) = B(\omega)H_B(\omega)H_{sc}(\omega) \tag{6.105}$$

In order to eliminate the strong coherent noise, we adopt (6.105) complex natural logarithm:

$$\log G(\omega) = \log B(\omega) + \log H_B(\omega) + \log H_{sc}(\omega) \tag{6.106}$$

Clearly, strong coherent noise becomes the additive noise component in $\log G(\omega)$. Based on the characteristics of strong coherent noise in the logarithmic domain, the new method can suppress the strong coherent noise generated during deviated drilling and at the same time strengthen the bit signal by summing the contributions of sufficient amounts of data $\log G(\omega)$. The key is to obtain the sum of the amount of data $\log G(\omega)$. In general, the larger the amount of data, the more effectively the strong coherent noise can be suppressed. So (6.106) can also be written as:

$$\log G(\omega) = \log B(\omega) + \log H_B(\omega) \tag{6.107}$$

There is no corresponding term of noise in Eq. (6.107). In order to obtain the bit signal in time domain, we first use the natural index of (6.107):

$$e^{\log G(\omega)} = e^{\log [B(\omega)+H_B(\omega)]} \tag{6.108}$$

Meanwhile, $G(\omega) = B(\omega)H_B(\omega)$, conventional processing techniques such as adaptive prediction filtering are used to enhance bit signals in the frequency domain, and then inverse Fourier transform was performed to obtain the time-domain characteristics of bit signals.

$$g(t) = b(t) * h_B(t) \tag{6.109}$$

Eq. (6.109) represents a common-source gather of each depth point, which contains the bit signals directly reached and reflected by the drill bit.

6.6 SWD Application

SWD technology has various applications in the oil and gas industry [33–35]. According to wave velocity and density, key basic parameters such as pore pressure [31], rock strength [32] and crustal stress that affect the stability of wellbore wall can be calculated directly or indirectly. With the appropriate velocity-pore pressure conversion, the pore pressure before drilling can be predicted from the seismic velocity. This requires high resolution of seismic velocity obtained. The porosity accuracy obtained by inversion of shear wave velocity should be lower than that of longitudinal wave. The analysis of logging data shows that the influence of porosity and argillaceous content on the longitudinal wave velocity is much greater than that of water saturation, and the longitudinal wave velocity can also be affected by the volume modules, which is influenced by rock matrix and porous medium. Therefore, for low porosity and low permeability rocks containing gas, the relationship between longitudinal wave velocity and water saturation should be treated differently with lithology.

6.6.1 Formation Velocity

The reflection coefficient and velocity of each layer are different due to the influence of formation properties, both are discrete random sequences. Based on the assumption that the direct wave of the SWD is vertically incident to the ground, the average velocity of the drilled formation can be obtained by the time that the direct wave propagates from the drill bit to the geophone and the depth of the drill bit, as shown in the following equation:

$$v_{av}(n) = \frac{\sum_{i=1}^{n} h_i}{\sum_{i=1}^{n} v_i} \tag{6.110}$$

Where $v_{av}(n)$ is the average velocity of the n layer, h_i is the thickness of layer i. v_i is the layer velocity of layer i.

According to the obtained data of formation velocity sequence and the mathematical model of formation velocity changing with depth, the formation velocity in front of the drill bit can be deduced.

6.6.2 Wave Impedance

According to the SWD observation data convolution forward model, each upstream wave after the SWD wave field separation can be regarded as the convolution of the downstream wave and the formation reflection coefficient sequence below the depth of the channel, as shown in the following equation:

$$\begin{aligned}\{S_f(t)\} &= k \cdot \{S_u(t)\} = k\sum r(f)S_d(t-f) = \sum r(f) \cdot k \cdot S_d(t-f) \\ &= \{r(t)\} * \{S_z(t)\}\end{aligned} \quad (6.111)$$

Where $\{S_f(t)\}$ is the reflected wave of SWD and $\{S_z(t)\}$ is the direct wave of SWD.

Seismic record is the convolution of seismic wavelet and formation reflection coefficient, and the emission coefficient is determined by wave velocity and density. There is also a correlation between speed and density. SWD data can be separated into direct wave $\{S_z(t)\}$ and reflected wave $\{S_f(t)\}$, and the convolution relation with reflection coefficient sequence$\{f(t)\}$ is formed:

$$S_f(t) = S_z(t) * f(t) \quad (6.112)$$

Therefore, the wave impedance profile can be obtained by inverting the reflection coefficient sequence with the least-square filter, so as to achieve the purpose of formation prediction in front of the drill bit. In this method, direct wave is directly used as auto-correlation function of deconvolution, and reflection coefficient sequence can be obtained by filtering only once, which avoids the steps of obtaining wavelet and inverse wavelet in traditional inversion, improves the inversion accuracy, and has the function of eliminating multiples interference and the advantage of not being constrained by the minimum phase of wavelet.

6.6.3 Seismic Wave Inversion Pore Pressure

Based on the correct velocity model, the spatial resolution of seismic velocity field can be improved by iteratively optimizing the model so as to obtain more reliable pore pressure estimation before drilling.

Based on the assumption that the elastic wave velocity depends on the pore pressure P and the total stress tensor S_{ij}, then,

$$\sigma_{ij} = S_{ij} - \alpha p \delta_{ij} \quad (6.113)$$

Where α is coefficient, $\delta_{ij} = 1$ (if $i \neq j$), $\delta_{ij} = 0$ (if $i = j$). The parameter σ_{ij} is called the effective stress tensor. Assuming parameter $\alpha = 1$, σ_{ij} is the effective stress:

$$\sigma_{ij} = S_{ij} - p\delta_{ij} \tag{6.114}$$

The combination of pore pressure P and total tensor S_{ij} is called differential sensor. For uniaxial compression, it is generally assumed that porosity and velocity depend only on the vertical component of differential stress, which is defined as follows:

$$\sigma = S - p \tag{6.115}$$

Where σ is the vertical component of the differential stress tensor defined in the Eq. (6.114). S is the vertical component of the total stress tensor S_{ij}.

Assume that the vertical component of the total stress S at any point is given by the total weight of the water column above the sea bed, the rock matrix, and the fluid in the pore space. This can be calculated by the density integral as follows:

$$S = g\int_0^z \rho(z)dz \tag{6.116}$$

Where ρ(z) is the density at depth z below the sea surface, and g is the acceleration of gravity.

Given the relation between elastic wave velocity and vertical diffusion stress, the vertical stress can be obtained by using (6.116), and the pore pressure can be obtained by (6.115) formula. In the absence of density logging, the sediment density can be estimated from the depth below the sea floor by using the empirical relation:

$$\overline{\rho}(h) = 16.3 + (h/3125)^{0.6} \tag{6.117}$$

Here, $\overline{\rho}(h)$ is the mud weight equivalent of the average sediment density (in pounds per gallon) between the sea floor and the depth h (feet) below the sea floor. The empirical formula is $\rho = av^b$, and sediment density can be measured by seismic velocity. a and b are coefficients related to formation properties. Eaton's method use seismic velocity v to estimate differential stress, and its relation is expressed as follows:

$$\sigma = \sigma_{Normal}(v/v_{Normal})^n \tag{6.118}$$

Where σ_{Normal} and v_{Normal} are the vertical component and the seismic velocity of differential stress respectively, and n is the quantity describing the sensitivity of velocity to differential stress.

v_{Normal} can be obtained by the following equation:

$$\log dt_{Normal} = a - bz \tag{6.119}$$

Where $dt_{Normal} = 1/v_{Normal}$, the relation between the velocity and depth of the simplest and earliest earthquake is linear, which can be expressed as:

$$v = v_0 + kz \tag{6.120}$$

Where v_0 is the velocity of seabed sediments, and k generally ranges between 0.6-1.

Some researchers believe that there is a quantitative relationship between longitudinal wave velocity and pore pressure, as shown in the following equation:

$$v_P = C_0 + C_1(\sigma_V - \alpha p_P) + C_2 \exp\left[C_3(\sigma_V - \alpha p_P)\right] \tag{6.121}$$

Where v_p is the velocity of longitudinal wave, km/s; σ_v is the overlying formation pressure, MPa; α is the effective stress coefficient; p_p is formation pore pressure, MPa; C_0, C_1, C_2 and C_3 are empirical parameters of the model.

Based on various relations, the seismic record can be expressed as a nonlinear function with pore pressure as the independent variable, i.e., *S=f(p),* where *S* represents the seismic record and *f(•)* represents the nonlinear function.

According to the nonlinear optimization principle, when the seismic information is known, the least square method can be used to directly invert the pore pressure according to this functional relation as follows:

$$P_{P,i+1} = P_{P,i} - \left[A(P_{P,i})^T A(P_{P,i}) + \gamma_i^2 I\right]^{-1} A(P_{P,i})^T \left[S(P_{P,i}) - S_0\right] \tag{6.122}$$

Where P_p is the solution vector of pore pressure obtained by the *t*th iteration; *t* is the number of iterative calculation and the damping factor calculated successively; *A* is the partial derivative matrix of nonlinear function; *I* is the unit matrix; S_0 is the original post-stack seismic record vector; *S* is the composite seismic record vector calculated based on the *t*th pressure iterative solution. The optimization process is completed when the synthetic seismic records obtained by iteration reach the specified accuracy. The inversion process of crustal stress and rock strength is similar to the above pore pressure seismic inversion, so the wellbore stability can be predicted by seismic inversion.

6.7 Looking Into the Future

Compared to conventional VSP, SWD has its own characteristics and unique advantages in that it uses bit vibration in the process of drilling as the source for seismic measurement, without interfering with the drilling or occupying drilling time, and without any risk to the hole, especially the bit can be predicted in real-time structure details of the formation in front of the bit through the field seismic imaging processing, with the main purpose of reducing drilling risks. The key of this technology is how to collect and recover the weak bit reflection signal under strong disturbance noise and make it the equivalent formation impulse response.

The SWD techniques are evolving at fast pace and the future seems quite bright for SWD as several new MWD Technologies are emerging.

- Memory tools with rugged high precision clock (application to Drill Bit-SWD and VSP-WD)
- High speed / high transmission rate wired drill string system, such as Intellipipe™ applicable to all seismic applications. Including, Drill-Bit seismic, VSP-WD and SONIC-WD.
- Through bit logging device.
- Downhole mechanical devices to enhance the drill-bit axial vibration or the near bit pressure seismic noise, such as hammer drill and Hydropulse Drilling.

The downhole technology to be developed for drill-bit walkaway SWD can be restrained to an electronic equipment in central position inside the drill collars, the precision and drift constrains on the downhole clock do not need to be very high, which lead to reasonable equipment cost. The preferable downhole technology to develop for VSP-WD is with an electronic equipment built in the crown of a drill collar, for all possible drill collar diameters, since the VSP-WD measurements are needed continuously in the deep half of vertical, deviated and horizontal wells. It is also expected that the real time availability of full-waveform data will increase the possibility of accurate well placement relative to a drilling target.

SWD technology is one of the important means to predict pre-drilling formation information, which can effectively guide drilling construction and reduce drilling risks. However, it is also a recognized worldwide problem. How to form perfect field acquisition method and corresponding data processing method is a long-term exploration, practice and summary process. This technology has considerable advantages over conventional VSP in terms of real-time prediction of formation structure details in front of the bit, reduction of drilling risks, improvement of exploration efficiency and reduction of exploration cost (especially for multi-azimuth, multi-arrangement and 3D observation). Therefore, it has good prospect in promotion and application, mainly including the following aspects:

a. Real-time determination of the position of the drill bit on the seismic profile.
b. Predict in advance the structural details of the formation around the wellbore and in front of the bit.
c. Predict pore pressure in formation in front of the bit by precise velocity measurement combined with LWD.
d. Provide comprehensive formation evaluation services and dynamic bottom-hole assembly monitoring in combination with MWD in bit position.
e. Obtain a full set of zero-offset, non-zero-offset, multi-offset, multi-azimuth VSP and three-dimensional VSP measurements.

References

1. Freire SLM, Ulrych TJ. Application of singular value decomposition to vertical seismic profiling[J]. Geophysics. 1988;53(6):778–85.
2. Chang WF, McMechan GA. Reverse-time migration of offset vertical seismic profiling data using the excitation-time imaging condition[J]. Geophysics. 1986;51(1):67–84.
3. Dubinsky V. Reference signal encoding for seismic while drilling measurement: U.S. Patent 6,078,868[P]. 2000-6-20.
4. Petronio L, Poletto F. Seismic-while-drilling by using tunnel boring machine noise [J]. Geophysics. 2002;67(6):1798–809.
5. Poletto F, Malusa M, Miranda F, et al. Seismic-while-drilling by using dual sensors in drill strings [J]. Geophysics. 2004;69(5):1261–71.
6. Masak P C, Malone D L. Inverse vertical seismic profiling using a measurement while drilling tool as a seismic source: U.S. Patent 6,094,401[P]. 2000-7-25.
7. James W. Rector III and Bob A. Hardage. Radiation pattern and seismic waves generated by a working roller-cone drill bit: Geophysics, 1992, 57, 1319-1333.
8. Mueller MC. Prediction of lateral variability in fracture intensity using multicomponent shear-wave surface seismic as a precursor to horizontal drilling in the Austin Chalk [J]. Geophys J Int. 1991;107(3):409–15.
9. Yilmaz Ö. Seismic data analysis: Processing, inversion, and interpretation of seismic data [M]. Soc Explor Geophys. 2001;
10. Claerbout JF. Synthesis of a layered medium from its acoustic transmission response [J]. Geophysics. 1968;33(2):264–9.
11. Schuster G T, Rickett J. Daylight imaging in V (x, y, z) media[J]. Utah tomography and modeling-migration project midyear report and stanford exploration project midyear reports, 2000: 55-66.
12. Haldorsen JBU, Miller DE, Walsh JJ. Multichannel Wiener deconvolution of vertical seismic profiles[J]. Geophysics. 1994;59(10):1500–11.
13. Larose E, Khan A, Nakamura Y, et al. Lunar subsurface investigated from correlation of seismic noise[J]. Geophys Res Lett. 2005;32(16)
14. Wu Z, Lu J, Han B. Study of residual stress distribution by a combined method of Moire interferometry and incremental hole drilling, Part I: Theory[J]. J Appl Mech. 1998;65 (4):837–43.
15. Wapenaar K, Slob E, Snieder R, et al. Tutorial on seismic interferometry: Part 2—Underlying theory and new advances[J]. Geophysics. 2010;75(5):75A211–27.
16. He R, Hornby B, Schuster G. 3D wave-equation interferometric migration of VSP free-surface multiples[J]. Geophysics. 2007;72(5):S195–203.
17. He R, Hornby B, Schuster G. 3D wave-equation interferometric migration of VSP multiples [M]//SEG Technical Program Expanded Abstracts. Soc Explor Geophys. 2006;2006:3442–6.
18. Yu J, Schuster GT. Crosscorrelogram migration of inverse vertical seismic profile data [J]. Geophysics. 2006;71(1):S1–S11.
19. Derode A, Tourin A, de Rosny J, et al. Taking advantage of multiple scattering to communicate with time-reversal antennas[J]. Phys Rev Lett. 2003;90(1):014301.
20. Schmidt R. Multiple emitter location and signal parameter estimation[J]. IEEE Trans Antennas Propag. 1986;34(3):276–80.
21. Asgedom EG, Gelius LJ, Tygel M. Seismic coherency measures in case of interfering events: A focus on the most promising candidates of higher-resolution algorithms[J]. IEEE Signal Process Mag. 2012;29(3):47–56.
22. Gelius LJ, Tygel M, Takahata AK, et al. High-resolution imaging of diffractions—A window-steered MUSIC approach[J]. Geophysics. 2013;78(6):S255–64.
23. Tewfik AH, Kim M. Correlation structure of the discrete wavelet coefficients of fractional Brownian motion[J]. IEEE Trans Inf Theory. 1992;38(2):904–9.

24. Vasconcelos I, Snieder R. Interferometry by deconvolution: Part 1—Theory for acoustic waves and numerical examples[J]. Geophysics. 2008;73(3):S115–28.
25. Vasconcelos I, Snieder R. Interferometry by deconvolution: part 2—theory for elastic waves and application to drill-bit seismic imaging[J]. Geophysics. 2008;73(3):S129–41.
26. Comon P. Independent component analysis, a new concept?[J]. Signal processing. 1994;36 (3):287–314.
27. Bernasconi G, Vassallo M. Efficient data compression for seismic-while-drilling applications [J]. IEEE Trans Geosci Remote Sens. 2003;41(3):687–96.
28. Gastpar M, Dragotti PL, Vetterli M. The distributed karhunen–loeve transform[J]. IEEE Trans Inf Theory. 2006;52(12):5177–96.
29. Kingsbury BED. Perceptually inspired signal-processing strategies for robust speech recognition in reverberant environments[M]. Berkeley: University of California; 1998.
30. Grinsted A, Moore JC, Jevrejeva S. Application of the cross wavelet transform and wavelet coherence to geophysical time series[J]. Nonlinear Process Geophys. 2004;11(5/6):561–6.
31. Cheng CH, Toksöz MN. Inversion of seismic velocities for the pore aspect ratio spectrum of a rock[J]. J Geophys Res Solid Earth. 1979;84(B13):7533–43.
32. Alber M, Fritschen R, Bischoff M, et al. Rock mechanical investigations of seismic events in a deep longwall coal mine[J]. Int J Rock Mech Min Sci. 2009;46(2):408–20.
33. Poletto F, Magnani P, Gelmi R, et al. Seismic while drilling (SWD) methodology in support to moon subsurface stratigraphy investigations[J]. Acta Astronaut. 2015;110:99–114.
34. Naville C, Serbutoviez S, Throo A, et al. Seismic while drilling (SWD) techniques with downhole measurements, introduced by IFP and its partners in 1990-2000[J]. Oil Gas Sci Technol. 2004;59(4):371–403.
35. Jaksch K, Giese R, Kopf M, et al. Seismic prediction while drilling (spwd): Looking ahead of the drill bit by application of phased array technology[J]. Sci Drill. 2010;9:41–4.

Chapter 7
Signal Processing in Logging While Drilling

Zizheng Wang

Abstract Logging while drilling (LWD) technology can be roughly divided into three generations. Correlation with construction and formation evaluation; logging while drilling ensures that the logging data required to determine capacity and economy and reduce drilling risks can be collected. The target location is found in conjunction with wellbore imaging, formation dip and density data. These advances have led to high success rates for drilling of many types of wells, especially high-angle, super-long and horizontal wells. So far, serialized and complete sets of logging while drilling equipment have been developed, including electric logging while drilling, sonic logging while drilling, nuclear logging while drilling, formation pressure logging while drilling, NMR logging while drilling and seismic logging while drilling, etc.

Keywords Logging while drilling (LWD) · Wellbore imaging · Electric logging · Sonic logging · Nuclear logging

7.1 Overview of Logging While Drilling

Simultaneous logging during the drilling process is referred to as logging while drilling. The downhole instrumentation of the logging while drilling in the logging while drilling system is basically the same as that of the conventional logging tool. The difference is that each instrument unit is installed in the drill collar, and these drill collars must be able to adapt to the normal mud circulation. First, we will calibrate all kinds of logging while drilling tools on the ground, then dock them together for overall inspection, then connect the logging while drilling tool to the bottom of the drill pipe, and finally connect the bottom drilling tool assembly and the drill bit [1].

Contributions by Zizhen Wang, China University of Petroleum (East) and Qilong Xue, China University of Geosciences (Beijing).

Q. Xue, *Data Analytics for Drilling Engineering*, Information Fusion and Data Science, https://doi.org/10.1007/978-3-030-34035-3_7

In the early 1930s, J. C. Karaher of Dallas Geophysical Company insulated the drill bit from the drill string with a length of 4–5 ft of insulated wire, embedded an insulating rod in each drill pipe, and passed a wire through the insulating rod. Passing to the ground, transmitting downhole signals through this wire, in this way, has obtained encouraging results, measuring a continuous resistivity curve. In 1938, the first LWD resistivity curve was acquired, which is the first LWD curve for transmitting data by electrical connection.

Logging while drilling technology can be roughly divided into three generations. It was the first generation before the late 1980s and provides basic azimuth measurements and formation evaluation measurements. It is used as "insurance" logging data for horizontal and high-angle wells, but its main application is to perform formation near the wellbore. Correlation with construction and formation evaluation; logging while drilling ensures that the logging data required to determine capacity and economy and reduce drilling risks can be collected. From the early 1990s to the mid-1990s, it belonged to the second generation. Azimuth measurement, borehole imaging, automatic steering motor and forward modeling software were successively launched to accurately determine the well trajectory through geosteering. The driller can use real-time azimuth measurement. The target location is found in conjunction with wellbore imaging, formation dip and density data. These advances have led to high success rates for drilling of many types of wells, especially high-angle, super-long and horizontal wells. From the mid-1990s to the present, the third generation, called Logging for Drilling, provides the data required to define the geological environment, the drilling process, and collect real-time information [2].

So far, serialized and complete sets of logging while drilling equipment have been developed, including electric logging while drilling, sonic logging while drilling, nuclear logging while drilling, formation pressure logging while drilling, NMR logging while drilling and seismic logging while drilling, etc. Schlumberger, Baker Hughes, Halliburton, Weatherford and other large oilfield technical service companies have developed a complete set of logging while drilling equipment, the degree of integration is further improved, both can provide neutron porosity, density, resistivity of multiple depths of detection, gamma, and parameters such as drilling orientation, well inclination and tool surface can basically meet the needs of formation evaluation, geosteering and drilling applications. According to the needs of oilfield companies and formation evaluation. Choose a different combination of instruments. The two most commonly used combinations are MWD + gamma + resistivity and MWD + gamma + resistivity + density + neutrons. The first one can provide geosteering services. The second one can provide geosteering and basic formation evaluation services.

7.2 Data Transmission of Logging While Drilling

There are two kinds of data transmission methods for logging while drilling. One is wired transmission mode, which realizes signal transmission through cable and optical fiber. The other is wireless transmission mode, which transmits signals through mud, electromagnetic and acoustic waves. The components of the tool include: (1) the power source of the operating system; (2) the sensor that measures the required information; (3) the transmitter that transmits the data to the ground in the form of code; and (4) the micro-functions of the coordination tool. Processor or control system.

7.2.1 Wired Transmission Mode

Wired transmission methods include cable transmission, special drill pipe transmission and fiber transmission [3].

7.2.1.1 Cable Transmission Method

The cable transmission method is to pass the electric wire inside the drill pipe. The type of the wire is similar to the cable in the electric test, and is an armored cable. As the well is deepened, cables and instruments must be placed when adding a single root, or the cable should be pre-fitted into the bore of the drill pipe [4].

In addition, there is a coaxial conductor system that can be used for signal transmission. It uses a coaxial wiring method with a copper tube center conductor with an insulating layer. The conductor is inserted into a conventional drill pipe, and the liquid pressure expands to seal against the inner wall of the drill pipe. When the drill string is connected, signals can be transmitted between the drill rods to form a high-speed two-way information network. The advantage of cable transmission is that the transmission rate is high, the information can be transmitted in both directions, and the power can be directly supplied from the ground to the downhole sensor. The bottom of the well does not need an additional power source; the disadvantage is that the manufacturing process is relatively complicated and often affects the normal drilling process.

7.2.1.2 Special Drill Pipe Transmission Method

This method attaches a continuous conductor to the drill pipe to make it part of the drill pipe. A special attachment device mounted in the joint allows the drill string to conduct electricity over the entire length. The sensor is housed in a special drill collar. An armored cable connects the drill collar to the lower end of the drill pipe. An insulating ring is attached to the top of the kelly, which is connected to the ground equipment.

The key to this transmission method is the drill pipe joint design. Drill pipe joint design methods mainly include induction method, wet joint method, Hall effect sensor method and wire docking method. The advantages of special drill pipe transmission mode are fast data transmission and simple two-way communication; the disadvantage is that special drill pipe is required, the cost is high, it is difficult to obtain continuous circuit at the joint, the reliability is poor, and it is difficult to realize the power transmission [5, 6].

7.2.1.3 Optical Fiber Transmission Method

A cheap fiber with a simple protective layer is run into the wellbore. The length of the fiber is the length of the entire drill string. The fiber can both circulate downhole from the ground in the axial direction and back from the bottom hole assembly to the ground [7]. The fiber optic cable used is very small, low cost, can be used for a short time, and finally wears away in the drilling mud and is washed away. Fiber optic telemetry technology can transmit data at approximately 1 Mbit/s, five orders of magnitude faster than other commercially available wireless LWD telemetry.

7.2.2 Wireless Transmission Mode

The wireless transmission mode is divided into three types: mud pulse, electromagnetic wave and sound wave according to the transmission channel.

7.2.2.1 Mud Pulse Transmission Method

The mud pulse method mainly uses a rotary valve to generate pressure waves in the drilling fluid column under the well. This rotary valve is called a decomposer. By changing the phase of the wave downhole and detecting these phase changes on the ground, the signal can be continuously transmitted to the ground.

In the mud pulse system, due to the limitations of pulse diffusion, speed regulation and other characteristics of the mud system, the data transmission speed is relatively slow, the propagation speed of the pressure wave in the mud is about 1200 m/s, and the data transmission rate is not high. The transmission signal is susceptible to noise. The advantage is that it does not require insulated cables and special drill pipes, but uses mud flow as a power source, reducing development costs.

7.2.2.2 Electromagnetic Transmission Method

Electromagnetic Measurement While Drilling is a new technology that entered industrial application in the 1980s. There are two methods of transmitting signals. One uses the formation as the transmission medium and the other uses the drill string as the transmission medium. The downhole instrument loads the measured data onto the carrier signal, which is transmitted from the electromagnetic wave transmitter to the surrounding with the carrier signal. The ground detector unloads, decodes, and calculates the measurement signal in the detected electromagnetic wave on the ground to obtain actual measurement data. This method is bidirectional transmission and can be transmitted up and down in the well without mud circulation. The advantage of EM transmission is that no mechanical receiving device is needed and the data transmission speed is faster. It is suitable for transmission orientation and geological data parameters in drilling construction such as ordinary mud, foam mud, air drilling and laser drilling. The disadvantage is that the EM measurement method is only suitable for use in shallow wells due to the rapid attenuation of the transmission signal, and the low electromagnetic wave frequency is close to the earth frequency, which is easily affected by the well field electrical equipment and the formation resistivity, so that the signal detection and reception become more difficult.

7.2.2.3 Acoustic Transmission Method

This transmission method uses sound waves or seismic waves to transmit signals through a drill pipe or a formation. The downhole data testing process is to test the instrument and the acoustic wireless transmission transmission system with the drill pipe or the oil pump. The test instrument converts various downhole parameters into digital information, and then encodes and temporarily stores the binary code pulse representing the downhole parameters. It is sent to the control circuit to transmit the acoustic vibration signal, which is transmitted to the ground along the drill string or oil pipe. It is received by the acoustic wave receiving probe installed at the wellhead, amplified and sent to the storage medium for recording, data processing and interpretation, and the current well is obtained. Formation evaluation or production dynamic data. Acoustic telemetry and electromagnetic wave telemetry do not require mud circulation, and the implementation method is simple and the investment is small. The disadvantage is that the attenuation is very fast, and the environmental interference is large. The low intensity signal generated by the wellbore and the acoustic noise generated by the drilling equipment make the detection signal very difficult.

7.3 Resistivity Logging While Drilling

Like wireline logging technology, electromagnetic wave resistivity logging while drilling technology is also divided into lateral and induction classes. Lateral types are suitable for use in conductive mud, high resistivity formations, and high resistivity intrusion environments. Inductive measure well in conductive formations and are suitable for conductive or non-conductive muds.

An electromagnetic wave resistivity logging instrument, using multiple transmitters and multiple receivers. The current in the transmitting antenna is the "wave source" of the measuring system. The electromagnetic wave emitted by it is propagating in all directions in the ground. The attenuation rate and phase shift of the wave are closely related to the conductivity of the formation, and the influence of the fluid in the well is relatively small. The formation conductivity can be calculated by receiving the antenna attenuation rate and phase difference. Different electromagnetic resistivity logging instruments have similar operating frequencies. Only a limited number of frequencies can be used to eliminate background effects such as drill collars, such as 20 kHz, 250 kHz, 400 kHz, 500 kHz, 2 MHz [8].

The electromagnetic wave resistivity logging while drilling tool has the following features:

a. Reduces the impact of the caliper.
b. Reduces the effects of fluids in the well and its intrusion.
c. Drilling fluid intrusion is small.
d. The impact of drilling fluid intrusion is small, and the measured value of resistivity is equivalent to the measured value of deep induction.
e. Has a good (higher) vertical (longitudinal) resolution, which is very beneficial for the thinning and interlayer subdivision

1. EWR While Drilling Resistivity Logging Tool

The EWR instrument consists of a transmitting coil and two receiving coils. As shown in Fig. 7.1, the operating frequency is 2 MHz, the receiving coil spacing is 6 inches, and the source spacing is 27 inches. A phase difference can be measured. Electromagnetic waves are emitted by the transmitting antenna through the wellbore and the surrounding formation, and are received by two receiving antennas located at different distances from the transmitting source. The difference in propagation time between the two receiving points can be measured. Since it is a fixed frequency transmission, this time difference can be expressed as a phase difference. This phase difference is a function of the conductivity of the formation [9].

2. DPR Logging Tool

Compared with EWR, DPR has different receiving coil pitch and source distance and performs amplitude ratio measurement. The operating frequency is 2 MHz, the

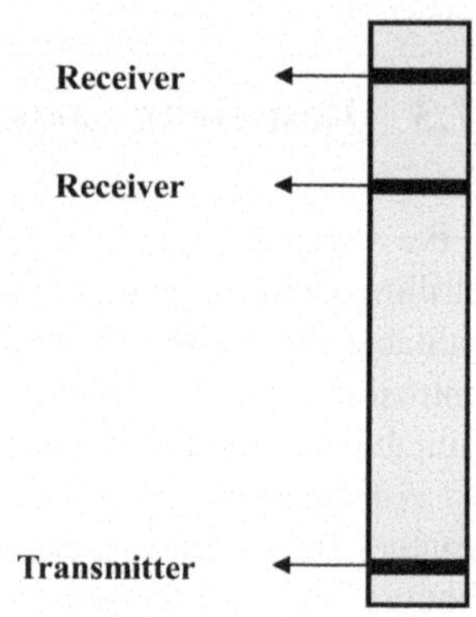

Fig. 7.1 Schematic diagram of the EWR instrument

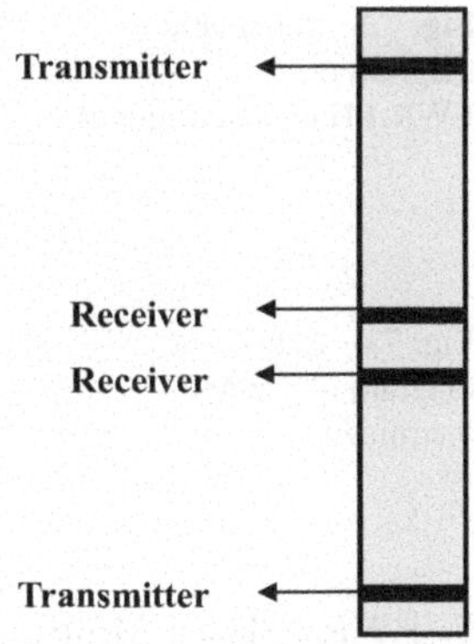

Fig. 7.2 Schematic diagram of the CDR instrument

spacing is 7 inches, and the source distance is 31 inches. The layering ability of phase difference is worse than the layering ability of amplitude ratio, but the detection depth is not as large as the amplitude ratio. The combination of the two can be used to judge the formation intrusion. Therefore, the increase of DPR can obtain the formation information more accurately than the measurement. The depth of detection of the instrument is still defined by the geometric factor theory. For typical formations during drilling, the depth of detection is 29.2–63.5 cm, and the depth of detection is 49.5–76 cm [10].

3. CDR Compensation Double Resistivity Logging Tool

It adopts double-shot and double-receiver measurement mode. As shown in Fig. 7.2, the upper and lower transmitters work alternately. The operating frequency is 2 MHz, the spacing is 6 inches, the source distance is 28 inches, and the measurement results are averaged to obtain the amplitude ratio resistivity and phase resistivity. The wellbore is compensated to eliminate the thin layer response due to collapse of the wellbore while obtaining a vertical symmetrical response curve to the formation. In general, the detection depth of the amplitude specific resistivity is 25 cm larger than the depth of the phase difference. Since the sensitivity of the amplitude specific resistivity and the phase difference resistivity to the change of the formation resistivity is different, when the intrusion is not large, the two intersect at the boundary of the formation, thereby providing a simple method for determining the thickness of the formation [11].

4. EWR-S Multi-detection Depth Logging Tool

The instrument consists of three transmitting coils and four receiving coils with a pitch of 6 inches and a source distance of 15, 27, 39 inches. There are two operating frequencies of 1 MHZ and 2 MHZ, which can obtain six kinds of measurement parameters with deep, medium and shallow depth of detection. (three phase differences and three amplitude ratios). The 1 MHz frequency is used for deep depth measurement and 2 MHz for medium and shallow depth measurements.

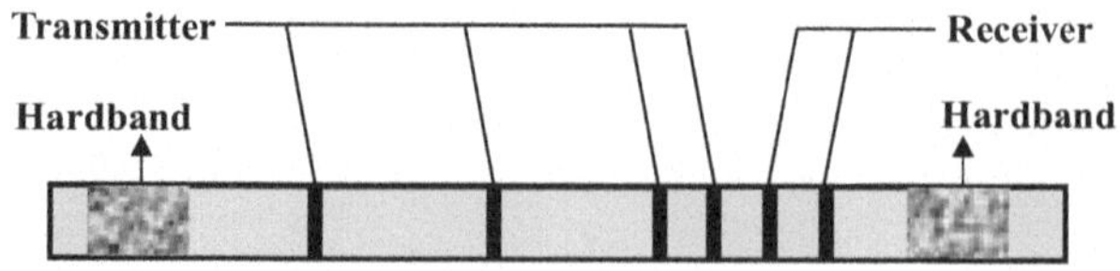

Fig. 7.3 Schematic diagram of the EWR-PHase4 instrument

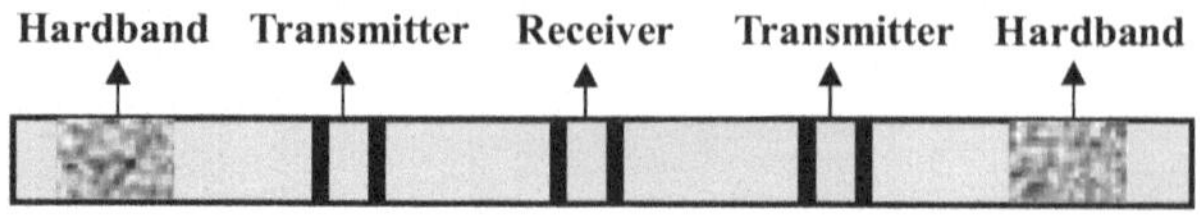

Fig. 7.4 Schematic diagram of the MPR instrument

5. EWR-PHase4 Multi-detection Depth Logging Tool

The EWR-PHase4 logging tool consists of four transmitting coils and two receiving coils, as shown in Fig. 7.3. The spacing is 6 inches, the source distance is 9, 15, 27, 39 inches, and there are two operating frequencies of 1 MHz and 2 MHz, The phase difference and amplitude ratio between each set of emitter and receiver are measured in total for eight measurement parameters (four phase differences and four amplitude ratios). The difference between this instrument and the conventional EWR is that it can measure two deep curves and one deeper curve. The formation conductivity is calculated as a function of the phase difference and amplitude ratio of the transmitted signal. The calculation method was based on the test in a test chamber made with different concentrations of brine and finite element simulation [12].

The phase difference combination curve is superior to the simple amplitude ratio curve. According to the solution of the single-phase difference or the amplitude ratio resistivity curve, the real resistivity and the intrusion band measurement are often misunderstood. After comparing all the measured data, the phase difference combination curve provides better real resistivity values and more information about formation intrusion zones.

6. MPR Small Borehole While Drilling Logging Tool

MPR technology is developed based on CDR technology. MPR belongs to the compensation electromagnetic wave propagation resistivity instrument. As shown in Fig. 7.4, it has two sets of compensation transmitting antennas. The receiver adopts the method of receiving the signals of the upper and lower symmetrical transmitters for compensation measurement. The spacing is 6 inches and the source distance is 23, 25 inches. The resistivity measurement is performed by the phase difference and the amplitude ratio of the long and short source distances. The operating frequencies are 400 kHz and 2 MHz. In order to improve the measurement accuracy of the coil symmetry setting, the 400 kHz operating frequency is selected to increase the detection depth and eliminate the influence of the dielectric constant under high water saturation conditions. A pair of receiving coils is added to increase the number of detection depths. The instrument provides eight different depths of measurement [13].

Advanced electronic technology and a complete antenna combination make up for many of the shortcomings of mechanical antennas. This method has many advantages over the method of collecting data by asymmetric transmission and then processing it by software technology. It uses array antennas to transmit electromagnetic waves of different frequencies to the ground. The frequencies are different and the depth of detection is different. Such instruments have a wide dynamic range from shallow to deep, and the deepest detection range far exceeds the depth reached by the mud filtrate.

The main features of MPR technology are: high precision, large detection range; multi-parameter measurement of intrusion profile; low borehole impact; reduced sensitivity to adverse effects of oil-based mud; improved longitudinal resolution; improved thin layer resistivity response; The combination of 2 MHz and 400 kHz signals improves the ability to divide the boundary of horizontal wells. The system can identify and correct environmental impacts and can perform dielectric parameter calculations. For high-angle wells, the system also could calculate horizontal resistivity and vertical resistivity values that represent anisotropy, with high-precision model support and strict quality control. The introduction of MPR technology has improved the accuracy of resistivity measurement, enhanced the ability of thin layer and its fluid interface division, and improved and comprehensive reservoir interpretation and detailed oil and gas analysis techniques.

7. RAB Drill Bit Resistivity Tool

RAB belongs to the lateral type of measurement while drilling resistivity instrument. The circular transmitting coil is only 1~2 inches away from the bottom of the instrument. The current flows through the drill bit into the formation (the drill bit serves as the power supply electrode) and returns to the hoop away from the drill bit. And establish a constant electric field near the drill bit, as shown in Fig. 7.5. Knowing the voltage, measuring the axial current flowing through the drill bit, the formation resistivity at the drill bit can be calculated using Ohm's law. This measurement is used to accurately indicate the location of the formation through which the drill bit passes, with a resolution of 2–6 inches [11].

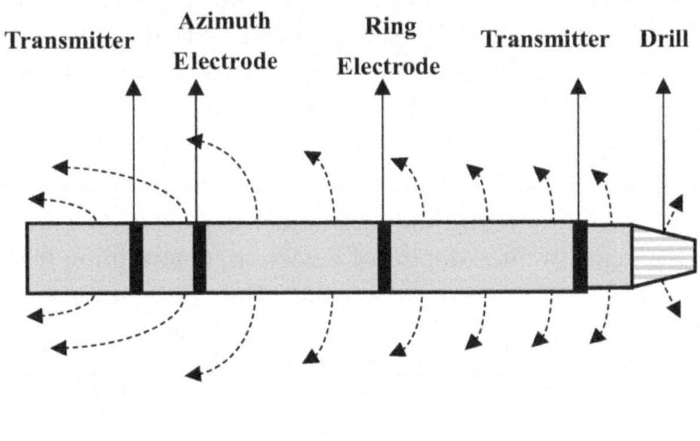

Fig. 7.5 Schematic diagram of the RAB instrument

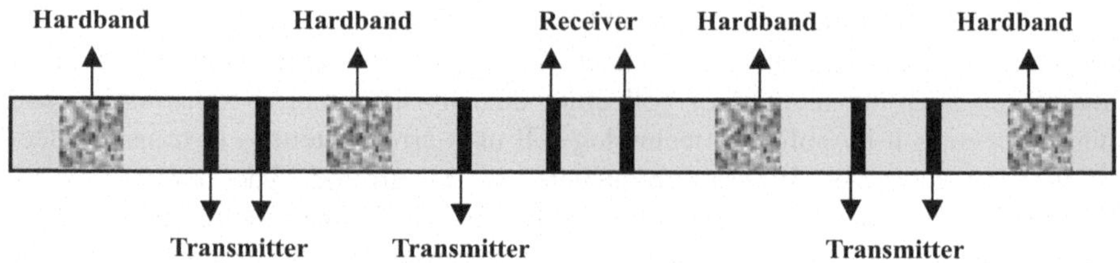

Fig. 7.6 Schematic diagram of the ARC5 instrument

8. ARC5 Compensation Type While Drilling Resistivity Logging Tool

Using two frequencies of 2 MHz and 400 kHz, the wellbore compensation removes the effects of pressure, temperature and vibration for high-resistivity formations. The most ideal frequency with high sensitivity is 2 MHz; the frequency of 400 kHz provides a deeper detection for the conductive formation depth and a smaller noise signal. The ARC5 instrument uses 5 transmitters to transmit 2 MHz of electromagnetic waves to the ground plane, as shown in Fig. 7.6, providing five raw phase difference measurements and five amplitude ratio measurements.

The ARC5 uses a unique borehole compensation technology. The standard wellbore compensation method combines the signals of two transmitters placed symmetrically around the receiver to obtain a compensated measurement. ARC5 relies on a linear combination of three sequentially arranged emitters for borehole compensation, a method known as hybrid wellbore compensation. The five phase shifts and attenuations after wellbore compensation can be converted into five scaled phase shift resistivity and five scaled amplitude specific resistivity. Since the depth of detection increases with the increase of the transmitter spacing, the five-phase shift resistivity represent the formation resistivity values of five different depths of detection with almost the same axial resolution. Similarly, the five-attenuation resistivity reflect five different depths of detection measurements.

9. GVR Drill Bit Resistivity Logging Tool

GVR belongs to the lateral type while drilling resistivity logging tool, as shown in Fig. 7.7. For conductive drilling mud, it provides 5 kinds of spacing lateral resistivity and 3 depths of borehole resistivity imaging data. The drill bit is used as a measuring electrode to truly measure near the bit. The measurement results directly reflect the formation information at the drill bit and are used to select the casing and scoring position in real time. The integrated columnar electrode provides a high-resolution lateral resistivity, called the ring resistivity, which has a small impact on the surrounding rock. The electric buckle has three azimuth focusing electrodes that provide detailed azimuth measurements. The three electric buckles are arranged longitudinally along the instrument axis, spaced 1 inch apart, thus providing measurements of three depths of detection, determining the layer boundary direction and quantizing the intrusion profile. The measurement results show a full borehole resistivity image of 3 depths of detection for thin layers, structural dip and crack analysis, and enhanced the accuracy of geosteering [14].

Fig. 7.7 Schematic diagram of GVR instrument

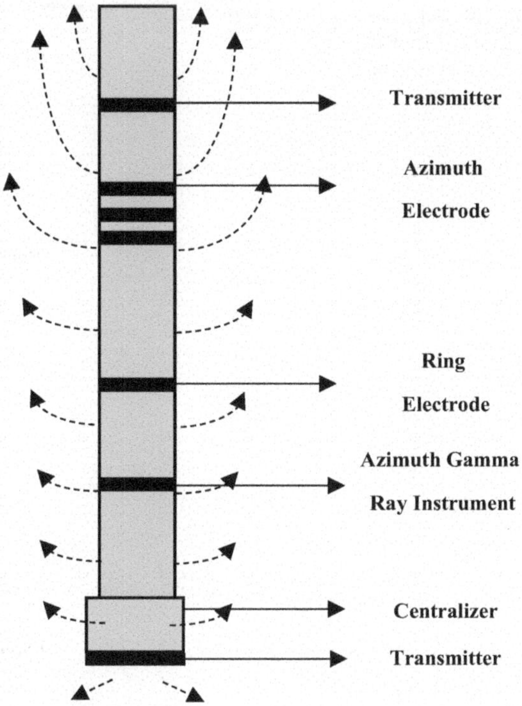

7.4 Acoustic Logging While Drilling

When sound waves pass through the medium in the wellbore, they propagate in different ways, such as longitudinal waves, transverse waves, Rayleigh waves, mud waves, and Stoneley waves, and the corresponding propagation speeds are also different, which can be identified from the acoustic signals. Longitudinal waves travel longer through formation rocks containing water, oil, and gas than through non-porous rocks. The change in propagation time is related to the volume of fluid in the pore space of the rock, which is a function of porosity. Therefore, by analyzing the logging acoustic wave information, the formation information can be obtained.

The information of acoustic logging while drilling is mainly used for: calculation of formation porosity, seismic data time-depth conversion and synthetic seismic record, rock mechanical characteristics analysis and wellbore stability prediction, lithology identification, gas layer identification, and conventional logging data correlation analysis and so on.

1. ISONIC Acoustic Logging Tool

The principle of ISONIC is very similar to that of acoustic logging in cable logging. A sonic generator and array receiver are mounted in the drill collar 12 m above the drill bit, as shown in Fig. 7.8. During drilling, the transmitter produces

Fig. 7.8 Schematic diagram of ISONIC instrument structure

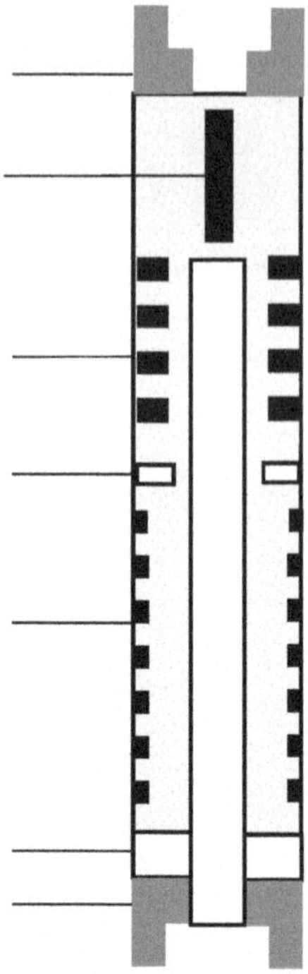

sonic pulses that travel through the mud and formation to the four receivers. The longitudinal wave time difference of the formation is extracted from the wave train recorded by the electronic collection part of the downhole instrument and can be sent to the ground in real time, and the waveform is recorded in the downhole memory.

2. CLSS Acoustic Logging Tool

The CLSS while drilling sonic logging tool is also called a compensating sonic logging tool, and its structure is shown in Fig. 7.9. Its working principle is basically similar to that of single-shot and dual-receiver instruments. The working frequency range of transmitter and receiver is 10~20,000 Hz. Since the transmitter and receiver are non-omnidirectional, they are mounted on one side of the barrel. This structure can eliminate the adverse effects on sound wave measurements such as caliper, rock thickness, and cycle jump. The CLSS records the average of the acoustic time difference and compensates for the effects of the wellbore factors. With high acoustic insulation, its unique isolator and high efficiency emitter and receiver ensure high signal-to-noise ratio and high anti-interference performance, as well as improved logging capability in soft formations [15].

3. BAT Dipole Acoustic Logging Tool

The BAT dipole acoustic logging tool has two opposing emitters and two sets of receivers consisting of seven receivers, as shown in Fig. 7.10, each receiver

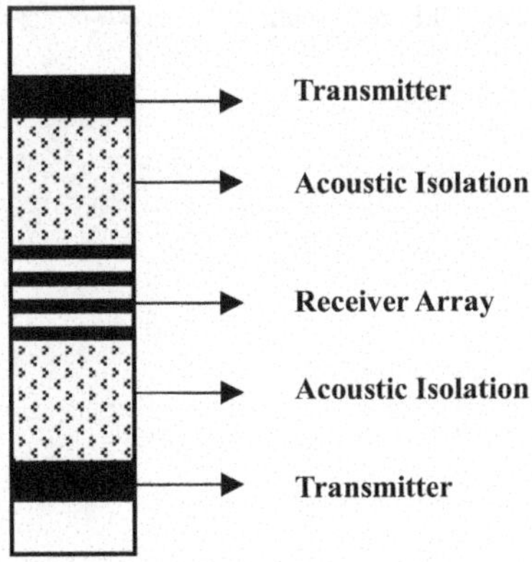

Fig. 7.9 Schematic diagram of the CLSS logging tool

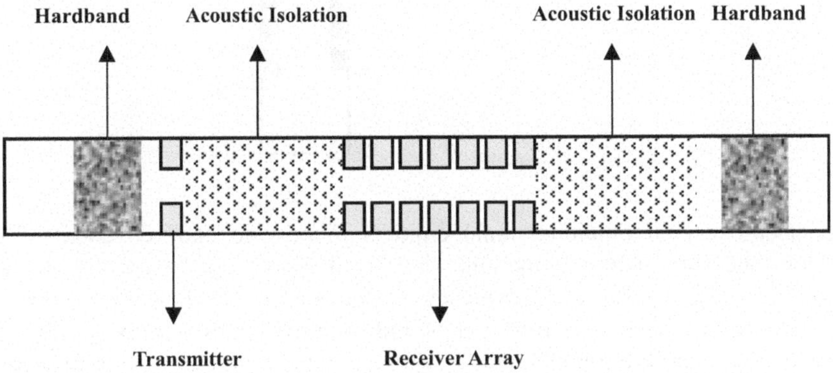

Fig. 7.10 Schematic diagram of BAT instrument structure

combination corresponding to a transmitter. It uses a variety of mechanical design methods to minimize the impact of drilling noise and reduce the direct coupling of acoustic signals. The transmitter excites the longitudinal and refracting transverse modes at a high frequency of 12–15 kHz, and then excites to the interface of the wellbore in a pseudo dipole mode of 6–8 kHz.

4. APX Acoustic Characteristic Parameter Detector

The APX instrument has a wide-band sound source and 24 receivers (6 groups of 4 receivers per group), as shown in Fig. 7.11. The sound source emits an acoustic energy pulse to the formation surrounding the wellbore at an optimal frequency. When the acoustic energy propagates along the wellbore, the receiver measures the wavefront. The velocity of the wavefront is directly affected by the environmental characteristics of the formation near the wellbore. The system uses advanced embedding technology to convert the received acoustic analog signal into a digital signal to obtain the formation acoustic time difference. Raw and preprocessed sonic waveform data is stored in high speed memory. APX uses a set of cylindrical

Fig. 7.11 Schematic diagram of the APX instrument

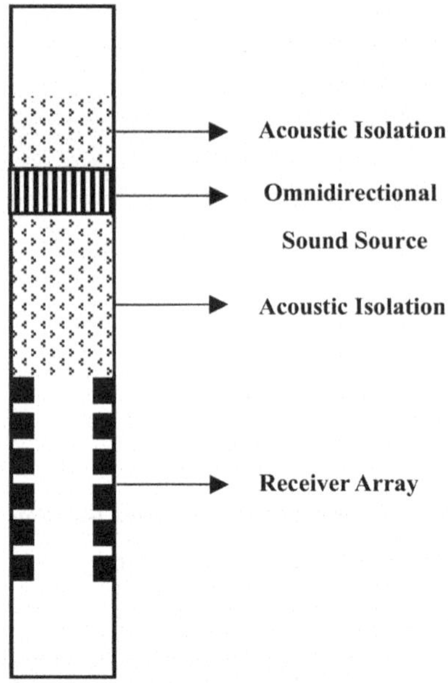

piezoelectric crystal omnidirectional emitters to provide 360° coverage of the wellbore and surrounding formation. The sound source can be transmitted with monopoles and dipoles and frequency modulated in the frequency range of 10–18 kHz. The APX uses a 6 × 4 omnidirectional receiver array to align the receiver array with the sound source to achieve radial multipole source excitation. From the aspect of frequency, signal amplitude and phase matching, the response of the matched receiver is very important for high-quality acoustic time difference measurement in the drilling noise environment [15].

Multiple receiver combinations and long and short source distance combinations provide high quality formation signals. Proper filtering techniques can reduce the effects of rig noise, bit runout and mud flow. Advanced sound insulation technology eliminates the interference of the body wave of the instrument. The acoustic velocity is directly affected by the formation near the wellbore, so this technique can be used to obtain accurate formation acoustic time differences. The measured information is stored in the downhole high-speed memory in addition to being transmitted to the ground in real time.

5. SonicVISION Acoustic Logging Tool

The signal transmitter array of the SonicVISION acoustic logging tool consists of an omnidirectional unipolar transmitter and two dipole transmitters. The unipolar signal transmitter has a frequency range of 1~12,000 Hz and a main frequency of 5~6000 Hz. This main frequency is 1/2 or 1/3 of the main frequency of the signal transmitted by other types of unipolar full-wave acoustic instruments, which can increase the detection range of shear wave and longitudinal wave. The instrument can transmit and receive wide-band acoustic signals in the frequency range of 3–19 kHz, and it is more likely to receive feedback information from most formations in this frequency range. The shear wave has a lower frequency than the longitudinal wave, which is difficult to obtain with a narrow-band instrument, while the frequency-optimized SonicVISION instrument uses a much wider frequency band than the previous test instrument to excite the formation, so when drilling in a formation with fast acoustic wave propagation speed Both shear and longitudinal wave measurements can be made. To more effectively and the formation width of a coupling acoustic energy, the power output is increased by 10 times the original [15].

It collects acoustic wave longitudinal and shear wave time difference data in real time. The full-wave waveform data stored in the downhole memory is played back for post processing. These data can be used for pore pressure prediction, gas reservoir detection, movable liquid estimation, and the like. The instrument uses automatic static measurements for quality control during drill pipe connection. In this short, noise-free environment, the instrument's static measurements are transmitted to the ground. These data can be used to predict pore pressure, detect gas layers, and estimate unbound liquids.

6. SonicScanner Acoustic Logging Tool

The SonicScanner instrument combines the long source distance method with the wellbore compensation TR short source distance method and also adds a circumferentially distributed receiver, as shown in Fig. 7.12. There are a total of 104 receivers with 13 axial receiving points on the receiver array and 8 receivers per receiving point. There are 5 emitters, one unipolar emitter at each end of the receiver array, and the other unipolar emitter and two orthogonal directional dipole emitters located further away from the lower part of the instrument. The SonicScanner's three

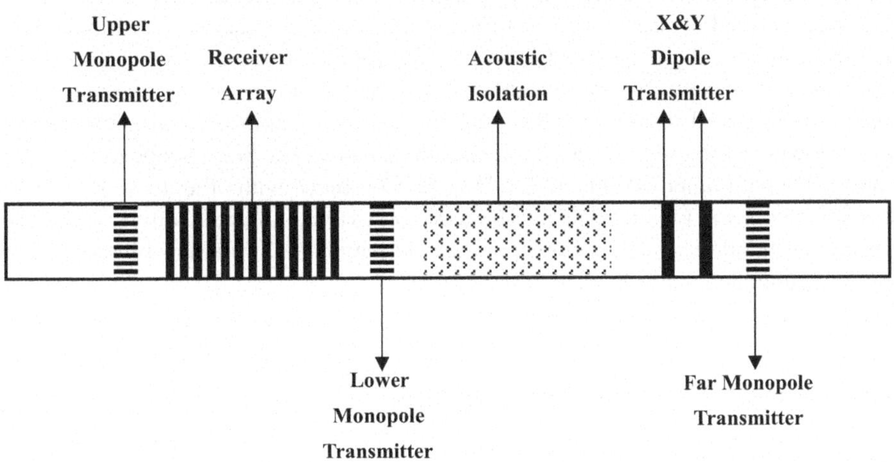

Fig. 7.12 Schematic diagram of the SonicScanner instrument structure

unipolar emitters produce a stronger pressure pulse than previous sonic instruments. These emitters produce clear longitudinal and shear waves, low-frequency Stoneley waves, and the high-frequency energy required for cementing evaluation. Both dipole emitters are a vibrating device consisting of an electromagnetic motor mounted on a cylinder suspended from the instrument. This mechanism produces a high voltage dipole signal that does not cause chattering of the instrument housing. The source can be driven in two modes: a pulse signal generated by a conventional dipole source, and a chirp signal generated by a frequency sweep. The two dipole sources are orthogonally oriented, one vibrating along the instrument reference axis and the other at 90° to the reference axis, producing a bending wave. The bending wave propagates up and down along the wellbore and passes to different depths of the formation depending on the frequency. Under all wellbore and formation conditions, the frequency components of the new chirped pulse dipole source (300Hz to 8 kHz) can produce curved waveforms, ensuring maximum signal to noise ratio.

7. SonicScope Acoustic Logging Tool

SonicScope is a quadrupole LWD tool with a wide range of applications for collecting downhole data in different modes. Acoustic waves are excited in the wellbore in both monopole and quadrupole modes using a powerful broadband transmitter, with frequencies ranging from 1 kHz to 20 kHz. There are 48 receivers in the protection groove on the outside of the instrument, 10 cm (4 inches) apart from each other, and the protection grooves are separated by a 90° angle. The receivers are arranged in four arrays providing 12 axial and 4 azimuth measurements. Each row of arrays contains 12 digitizers, one for each sensor. The signal to noise ratio is maximized by optimizing the distance from the transmitter to the receiver. The 1G memory is configured in the logging tool to store data recorded in all modes even when the recording speed is once per second. Typically, the SonicScope tool is programmed in the field to record high frequency monopole data, obtain longitudinal and shear wave slowness in the fast formation, record low frequency monopole data and quadrupole data, respectively for the slow formation stoneley wave and shear wave. Through sonic dispersion analysis (using the inversion algorithm to best match the recorded results to the simulated response), engineers can extract shear wave slowness values below 2000 microseconds/meter. The SonicScope tool is fully compatible with other MWD and LWD tools. Combined with other records such as density data, sonic logging data can be used for a variety of applications, including seismic correlation analysis, pore pressure determination, complex lithologic log interpretation, and analysis of rock mechanics.

7.5 Nuclear Logging While Drilling

Nuclear logging while drilling includes: natural gamma logging while drilling, density while drilling – neutron logging, and nuclear magnetic resonance logging while drilling.

7.5.1 Natural Gamma While Drilling Logging

Natural rocks contain natural radioactive elements, including potassium (40 K), strontium (232Th) and uranium (238 U) elements, which produce gamma rays of different energies when the radioactive elements naturally decay. The content of radioactive elements in different lithologies is different and the types are also different. The principle of natural gamma logging is to measure the total natural gamma ray intensity in the formation using a gamma ray detector to divide the formation lithology and permeable layer, determine the mudstone content, and provide an accurate indication of shale.

The natural gamma-while-drilling logging tool uses two sets of Geiger-Miller tube (GM tube) gas detectors for the detection of gamma rays. Each group of 8 and two sets of detectors are circumferentially distributed, as shown in Fig. 7.13. Show. The gamma ray detection method uses the incident gamma photon to interact with the target atoms in the detector, and the secondary electrons are generated by the three effects of photoelectric, Compton scattering and electrons, thereby inducing gas ionization or crystal excitation to detect.

The measurement error of natural gamma measurement is related to instrument design, ROP and sampling rate. The estimation formula can be as follows:

$$\frac{\delta_{GR}\sqrt{N}}{GR\,N} = \sqrt{\frac{k \times ROP}{GR \times d}} \tag{7.1}$$

Where N: gamma count value, 1/s; k: scale factor, gAPS/cps; d: longitudinal sampling interval, m; ROP: mechanical drilling rate, m/s; GR: gamma measurement, gAPI.

For the natural gamma logging tool while drilling, according to the instrument scale coefficient $k = 2$gAPI/cps; and assume: $ROP = 50$ m/h, $d = 0.1$ m, $GR = 50$gAPI, calculate the measurement error of the instrument, and get:

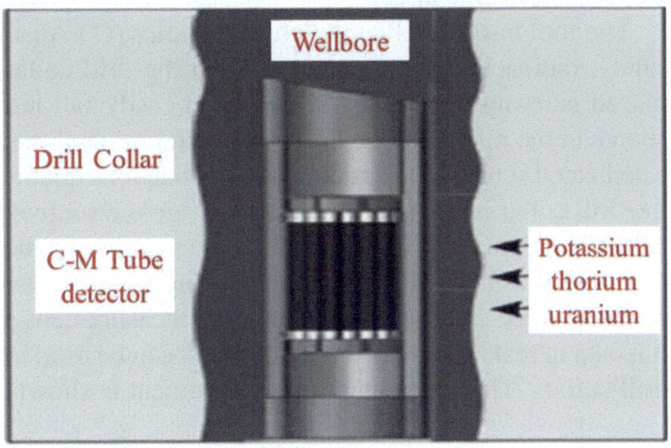

Fig. 7.13 Schematic diagram of the detector structure

$$\frac{\delta_{GR}}{GR} = \sqrt{\frac{2 \times 50}{50 \times 0.1 \times 3600}} = 7.45\% \tag{7.2}$$

which is: GR = 50 ± 3.7gAPI。

It can be concluded that in the instrument design from the above analysis, the *k* value is reduced as much as possible, that is, the detection efficiency of the gamma ray is improved, which is the main means to improve the measurement accuracy. In addition, proper reduction of *ROP* and sampling rate is also conducive to improve measurement accuracy, but it will directly limit the efficiency of drilling engineering operations for logging while drilling tools [16].

There are many factors affecting the error of natural gamma logging while drilling, including: formation thickness, environment in the wellbore, radioactive fluctuation error and vibration, etc. [17]

7.5.2 Neutron-Density Logging While Drilling

The high-energy fast neutron generated by the neutron source enters the formation and inelastically scatters with the nucleus of the formation element and releases the corresponding characteristic gamma ray; as the neutron energy decreases, the neutron and the nucleus elastically scatter; When further reduced to become a thermal neutron, a radiation capture reaction with the nucleus will occur, releasing the captured gamma ray. Using inelastic scattering gamma rays generated by neutrons and formations as gamma sources, density logging can be achieved by detecting secondary gamma rays. Instrument introduction: including compensated neutron density (CDN), azimuth neutron density (AND), azimuth density while drilling – middle word logging tool (AND) and density – neutron gap caliper.

1. Compensated neutron density (CDN):

It consists of 2 neutron sources, 1 neutron detector, 1 density detector, 1 centralizer and electronic circuit.

The tool has the following characteristics: (1) Anadrill special emphasis on tool safety, radioactive source placement In the drill collar, while other companies are placed sideways, sealed by screws, and easily fall into the well, Anadrill puts the source in the middle of the drill collar and has a unique installation method. There is a dedicated source installation platform that does not harm the surrounding people. If the drill collar is stuck, it is easy to salvage the source; (2) the purpose of using two detectors is to compensate for the impact of the wellbore; (3) compensating for the thermal neutron density; (4) compensating for the density of the rock; (5) A full-size or under-sized centralizer; (6) can store measurement data downhole; (7) can transmit data in real time with MWD tools; (8) can be used in φ165.1 mm and φ203.2 mm drill collars. The schematic of the instrument is shown in the Fig. 7.14.

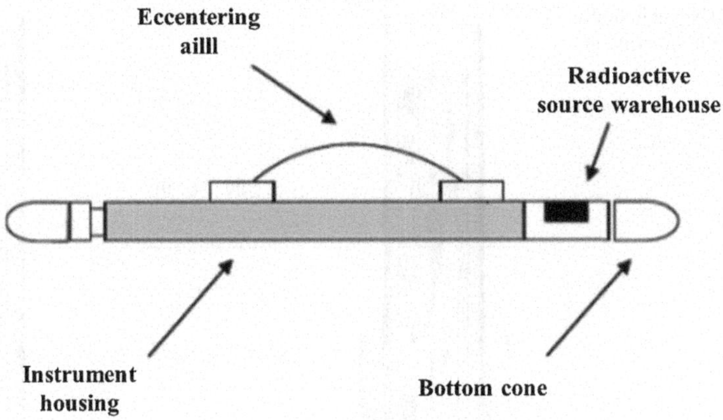

Fig. 7.14 Compensated neutron density logging tool

2. Azimuthal Density-Neutron Tool (AND):

Schlumberger's AND (Azimuthal Density-Neutron Tool) is a new generation of wellbore compensation formation density-neutron drilling technology based on CDN (Compensated Density-Neutron Tool) technology. In order to improve the measurement accuracy, the instrument uses the wellbore compensation and gap correction technology to measure the photoelectric absorption index. Compared with CDN, AND technology is characterized by multi-directional real-time neutron porosity, formation volume density and photoelectric absorption index measurement during drilling, describing formation lithology and porosity characteristics. 360° imaging of the density and porosity results measured by the instrument's rotating probe in four quadrants. In addition to the orientation parameters, the average of each parameter is also recorded. In order to avoid affecting the measurement of the direction and inclination of the drilling, the upper part of the instrument is neutron measurement and the lower part is the density measurement part. The ultrasonic transducer is located below the density measurement section (Fig. 7.15).

The density, neutron porosity detector and ultrasonic transducer are located on one side of the drill and are in one. The improvement of the instrument is that the instrument collects the quadrant measurements around the wellbore according to the azimuth distribution as the instrument rotates. Ultrasonic detectors are used to provide gap measurements for each quadrant instrument to ensure density and neutron data quality, indicating wellbore size and shape. Improved understanding of geological heterogeneity and stratigraphic boundaries has been improved. Azimuth measurements provide a means to quantitatively evaluate formation porosity and lithology heterogeneity. The ADN instrument uses different sizes of stabilizers for logging, or in flexible mode, without stabilizers. The new instrument series is a video density neutron meter (VDN) with a density data of 16 points per well, which improves the azimuth resolution.

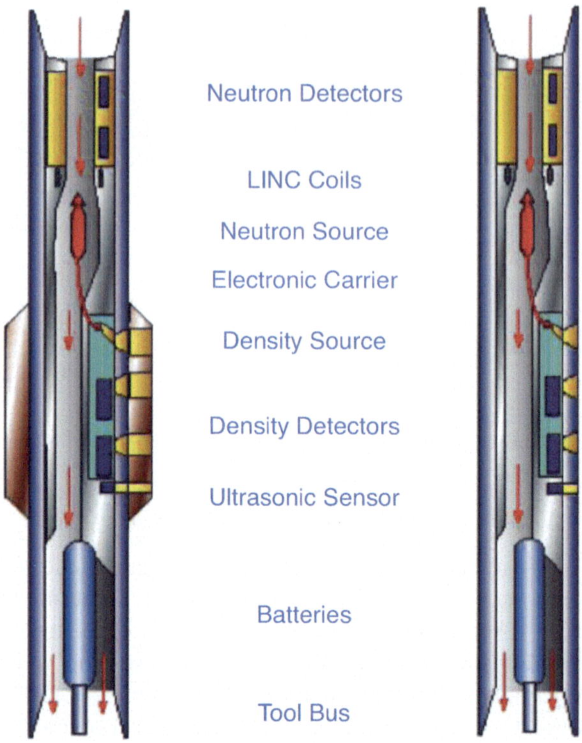

Fig. 7.15 AND schematic diagram of the instrument structure

3. Density-Neutron Gap Caliper:

Halliburton's density neutron gap well caliper (DNSC) is part of its Path Finder LWD system. The density measurement section is located at the bottom of the instrument and the detector is at the top of the source. The neutron measurement section is above the density measurement section and the detector is below the source. This arrangement prevents gamma rays generated by neutron interaction from reaching the density detector. The difference is that the combination of three ultrasonic detector measurements produces caliper measurements to correct for density and neutron measurements. The gap ultrasonic measurement is provided by the gap ultrasonic transducer, and the gap correction is performed according to the small gap weighting over the large gap weighting. The primary purpose of measuring the deviation gap and caliper is to correct the effect of the offset gap on the neutron porosity measurement and improve the quality of all nuclear measurement data. The DNSC instrument has 6.75in and 8 in.

In recent years, the nuclear logging tools have been greatly improved. The main improvements on the hardware are: (1) to strengthen the source's robustness, and to salvage when stuck; (2) use ultrasonic transducer to measure the gap and caliper; (3) The bottom side of the directional well is identified by a magnetometer; (4) The nuclear logging tools of the same drilling method adopt two measurement

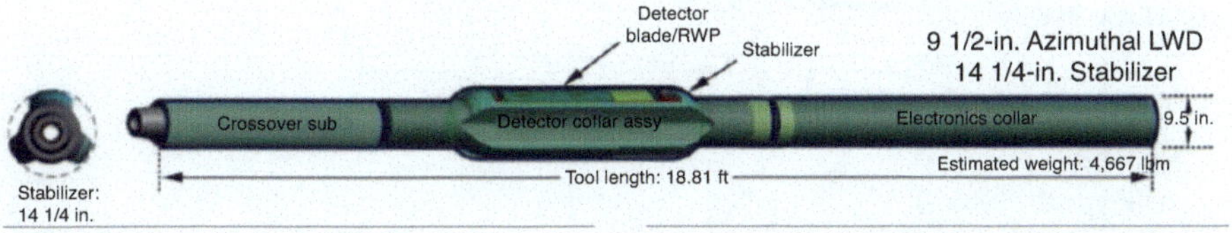

Fig. 7.16 Drilling density imaging tool

combinations to solve the physical combination in the same drilling and the instrument length is minimized. The improvement in the processing method mainly includes: (1) using the quadrant data to obtain a large amount of information content to improve the accuracy of the measurement. And use the average counting rate to equalize the effect of dynamic gap; (2) use energy spectrum-based algorithm and fast sampling technology; (3) use weighting technology to correct small gap data; (4) Monte Carlo technology improves the response of instruments in horizontal wells The understanding of the characteristics, especially the change of the gap in the elliptical wellbore, is relatively accurate, and the calibration plate for the field application can be developed [18].

4. Drilling Density Imaging Logging Tool:

Larger sizes of wellbore are often encountered in formation evaluation. This unique size logging while drilling tool can measure large size wellbores and the tool is capable of accurately providing density and optoelectronic data. The Fig. 7.16 shows the schematic of the tool.

This is a unique 9.5-inch Drilling Density Logging (LWD) tool that addresses the challenges of large bore sizes. New tools provide density and optoelectronic measurements in large diameter boreholes. It also includes an ultrasonic sensor that provides accurate borehole geometry information, which helps identify stress-related breakthroughs and provides accurate borehole volume estimates for later cement placement for interlayer isolation [19].

7.5.3 NMR Logging While Drilling

NMR logging is the response of formation hydrogen nuclei to external magnetic field [20]. An NMR probe equipped with a permanent magnet is passed through the formation to magnetize randomly arranged protons in the formation fluid. The permanent magnetic field causes the proton magnetization to generate a macroscopic magnetization vector, and then the antenna emits an alternating excitation pulse to generate an alternating magnetic field according to the principle of electromagnetic induction. The frequency of the variable magnetic field is determined by the frequency of the excitation pulse emitted. When the frequency of the transmitted pulse coincides with the natural frequency of the protons of the formation, nuclear

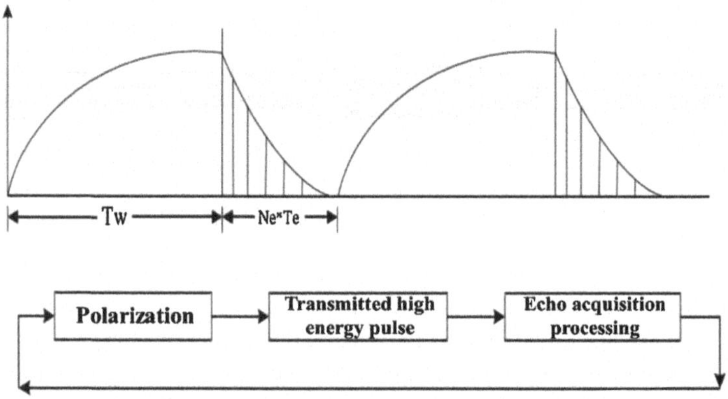

Fig. 7.17 NMR logging tool logging diagram

magnetic resonance occurs. When the alternating magnetic field is cancelled, the proton will gradually return to the state before the polarization due to the relaxation process, and a set of spin echo signals will be generated. If the antenna probe is tuned to the specific frequency, the original spin echo signal can be received. The peak set of spin echo signals is the spin echo train. The measurement of the spin echo train takes only a few hundred milliseconds or less, the instrument moves only a short distance in the well, and the information of the recorded echo train can completely represent the information of the fixed depth point. Thus, by transmitting a set of excitation pulse signals at each depth, an echo train of a corresponding depth can be acquired. The amplitude and attenuation information of the echo train is related to the acquisition parameters set by the instrument on the one hand and the fluid characteristics in the corresponding pore region on the other hand. The parameters of the instrument work are set by the ground staff. There are mainly echo interval time TE, polarization time TW, number of echoes NE, etc. By adjusting the length of TE and TW, and combining various modes, we can Measuring different information about the fluid [21] (Fig. 7.17).

The MWD logging tool measures rock formation parameters during drilling. There are three drilling modes: stationary, sliding and rotating. Ground engineers can adjust the sequence of measurements based on actual conditions at the site and adjust the drilling mode according to the characteristics of the formation. It can also automatically switch the drilling status according to the drilling conditions. At the same time, there are three measurement modes of the instrument: measuring T1, measuring T2, or measuring both T1 and T2. The T1 mode is unaffected by vibration during drilling, while the T2 mode has short measurement times and higher vertical resolution and repeatability. Both T1 and T2 measurements are processes that reflect changes in formation fluid properties over time. The MWD instrument is installed in the drill collar. The instrument consists of two main components: the bottom probe and the upper electronic circuit. The probe includes a permanent magnet, a flow tube, and an antenna; the electronic circuit portion includes a battery, a mud generator, a power controller, a mud pulse generator, a data memory, an azimuth probe, and the like [22].

Commonly used NMR logging tools are Halliburton's MRIL-WDTM, Schlumberger's LWD-NMR, BakerAtlas' MagTrack, and American Petroleum's provision.

1. MRIL-WDTM

The instrument measures the vibration insensitive T1 and the instrument sliding phase measurement T2 while drilling to achieve the purpose of measuring formation parameters [23]. The instrument is measured in a medium-sized single-frequency measurement. The upper part of the instrument is an electronic short section, and the lower part is a probe, which is mainly composed of a permanent magnet, an antenna and a flow tube. The static magnetic field design is the same as the magnetic field design of MRIL. The uniform magnetic field formed by the magnet at the round shell has an intensity of 120 Gs, the magnetic field gradient is 14 Gs/cm, and the frequency of the RF pulse is about 500 kHz. The antenna is wrapped in fiberglass and rubber sleeves, and the longer antenna ensures sufficient signal strength and signal-to-noise ratio. The high magnetic field strength and resonant frequency of MRIL-WDTM guarantee high signal strength and signal to noise ratio. According to the drilling mode, there are two main data acquisition modes for the nuclear magnetic while drilling: Reconnaissance Logging and LOG (Evaluation Logging) (Fig. 7.18)

2. LWD-NMR

Both ends of the instrument are real-time LWD connectors that are compatible with any part of the BHA [24]. The instrument's operating rating is similar to that of other commercial 6.75inLWD instruments: maximum temperature resistance is 300 °F, maximum withstand voltage is 2000ib/in2, maximum dog leg severity is 8°/100 ft. (1 ft. = 305 mm) when rotating, maximum during sliding The dog leg has a severity of 16°/100 ft., which satisfies the seismic performance of the logging while drilling. The permanent magnets of the LWD-NMR probe are cylindrical and mounted vertically. The opposite poles create a gradient magnetic field around the magnet. The resonant region is a cylindrical shell with a diameter of 14 in. and a length of 6 in. The magnetic field strength in the resonance region is 61 Gs, and the

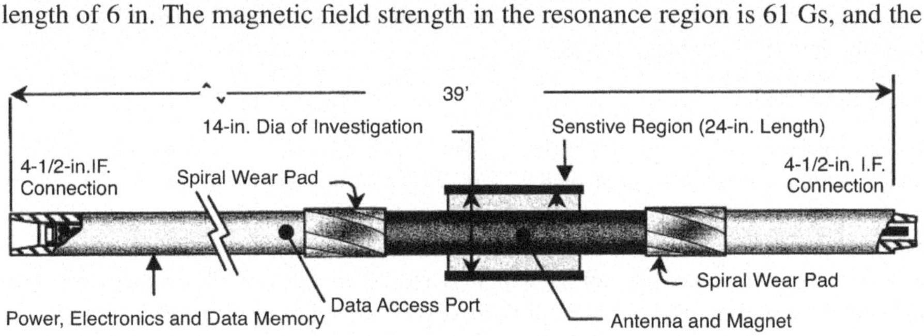

Fig. 7.18 Schematic diagram of MRIL-WDTM logging tool

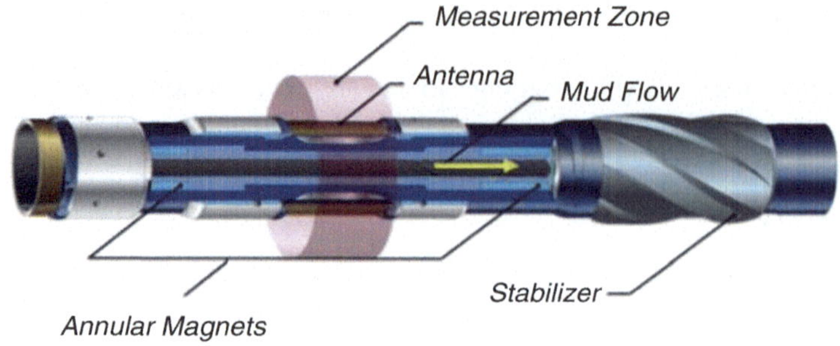

Fig. 7.19 Schematic diagram of LWD-NMR instrument structure

Table 7.1 Data of LWD-NMR pulse signal synthesis

Sequence	Tw(ms)	Te(ms)	Ne	Noise(pu)
A	8422	0.9	500	1
B	2200	0.9	20	1
C	300	0.6	10	1
D	100	0.6	10	1
E	30	0.6	10	1
F	10	0.6	10	1

gradient of the magnetic field is 3 Gs/cm. The high gradient static magnetic field produces a thinner thickness of the round shell, while the LWD-NMR uses a low magnetic field gradient design with a thicker measuring housing that is insensitive to instrument motion. The depth of detection of the instrument depends on the size of the wellbore and the eccentricity of the instrument. For a 8.5in wellbore, if the instrument is centered, the depth of detection is 2.75in; if the instrument is eccentrically measured, the depth of detection is between 2.375in and 3.125 in. For larger boreholes, the stabilizer can be used to position the instrument to ensure that the measured information comes from the formation, not the mud signal. When the size of the wellbore exceeds 10.625 in, the measured signal may come from the mud signal of the wellbore. The Fig. 7.19 shows the structure of the instrument.

LWD-NMR data inversion processing table, as shown in the following table, the A-F pulse sequence is subjected to inversion processing to synthesize T2 spectrum logging data [25] (Table 7.1).

3. MagTrack

The instrument has a total length of 32.8 (1 ft. = 305 mm) ft, a diameter of 6.75in, and a minimum measurement of 8.375 in. For ease of operation, the MagTrack is pre-programmed for a variety of formation and fluid properties. For each selected mode, the raw data is processed downhole and the calculated parameters are passed to the surface in real time, all of which is stored in the downhole memory. The instrument has the following features: it can be compatible with other downhole instruments; vibration has less influence on measurement; real-time measurement data is obtained by mud pulse; has higher longitudinal resolution; can be measured in

Table 7.2 ProVISION instrument pulse sequence parameters

Expected output	Waiting time (seconds)	Repeat times	Number of echoes
Oil	6.00	2	500
	0.60	2	300
	0.04	40	20
Oil and gas	13.00	2	500
	0.60	2	300
	0.04	40	20

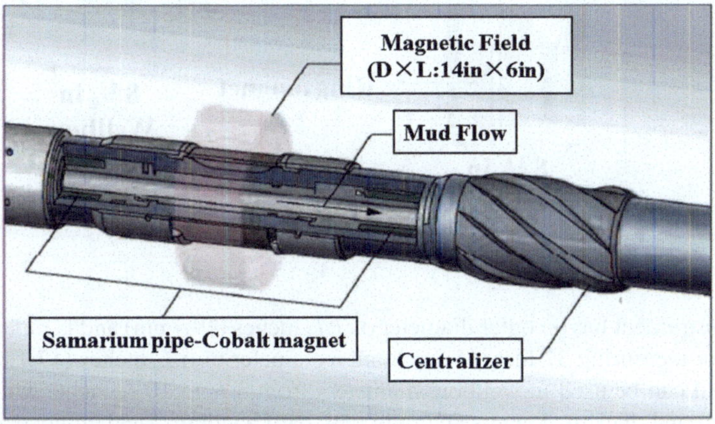

Fig. 7.20 Schematic diagram of ProVISION instrument design

horizontal wells; less occupied drill floor Time can save money; real-time data accuracy is high, which is conducive to geosteering [24].

4. ProVISION

ProVISION is a real-time reservoir-oriented service system introduced by American Petroleum Corporation in 2001. The LWD tool can perform accurate high-resolution NMR measurements under common harsh drilling conditions. The ProVISION NMR magnetic logging tool measures lithology-independent porosity, bound fluid volume (BFV), free fluid volume (FFV), permeability and T2 distribution, and hydrocarbon detection (Table 7.2).

This table shows the three different waiting time parameter acquisition sequences used in the evaluation of the oil production interval (the upper group) and the oil production/gas layer (the lower group) in the Deepwater oil and gas wells in the US Gulf of Mexico, indicating that the instrument has very Good programmable features. The instrument can be controlled by manual programming or set to an automatic state to automatically switch states according to drilling conditions (Fig. 7.20).

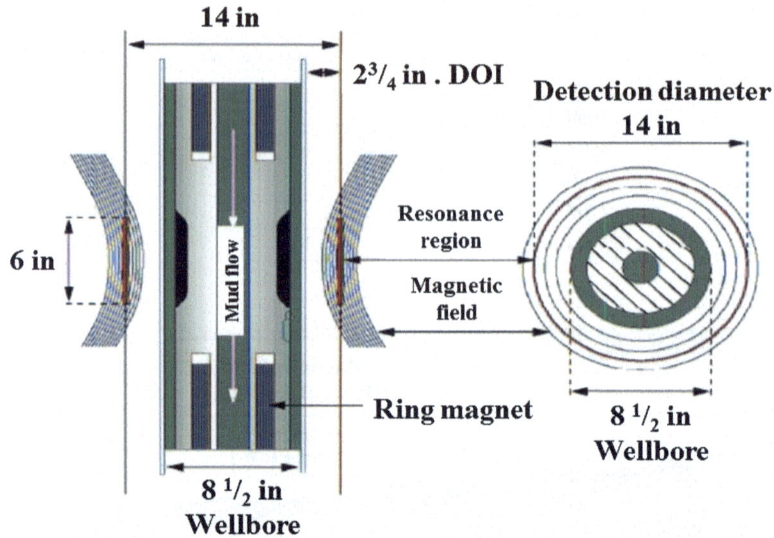

Fig. 7.21 ProVISION instrument profile

The instrument has an outer diameter of $7^3/_4$ inches (19.7 cm) and is mounted in a drill collar measuring 37 feet (11.3 m) and a diameter of $6^3/_4$ inches (17.1 cm). The instrument can be used in wellbore diameters from $8^3/_8$ to $10^5/_8$ inches without the use of external thickened layers and only with hard banding. Field engineers can use bolted dome centralizers to reduce lateral movement of the instrument and center it within the wellbore. The telemetry connections on both ends of the instrument allow it to be mounted anywhere in the bottom hole assembly (BHA). The instrument is powered by a turbine instead of a battery and is suitable for flow rates from 300 to 800 gallons per minute (1136–3028 liters per minute) (Fig. 7.21).

The axial cross-section of the antenna (left) shows the symmetrical design of the instrument, the dark blue strip is a hollow cylindrical magnet, and the equal field strength line (blue) indicates that the gradient magnetic field strength decreases as the distance from the instrument increases.. Black shows the cross section of the coaxially wound antenna coil. The interaction between the antenna and the magnet produces a cylindrical resonant housing (red strip) that is 6 inches (15 cm) long and 0.4 inches (10 cm) thick. The diameter is 14 inches (36 cm). The cross section (right) of the coaxially wound antenna coil shows an axisymmetric resonant housing (red). The resonant housing is the only area where measurements can be made, and measurements cannot be made between the instrument and the resonator and from the resonant shell to the more distant layers. Formation depth (DOI) in a $8^1/_2$ inches (21.5 cm) borehole is $2^3/_4$ inches (7 cm).

The MWD logging technology can be influenced by factors such as rock skeleton and muddyness of the formation, and intuitively and accurately provide parameters such as effective porosity, permeability, and irreducible water saturation of the reservoir. NMR logging technology can distinguish the movable fluid volume and

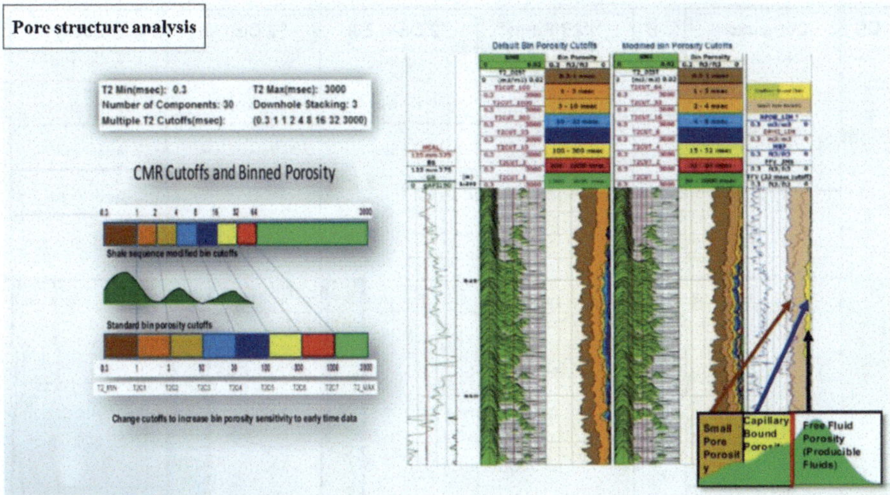

Fig. 7.22 Interpretation and application of T2 spectrum of NMR logging while drilling

the capillary bound fluid volume by T2 cutoff value, and further reflect the pore structure of the reservoir [26]. The pore structure analysis was performed using the T2 spectrum of the NMR logging while drilling as shown in the Fig. 7.22.

While drilling NMR logging is a new logging method that uses new techniques to evaluate reservoir porosity, completely different from conventional radioactive logging to find pores. Because it does not use radioactive sources, it ensures the safety of underground wells. At the same time, it can be used while drilling and can be used for drilling and testing, providing more solutions for drilling and logging work [27]. The following Fig. 7.23 shows the data of NMR logging while drilling [28].

7.6 Seismic Logging While Drilling

The seismic logging while drilling technology is a technique that uses a seismic while drilling instrument to measure the time when seismic waves propagate from the surface to the downhole receiver while recording the 4-component waveform data to improve the interpretation accuracy(Fig. 7.24).

To ensure the synchronization of surface excitation and downhole reception, the downhole tool uses a high-precision synchronous clock (UHPC) to synchronize with the GPS time of the ground to obtain an accurate time-depth relationship. The figure below shows the working flow chart of the seismic logging while drilling.

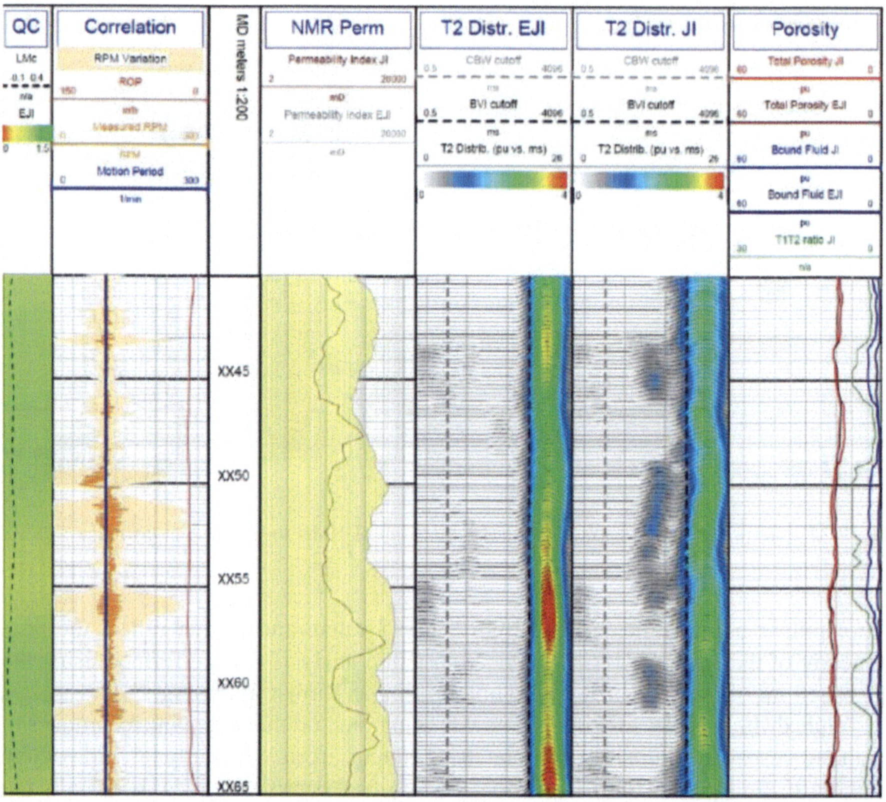

Fig. 7.23 NMR data while drilling

1. SeismicVISION B

Seismic data acquisition while drilling uses Schlumberger's second-generation seismic logging tool, SeismicVISION B, which has four sensors, one downhole processor and one memory. The four sensors include three orthogonal component detectors and one hydrophone. The following Fig. 7.25 shows the use of SeismicVISION B seismic logging tool [29].

The collected raw data is stored in the tool memory, and the tool also has real-time processing function, pre-processing and superimposing the original waveform, and using the drilling fluid pulse signal after the pump is turned on, the superimposed waveform data is transmitted to the ground in real time to the downhole. Data is collected for quality monitoring and real-time waveform data is further processed and interpreted as needed to provide pre-drill prediction information.

2. Earthquake technology while drilling

The drill bit while drilling earthquake uses the vibration generated by the bit breaking rock as the downhole source. The vibration signal transmitted by the drill

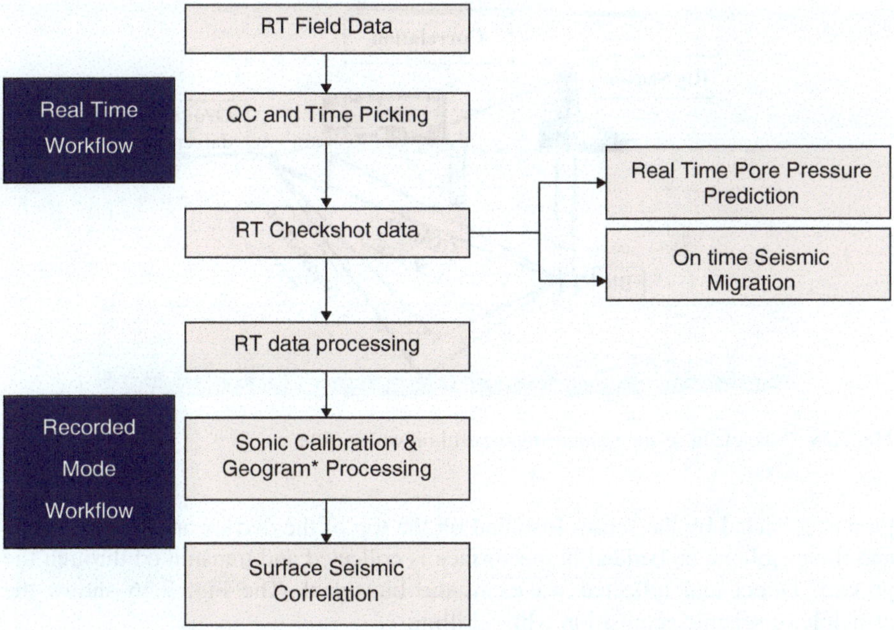

Fig. 7.24 Seismic logging workflow while drilling

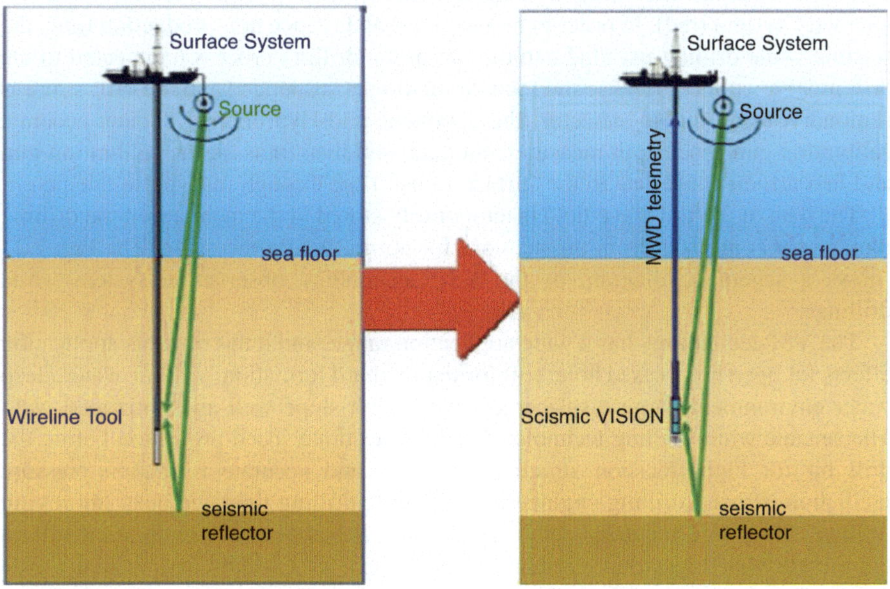

Fig. 7.25 Schematic diagram of seismic logging while drilling

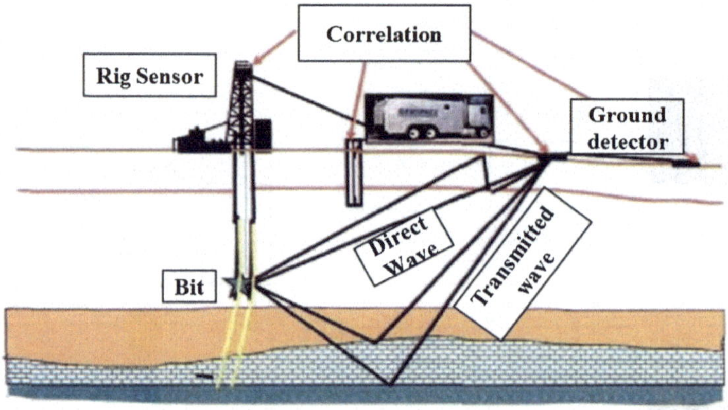

Fig. 7.26 Principle of seismic acquisition of drill bit while drilling

pipe is collected by the sensor installed on the top of the derrick and the drill pipe, and the geophone embedded in the surface is collected and transmitted through the ground. Direct and reflected waves of the bit signal. The Fig. 7.26 shows the principle of seismic acquisition while drilling.

The VSP technology uses the conventional VSP observation system to perform source excitation on the ground or the sea surface. The first wave (green) and the reflected wave (yellow) under the drill bit are detected by the seismic while drilling tool installed in the drill string. The signal is transmitted to the surface via the MWD telemetry system (red). In order to reduce noise and reduce non-production time, the seismic while drilling uses the natural gap of the drilling process (connected to the drill pipe) to collect, because the mud circulation is suspended and the drill string is stationary, using the 4C detector (the geophone and Hydrophone) Obtain accurate calibration gun time/depth measurement data, and then transmit the calibration gun and first arrival wave data to the surface in real time through mud-pulse telemetry.

The data of the full wave field is temporarily stored in the memory on the drilling tool, and is copied out from the memory for normal VSP processing. The Fig. 7.27 shows a schematic diagram of the VSP technology observation system while drilling.

The VSP technology has a wide application range, and it has obvious application effects for open hole, cased hole, soft formation, hard formation, shallow water, deep water environment, large displacement well, high slope well and horizontal well. The seismic while drilling technology enables accurate speed prediction before the drill bit for high-precision structural imaging and accurate formation pressure prediction, serving drilling engineering, reducing drilling time and cost, improving drilling safety, and reducing depth uncertainty. It has played an important role in optimizing well trajectories, reducing the number of casings, and reducing side drilling and directional drilling [30].

Fig. 7.27 Schematic diagram of the VSP technology observation system while drilling

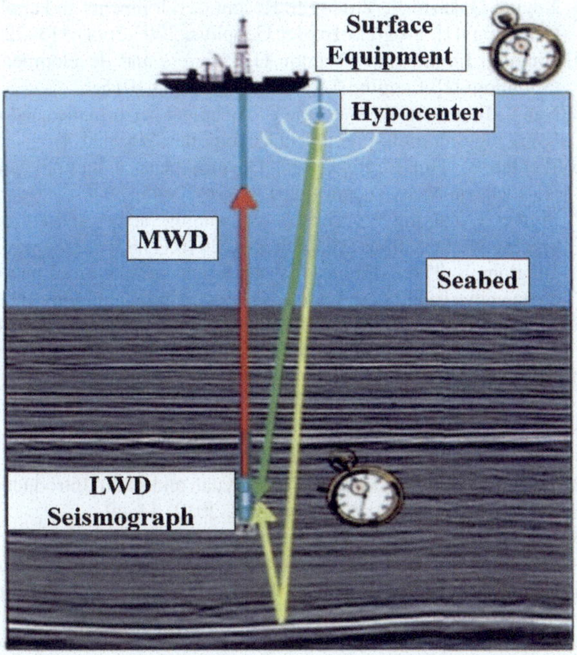

References

1. Zhihong B, Ganneng R, Le C. Logging while drilling technology[J]. Fault Block Oil Gas Field. 2001;08(4):22–4.
2. Xinyi Z, Yanjun G, Jingnong W. Yesterday, today and tomorrow of logging while drilling [J]. Logging Technol. 2006;30(6):487–92.
3. Xinping L, Jun F, Youhai J. Application status and prospect of data transmission technology while logging[J]. Logging Technol. 2008;32(3):249–53.
4. Xuanchao L, Shaozhen Z. Research on intelligent drill string information and power transmission system[J]. Pet Drill Tech. 2006;34(5)
5. Shihong X, Zheng L. Development status and prospect of rotary steering drilling technology [J]. Pet Mach. 2006;34(4):66–70.
6. Yang X, Wei J, Binwei Z. New technology of data transmission while drilling[J]. Pet Instrum. 2004;18(6):26–7.
7. Yong DU, Jian-bin HU, Yan-ping LI, et al. Application of acoustic transmission test Technology in Oilfield[J]. Meas Control Technol. 2005;24(11):76–8.
8. Linlin N, Xiaobing Z, Hailong L, et al. Current status and prospects of logging while drilling [J]. Foreign Well Logging Technol. 2008;7–12(6)
9. Rodney PF, Wisler MMM. Electromagnetic wave resistivity MWD tool[J]. SPE Drill Eng. 1986;1(05):337–46.
10. Meyer WH. Inversion of 2 Mhz propagation resistivity logs[J]. J Nutr. 1992;125(2):155–60.
11. Yonghao D, Zhoubo L, Hongyu M. Development of logging while drilling technology [J]. World Geol. 2004;23(3)
12. Haibo W. Research on structure and principle of EWR_P4 type resistivity resistor sensor[J]. W Prospect Eng. 2011;23(7):67–9.

13. Xuying Q, Lizhi X, Yufeng S. Recent developments and applications of logging while drilling technology[J]. Progress Explor Geophys. 2003;26(4):313–22.
14. Xinyi Z, Jingnong W, Yanjun G. Progress and development trend of logging while drilling technology[J]. Logging Technol. 2006;30(1):10–5.
15. Nan L, Jingmeng W, Wuchen Y. Technical performance and application examples of the latest LWD logging tool[J]. Pet Drill Techniques. 2006;34(4)
16. Wenbin Y, Hui L, Jie S, etc. Development of natural gamma logging tool while drilling [J]. Electron Meas Technol, 2013 (36): 42–45.
17. Liyang S, Zaixing W, Xiao Q. Influencing factors and correction methods of natural gamma logging while drilling [J]. Tech Res. 2016;(1):97.
18. Qinglong S. Three-dimensional numerical simulation of new natural gamma log response and data synthesis and high resolution processing technology [D]. Jilin University, 2013.
19. Michel F, Parker T, Rennie A, et al. A new logging-while-drilling Azimuthal density sensor for large borehole applications——Gulf of Mexico Case Studies[R].OTC-29052-MS, 2018.
20. Lei W, Jingjing C, Li K, etc. Nuclear magnetic resonance logging instrument pulse scale design [J]. Instrument Technol Sensors, 2011 (7): 21–23.
21. Zhixing X. Research on key technologies of NMR logging while drilling [D]. Huazhong University Science and Technology, 2013.
22. Yu Huiyuan, Song Gongpu, Cai Chiyuan and so on. Introduction to nuclear magnetic resonance logging instruments [J]. Pet Instrum, 2012: 44–49.
23. Donovan GG, Ugueto GC, Pellerin NN. Wireline and LWD NMR Applications in Undersaturated Oil Sands in Deepwater Us Gulf of Mexico[R].SPWLA, 2001.
24. Heaton NJ, Jain V, Boling B, et al. New generation magnetic resonance While drilling [R]. SPE,2012.
25. Akkurt R, Kersey DG, Zainalabedin K et al. Challenge for everyday NMR: an operator perspective[R].SPE,2006.
26. Enlong F, Le P, Jun C, etc. The application of the NMR logging tool magtrak in bohai oilfield [J]. Sci Manag, 2018: 333–334.
27. Coman R, Thern et al. Lateral-motion correction of NMR logging-While-drilling data [R].SPE, 2018.
28. Hongxiang S, Hui L, Duoming Z, etc. Seismic-steering drilling technology based on seismic logging while drilling—Taking the fracture-cavity reservoir in the Harahatang block of Tarim Oilfield as an example [J]. 2016 (43): 662–668.
29. Dotiwala F, Kumar R, Lim TK et al. Reducing target uncertainties and guided drilling using seismic while drilling technology, a novel approach in Andaman Sea deep water[R]. SPE165834-MS, 2013.
30. Xiaohui Z, Guiqiao S, Weihua Z, Yao N, Jiexiong C. Earthquake technology while drilling and its new development[J]. Geophys Prospect Pet. 2016;55(06):913–23.

Chapter 8
Prospect of Big Data Application in Drilling Engineering

Abstract The big data is often called the petroleum in the information era, but the connection between petroleum and big data is not limited to it. The big data technology and petroleum and natural gas industry are closely related. In the economic environmental background where the global energy market is gloomy, the petroleum and natural gas companies increasingly obviously pay close attention to the big data. Not merely the big data technology will generate influences on the oil & gas industry. The progresses obtained in computing technology, Internet of Things, cloud computing, mobile communication technology, robot technology and artificial intelligence bring new innovations for the oil & gas industry. Integrating the traditional production mode in the oil & gas industry with the rapidly developing Internet industry will definitely make the oil & gas industry glow the new vitality.

Keywords Big data · Oil & Gas · Fault identification · Optimization of drilling · Drilling safety

8.1 Introduction

"Data have become a torrent flowing into every area of the global economy [1]". According to IDC's estimation, the data increases by 50% annually all the time. Much information is contained in plenty of data. The big data technology refers to "obtaining the regularities and relevance of a phenomenon through concluding, analyzing and processing plenty of essential data". The big data pays no attention to the effect generation process and cause between two factors, but pays attention to the result of effect on another factor due to the change of a factor. In other words, the big data knows what but does not have to know why. By virtue of big data processing, the regularities between them are obtained, and no attention is paid to the intrinsic connection of phenomenon. The larger the essential data volume, the

Contributions by Fangtao Li and Qilong Xue.

Q. Xue, *Data Analytics for Drilling Engineering*, Information Fusion and Data Science, https://doi.org/10.1007/978-3-030-34035-3_8

more accurate the relation analyzed. Mathematically, in the big data technology, based on fitting of discrete data, the relation of one to another is obtained. If the data volume is small, the fitting result may be linear. However, the more data, the more accurate data obtained. In the big data era, data is changed from static to dynamic. In the traditional situation, the dead outdated data is revitalized.

The most primitive definition of big data is presented by Meta Group [2], and there are three distinct features, namely, volume, velocity and variety.

Firstly, Volume-large data volume. The required data leaps from TB level to PB level;

Secondly, Variety- various data types, and wide heterogeneous data sources.

Thirdly, Velocity-fast processing. The big data calculation requires extremely high velocity, and the huge lengthy data should be fulfilled within second. The big data technology has an essential difference from the traditional data mining technology. The big data technology has to more quickly meet the real-time demand. At present, the requirement for smart and real-time data is increasingly tougher. The information exchange and interaction between human and human and between human and machine inevitably bring the data exchange. The key to data exchange is reducing the delay, and the data is presented to the user in an almost real-time way. IBM has put forward the fourth characteristic, namely, accuracy, for the data quality. In the big data, the data should accurately describe the characteristic, so as to figure out the definite association. Accuracy refers to quality and validity of data provided for analysis and decision-making. It is about distinguishing the clean data and dirty data. This point is very important, for the dirty data will leave obvious effects on the velocity and accuracy of data analysis. The original data should undergo the professional and effective processing and filtration, for the purpose of data analysis. Otherwise, the result will be unreliable. Oracle stresses the value of data, and increases the fifth definition, namely, value. If the value density is low, the commercial value is high. As the value density of data is low, the big data has to extract the useful data from the mass basic information, and value is an important characteristic of big data. The return value of big data infrastructure investment is very important. The big data analyzes the huge data set, so as to reveal the potential trend, and helps the engineers predict the potential problem. The future performance of equipment used in the operation process is known, and the failure is discovered so that the company has the competitive advantages and the value is brought to the company.

To be specific, the big data processing may be divided into three typical steps, namely, data extraction and integration, data analysis and data interpretation.

1. There are miscellaneous big data sources and various data types. At first, the data of required data source must be extracted and integrated, and relation and entity are extracted. After association and aggregation, such data is stored by the uniformly defined structure. In the data integration and extraction, the data should be cleaned, so as to guarantee the data quality and credibility.
2. The data analysis is a key step of big data technology. The extracted and integrated data constitutes the big data era of original data of big data analysis.

The data volume will not directly lead to the increase of useful information. The increase of data volume will also lead to the increase of interference data, and the difficult extraction of valuable information. The traditional analysis technology contains data mining, machine learning and statistical analysis. In the big data background, these technologies should be adjusted to adapt to the real-time property of big data technology. Due to the complicated and various types of original data of big data, the difficulty in the design of data analysis algorithm is caused. If the feature of overall data cannot be mastered, the analysis result will be inaccurate.

3. The final result of data analysis is obtaining the data interpretation so that it is provided to the user. By virtue of visual technology and enabling the users to participate in the data processing, the data interpretation is more convincing.
4. The big data is often called the petroleum in the information era, but the connection between petroleum and big data is not limited to it. The big data technology and petroleum and natural gas industry are closely related. In the economic environmental background where the global energy market is gloomy, the petroleum and natural gas companies increasingly obviously pay close attention to the big data.

8.2 Oil & Gas Engineering and Big Data

8.2.1 Prospect of Developing the Oil & Gas Industry by Big Data Technology [3–5]

In recent years, the oil & gas industry is moving ahead in the face of various challenges. Although the oil & gas development company and oil & gas service company are always committed to reducing the costs in the oil & gas development process, the gloomy petroleum price and constantly rising operating costs all the time still pose a huge impact on the oil & gas industry. The employees in the new situation must have courage, wisdom and technology to face the current difficult environment. The big data technology is a data processing tool or technology, and it is used for storing and processing mass data of poor structure. The big data has thoroughly changed every industry in the world, and the petroleum and petrochemical industry is not an exception. The petroleum and petrochemical industry (P&P) are an industry of the largest trading volume in the world. The oil & gas field, exploration, drilling and production processes, petroleum and petroleum product marketing and operation of oil & gas companies in the market will generate lots of data. At the same time, with the development of digital technology, the data generated in the upstream, mid-stream and downstream processes of oil & gas industry constantly increases progressively.

Such data contains rich information and many hidden values, and they are like the industrial raw materials. Like crude oil, the original data must be refined so that their real value is brought into full play. The traditional data processing technology fails to meet the requirements for velocity and accuracy in the oil & gas industry. Meanwhile, there are shortcomings in the processing of a large amount of heterogeneous data. The big data enjoy exceptional advantages in the oil & gas industry. In recent years, the big data technology has been increasingly widely applied in the oil & gas industry. With the help from progress in such aspects as high performance computing, network, storage, and machine learning, the operator and the service company install the infrastructure and compile the necessary algorithm, so as to mine the data and detail them into the operable steps.

In the interview, Darryl Willis, Vice President for Petroleum, Natural Gas and Energy of Google Cloud said that: "*The industry has seized the opportunity, but the pace at which it's been able to pull that opportunity forward and leverage it has not been at the right pace. We have to pick up the pace of transformation and change. Everyone is using the right buzzwords -artificial intelligence, machine learning, digitalization – but truly leveraging it is taking too long*"

It is estimated that only 5% of data collected in the oil & gas industry is used, which will increase remarkably as the petroleum and natural gas companies continue the digital transformation. In the Report on "2018 Chief Information Officer Agenda: Insight into Petroleum and Natural Gas Industry", Gartner said that 54% petroleum and natural gas companies were carrying out the digitalization. According to the Study "IDC Future Escape: Worldwide Oil & Gas 2018 Predictions" by IDC Energy, 25% main operators make investment in the asset performance management, and 75% oil & gas companies have at least one comprehensively operating digital transformation plan.

The hierarchical decision-making in the petroleum and natural gas industry is based on the processing and analysis results of such data. Meanwhile, the oil & gas reservoir exploration and development, investment decision-making and oil & gas production development depend on the processing and modeling of large data. To shorten the data processing time and provide the real-time and effective data processing results, it is required to establish the high-performance computing (HPC) system. The big data has provided the method and technology in this aspect. In addition, the big data technology may dig the potential value of data. More attention is paid to the big data, predictive analysis, data science and machine learning in the petroleum and natural gas field.

The big data technology has developed the prospect in the petroleum and natural gas industry. The big data has provided an approach to obtain the new high-quality knowledge and opportunity, and it not only provides the competitive advantage for the enterprise in the market but also develops the whole industry through revealing its potential. The exploration and development of petroleum and natural gas are a large comprehensive complicated project, and the investment in the earlier stage is very huge. In order to establish the competitive advantage, the petroleum and natural gas companies will formulate a 10-year or a longer time strategy. When the big data strategy in the petroleum industry is formulated, the features of specific company and specific business should be considered. In the strategic direction, the overall corporate strategy must be considered in the hierarchical structure, so as to guide the big data technology to develop towards the implementation direction of business solution with better profitability. In addition, the implementation of main objective must be provided. In this aspect, the big data technology should not meet the requirement of scientific project but should focus on the business world. In some management aspects, the formulation and implementation of big data strategy include the decision-making of order of priority of application of these technologies, description about explicit and precise objective, establishment of proper structure and implementation of investment plan and effective feedback and assessment.

For a long time, technology is the key driving force for success in the petroleum and natural gas industry all the time. By virtue of digital technology, changing the business model and providing a new income and value creation opportunity may push the industry to a new level. In the era of economic recession, the company or the individual adopts the bottom-line thinking, and seeks the solution to maintain the low cost without affecting the workplace safety. Big data technology, machine learning, AI and other technologies can meet such demand, thus drawing great attention in the oil & gas industry. These technologies are increasingly widely applied in the oil & gas industry.

8.2.2 Common Big Data Algorithms in Oil & Gas Engineering

The big data mining technology may be used for reservoir evaluation, drilling fluid optimization, selection of measure operation way, prediction of production index and fault diagnosis. Common algorithms in the oil & gas engineering include: support vector machine algorithm, artificial neural network, cluster analysis, neural network, grey correlation, decision tree, rough set and the like. Different algorithms have their own advantages and disadvantages and scope of application. In the specific big data mining application, based on the specific data set, the algorithm is selected. Common algorithms are simply introduced as follows:

1. Artificial Neural Network

BP neural network [6], namely, error back propagation neural network, is the most widely used in the neural network model. It is composed of input layer, hidden layer and output layer. Suppose each layer of BP neural network has N nodes, the action function is non-linear Sigmoid function, $f(x) = 1/(1 + e^{-x})$ is adopted generally, the learning set includes M sample mode (X_p, Y_p). For P learning sample $(P = 1, 2, \ldots, M)$, sum of input of node j is recorded as net_{pj}, output is recorded as O_{pj}, $net_{pj} = \sum_{j=0}^{N} W_{ji} O_{pj}$, $O_{pj} = f(net_{pj})$.

If the network initial weight value is set at will, for each input sample P, error between network output and expected output d_{pi} may be expressed as:

$$\mathrm{E} = \sum E_p = \left(\sum_j (d_{pi} - O_{pj})^2\right)/2 \quad (8.1)$$

Correction formula of weight value of network

$$W_{ji} = W_{ji}(t) + \eta\delta_{pi}O_{pj} \quad (8.2)$$

$$\delta_{pj} = \begin{cases} f(net_{pj})(d_{pi} - O_{pj}), & \text{output} \\ f(net_{pj})\sum_k \delta_{pk}W_{kj}, & \text{input} \end{cases} \quad (8.3)$$

In the above mentioned formula, learning rate η is introduced to accelerate the rate of convergence of network. Generally, in the correction formula of weight value, an inertial parameter should be added to rewrite the formula (8.2) into:

$$W_{ji} = W_{ji}(t) + \eta\delta_{pj}O_{pj} + \alpha(W_{ji}(t) - W_{ji}(t-1)) \quad (8.4)$$

Where, α is a constant, which decides the effect of previous weight value on this weight value.

2. Decision Tree Algorithm

The decision tree algorithm based on information theory [7] includes ID3, CART and C4.5 algorithms. C4.5 and CART algorithms derivate from ID3 algorithm. ID3 algorithm principle is as follows:

1. Information entropy of classification system

Suppose the sample space (D,Y) of a classification system, D stands for sample (m characteristic), Y stands for n category, the possible value is C_1, C_2, $\cdots$, C_n. The occurrence probability of each category is $P(C_1)$, $P(C_2)$, $\cdots$, $P(C_n)$. The entropy of such classification system is:

$$H(C) = \sum_{i1}^{n} P(C_i) \cdot \log_2 P(C_i) \quad (8.5)$$

In the discrete distribution, the Occurrence Probability $P(C_i)$ of Category C_i may be obtained through the occurrence time of such category dividing the total number of sample. For the continuous distribution, the discretization per zone is often required.

2. Conditional Entropy

According to the definition of conditional entropy, the conditional entropy in the classification system refers to the information entropy when a Characteristic X of sample is fixed. Since the possible value of such Characteristic X may include

$x_1, x_2, \cdots, x_n$, if it is required to fix it in the calculation of conditional entropy, each possible value should be fixed. After that, the statistical expectation is figured out.

Therefore, the probability of sample Characteristic X value of x_i is P_i, the conditional information entropy when such characteristic is fixed as Value x_i is H (C|X = x_i), H(C|X) is the conditional entropy when Characteristic X is fixed in the classification system ($X = (x_1, x_2, \cdots, x_n)$):

$$\begin{aligned} H(C|X) &= P_1 H(C|X = x_1) + P_2 H(C|X = x_2) + \ldots \mathrm{P}_n H(C|X = x_n) \\ &= \sum_{i=1}^{n} P_i(C|X = x_i) \end{aligned} \tag{8.6}$$

3. Information Gain

The information gain of Characteristic X in the classification system is: Gain(D, X) = H(C)-H(C|X). The information gain is for each characteristic. The difference value between the system information volume with Characteristic X and the system information volume without Characteristic X is the information gain brought by such characteristic to the system. For the selection of characteristic each time, the information gain after division of data set for each characteristic value is calculated, and then the characteristic whose information gain is the highest is selected.

If the characteristic value is binary, the information gain brought by characteristic T to the system may be the difference between the original entropy of the system and the conditional entropy after fixed Characteristic T:

$$\begin{aligned} \mathrm{IG(T)} = &- \sum_{i=1}^{n} P(C_i) \log_2 \mathrm{P}(C_i) + P(t) \sum_{i=1}^{n} P(C_i|t) \log_2 \mathrm{P}(C_i|t) + P(\bar{t}) \\ &\times \sum_{i=1}^{n} P(C_i|\bar{t}) \log_2 \mathrm{P}(C_i|\bar{t}) \end{aligned} \tag{8.7}$$

After a round of information gain calculation above, a characteristic will be obtained as the root node of decision tree. Such characteristic has several values, the root node will have several branches, each branch will generate a new data sub-set D_k, the remaining recursive process is repetition of abovementioned process for each D_k till the sub-dataset belongs to the same class.

In the decision tree structure process, the following situation may occur: all characteristics are used up as split characteristic, but the sub-set is not the pure set (the elements in the set do not belong to the same category). In this case, since no more information may be used, "decision by majority" is carried out for these sub-sets. In other words, the category of the highest occurrence frequency in such sub-set is regarded as the node category. After that, such node is regarded as the leaf node.

Although the abovementioned decision tree algorithm is effective, such algorithm is often inclined to the attribute which selects more values. However, in reality, the attribute which selects many values is not optimal sometimes. To improve ID3 algorithm, top priority is given to optimizing the selection standard of attribute. Through the formula weighting of information entropy, the importance of attributes is increased.

4. Support Vector Machine Algorithm

Since the gradient algorithm convergence is slow and the local minimum problem exists simultaneously, in the solution process, the artificial neural network is largely affected by the initialization condition. The initialization parameter may decide whether the optimal solution is convergent. Besides, the initialization parameter can decide whether the algorithm is convergent. Therefore, the initialization parameter leaves great effects on the neural network solution. In addition, in the optimization solution process, the artificial neural network will encounter such problems as gradient explosion/disappearance and convergence to the local minimum value. As a result, the optimal solution of neural network cannot be figured out. Meanwhile, for the special field in the drilling industry, for example, fault diagnosis, the most important thing is knowledge acquisition and experience utilization, but it is difficult to obtain the training sample of neural network.

The support vector machine [8] favorably avoids the defects in the optimization solution of artificial neural network. Based on the structural risk minimization principle, the support vector machine guarantees that the learning model has good generalization capacity. Besides, the optimization algorithm of support vector machine may finally be converted into the convex optimization problem. Therefore, the global optimality of algorithm may be guaranteed, and the local minimum problem in the neural network is avoided. In addition, the support vector machine has a strict theoretical and mathematical foundation, which avoids the experiential element in the neural network.

In the practical application, the support vector machine (SVM) has an outstanding problem: how to select the key parameter in the algorithm, for different parameter combinations decide the learning performance and promotion capacity. At present, most solutions are integrating other algorithms to select the parameter. In essence, SVM is the classification algorithm. The effect depends on the classification goal boundary. However, for the vague and changeable objective function in the drilling engineering, applicability should be further researched.

SVM is a binary classification mode. Its basic model is the linear classifier of the largest interval defined in the characteristic space. SVM also includes kernel function, which enables it to become the non-linear classifier in essence.

Suppose the training dataset of a characteristic space is given:

$$\begin{aligned} D &= \{x_{1,}y_1, \ldots, x_k, y_k, \ldots, x_N, y_N\}, x_k \in R^n, y_k \in R \\ f(x) &= w^T \varphi(x) + b \end{aligned} \tag{8.8}$$

Where: Example x_k is input space, the corresponding y_k is the corresponding label value; *f*(*x*) is non-linear support vector machine, *w* is weight matrix; $\varphi(x)$ is kernel function, the input data is mapped to high-dimensional characteristic space; the non-linear problem is changed into the linear problem, and the original non-linear problem is solved through the method for linear problem after conversion; *b* is offset.

The objective function for SVM algorithm solution is:

$$minJ(w, b, \alpha) = \frac{1}{2}w^T w + \sum_{k=1}^{m} \alpha_k \left[1 - y_k \left(w^T x_k + b\right)\right] \tag{8.9}$$

Where: α_k is Lagrange multiplier, if partial derivative of $J(w, b, \alpha)$ to w and b is 0;

$$\mathrm{w} = \sum_{k=1}^{m} \alpha_k y_k x_k; \quad 0 = \sum_{k=1}^{m} \alpha_k y_k \tag{8.10}$$

After α is solved, w and b are figured out.

8.2.3 Big Data Architecture in Oil & Gas Engineering

The big data analysis can be applied into many complicated scenes, so it is necessary to conduct rapid and exact processing to data. The frequently used data processing tools include Apache Hadoop, Mango DB and Cassandra; wherein, the most extensively applied one in the oil & gas industry.

From mass of data, the big data technology finds out, analyzes and extracts valuable information through real-time recording, and its technological architecture comes from Apache Hadoop basically.

Hadoop comes from the open source item of Apache, and it is an open source, extension type and distributed big data analysis software, and is a kind of software architecture conducting distributed processing to mass of data. This platform is compiled in Java, with strong transportability, and is featured by linear expansion. Hadoop is the most popular big data processing platform at present. Now, the Hadoop has become a complete Eeosystem including the HDFS (Hadoop Distributed File System) [9], distributed database Hbase (HBase and Cassandra), data analysis processing MapReduce, and other functional modules (Table 8.1).

The early version doesn't realize the interface with other high-level languages. However, along with the emerging of Python, R and other scripting languages in the scientific computation field and the development of machine learning, there are simple tools and platforms such as RHadoop, Oryx and OxData, which can be used easily. Now, the Hadoop has been developed to the most popular big data processing platform. To some extent, the Hadoop has become the actual standard of

Table 8.1 Processing platform of main stream big data

Category		Examples
Platform	Local	Hadoop, MapR, Cloudera, Hortonworks, Info sphere big insights, ASTERIX
	Cloud	AWS, Google compute engine, Azure
Database	SQL	Greenplum, Aster data, Vertica
	NoSQL	HBase, Cassandra, MongoDB, Redis
	NewSQL	Spanner, Megastore, Fl
Data warehouse		Hive, Hadoop DB, Hadapt
Data processing	Batch	MapReduce, Dryad
	Stream	Storm, S4, Kafka
Query language		HiveAl, Pig Latin, DryadLINQ, MRQL, SCOPE
Statistic and machine learning		Mahout, Weka, R
Log processing		Splunk, Loggly

big data processing tool. The big data processing which improves the Hadoop and applies it into various scenes has become a new research hot spot. For the Hadoop doesn't require the "schema on load", the Petroleum and Natural Gas Companies can store the data in original way and all formats: PDF file, video, sensor data or structuralized ERP data. Then, it is only necessary to inquire relevant information of each report. This multi-function performance makes the companies able to respond to regulation change or information request flexibly.

The Hadoop is very powerful in the aspect of data warehouse. The Hadoop data is stored and backup at three nodes at least, and the accidents which may occur on node performance will not cause influence on the server, and the data even can be visit without passing the User's permit. However, the Hadoop has obvious insufficiency in dataset and real-time analysis. Now, a good solution is to set up the data warehouse of Hadoop.

At present, some main stream big data processing platforms in the market are complete processing platforms, while some are specialized in specific big data processing applications. But, all of them are expanded based on functions of Hadoop, or provide data interfaces with Hadoop.

The Apache Hadoop is also the most extensively applied data architecture [10, 11] in the oil & gas industry, and no matter in upstream, midstream, downstream or oilfield service, Apache Hadoop can slow down the attenuation of oil & gas output through optimizing production parameters, and it can monitor abnormal fluctuation, utilize parameters to forecast and optimize using prices of exploration and drilling rights, help the oil & gas companies to avoid risks, thus producing petroleum and natural gas in a faster and more economic and efficient way, enhancing output and production benefit. Figure 8.1 refers to the application architecture flow of Apache Hadoop in the petroleum and natural gas industry.

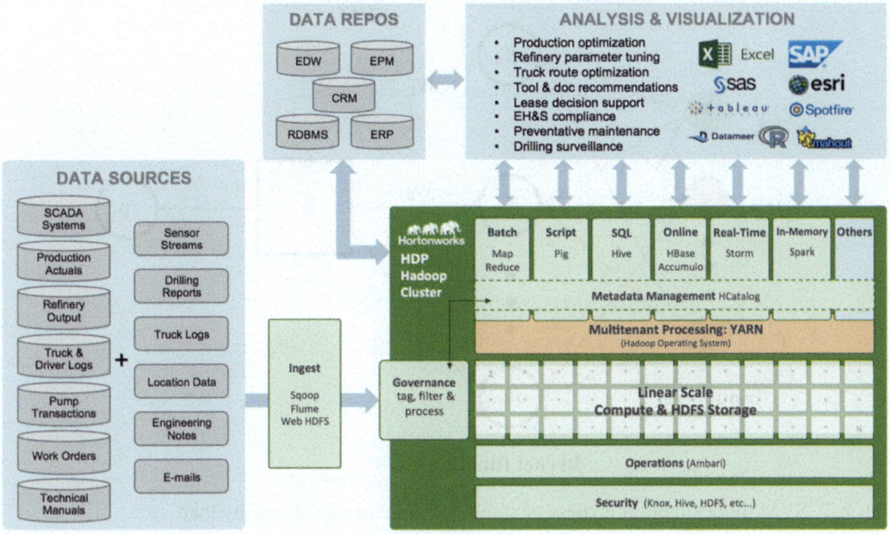

Fig. 8.1 Application of apache hadoop in the petroleum and natural gas industry

8.2.4 Big Data Technology for Reservoir Identification

The oil & gas company integrates the real-time of the drilling platform and office into the earth model, thus identifying reservoir better during the drilling process, so as to help the company formulate future strategies based on existing petroleum economy. The company can comprehend the down-hole formation better with big data tool, thus conducting detailed engineering research to the oil & gas well, optimizing the quantity of well heads required by efficient oil production, optimizing drilling resources, and reducing waste of drilling and well exploration.

1. Natural Gas Hydrate

In the research on the natural gas hydrate, the stability of hydrate and the thickness of the stabilized zone are related to the seawater ionic concentration, gas component, ground temperature gradient, seawater temperature and pressure. At present, there have been many kinds of calculation methods for the thickness of stabilized zone of natural gas hydrate, mainly referring to the phase equilibrium curve and warm pressing equation of natural gas hydrate. In various warm pressing equations, some secondary factors are neglected usually, so the obtained equation is just an approximate result. The machine learning algorithm can obtain the influence of all factors on the stabilized zone of hydrate, thus obtaining precise forecast to the thickness of stabilized zone of natural gas hydrate.

Utilize the phase equilibrium data formed by natural gas hydrate to establish SVM model [12], as shown in Fig. 8.2, and conduct optimization and solution

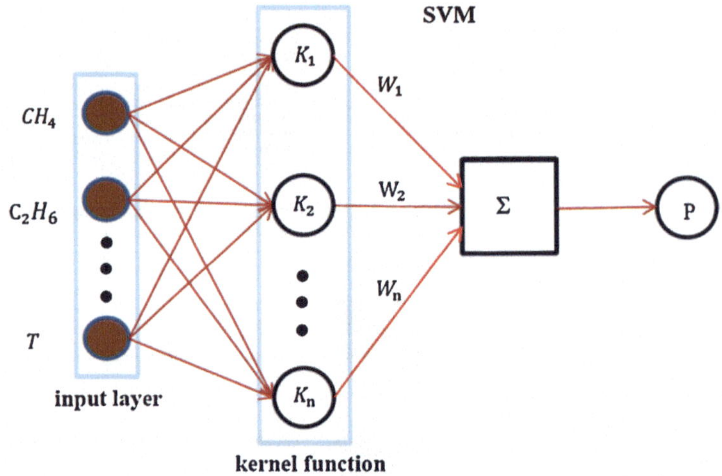

Fig. 8.2 SVM model about thickness of stabilized zone of natural gas hydrate

through the gradient descent method, thus obtaining the required SVM model. Next, the phase equilibrium conditions formed by natural gas hydrate can be forecasted, thus obtaining the pressure value formed by natural gas hydrate at some temperature.

2. Identification of Shale Gas Reservoir

For the geologic structure generated by shale gas is complicated, compared with conventional drilling of oil & gas, the quantity of drilling tasks required by production of shale gas is more. The modeling and analysis on reservoir of shale gas also need mass of dataset. The fracturing technology is a required technology for shale gas production. Although the fracturing technology has obtained remarkable progress, the cost of fracturing is still very high. Seek the most suitable well sections for fracturing, so as to get through the channel of flowing of hydrocarbon, thus reducing the cost in development of shale gas. To be successful in the shale gas exploitation, it is required to identify the reservoir region with the highest productivity, such as the optimal positions of shale gas development and hydraulic fracturing. The data measured on the site is adopted as basic dataset, and through adopting the data mining and machine learning technologies, the optimal position of shale reservoir can be determined, i.e. the region containing high total organic carbon (TOC) and breakable brittle rocks, thus enhancing the fracturing efficiency and recovery rate of shale gas, so as to achieve the purpose of reducing cost and enhancing drilling benefit. When developing the dense shale reservoir, it is required to formulate the optimal drilling strategies in case of utilizing the big data technology, thus reducing the drilling risk in the region where the well may be of low productivity to the greatest extent.

8.3 Big Data in Drilling Process

Along with the constant development of drilling technology, various kinds of measurement technologies and sensors are applied into the drilling process [13], logging and mud logging, seismic (Seismic While Drilling), reservoir parameter, formation information (porosity and pore pressure), etc. all will generate mass of data. Along with the promotion of drilling engineering information, more and more production and management data, field real-time data and tool specification data are acquired into the drilling database, so as to conduct further mining of data, thus forming an achievement which can provide support for decision of the next step finally.

Due to long-term development course, the whole department of the conventional drilling company will adopt different types of databases usually, and different types of databases usually cause the summary and analysis of data unchanged, and also cause difficulty of data inquiry. Targeting the mass of isomeric non-homologous, the analysis on relevant data shall be realized, thus obtaining the implicit relationship among data will be obtained, and the surplus value of data will be mined to guide the drilling operation, and this is a problem which the drilling engineering must be confronted with. The big data technology can be used to analyze the seismic, mud logging & logging data, as well as various kinds of relevant data generated during the drilling process, forecast reservoir characteristics, simulate and shorten drilling period, and enhance drilling safety.

8.3.1 Source of Big Data of Drilling Engineering

The development of computer and communication technology brings new development of petroleum and natural gas industry. Since the 1980–1990s, the digital and intelligent degree of petroleum and natural gas industry is improved constantly. The oilfield digital and intelligent [14] development enhances the exactness and efficiency of decision making, and the correlation among functional units of various well sites is closer, which enhances the petroleum output of petroleum to the greatest extent, reducing the cost of operation and maintenance. Generally, such circumstance of the petroleum and natural gas industry is called "digital revolution".

The innovation of most of software which plays a critical role during the digitalization process of oilfield is provided by the petroleum service suppliers (Halliburton and Schlumberger) and large-sized IT service suppliers (HP, Oracle, Microsoft and IBM). Main global participants of the big data market of the petroleum and natural gas industry are HP, Hitachi data system, IBM, Oracle, Cloudera, EMC, Mapr Technologies and SAP.

The digitalized and intelligent trend of oilfield will cause complexity of data acquisition and processing problems. The data analysis tool must be powerful enough and have high flexibility, so as to make full use of any extra engineering

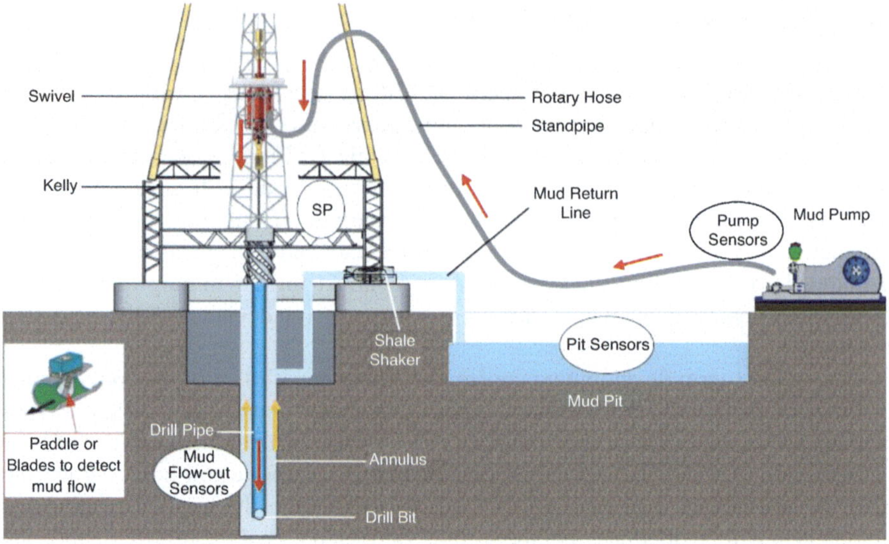

Fig. 8.3 Drilling operations and locations of sensors [13]

and geologic data, thus making the operator get to know the influence of these factors on oil well performance deeply. For these problems, the big data tool may be very useful. The digitalization and intelligence of oilfield rely on the sensor arranged on the well site, as shown in Fig. 8.3, shows drilling operations and possible sensor locations. This is also main source of big data of petroleum engineering.

The petroleum and natural gas industry are the largest consumer of different types of sensors all over the world [13]. The digitalized process of oilfield, including remote monitoring, modeling and management system, and real-time optimization and control of drilling, needs a large number of sensors. There have been tens of thousands of sensors monitors on the site. The sensors located down-hole, installed at the well head, along with stream line or inside the technological equipment such as those in the oil & gas exploitation, processing and transportation system will transmit a series of uninterrupted data. The operator will receive real-time, discontinuous and discrete field data, and extract the temperature, pressure, flow velocity or other measured values, so as to determine the states of down-hole and ground systems. The data of sensor is generated continuously, and it is a real-time data source. In addition, new data will be obtained constantly along with the production of oil well.

For example, the optical fiber sensor [15–17] in the oil well can be used to measure the temperature, pressure and other parameters of each oil well. The sensor can be utilized to control the performance and status of pipeline in a real-time way. In practice, the most frequently used tool is pipeline visualization instrument, which is used to monitor the states of mechanical and physical characteristics in various points of pipeline. The daily data volume of single well received from such type of sensor can reach several megabytes. With the appearance of down-hole and ground sensors and instruments for optimizing system performances, the operation personnel are confronted with processing and management of mass of dataflow generated by these systems. Meanwhile, in the past few years, with the rapid development of non-conventional oil & gas reservoir, the quantity of drilled wells is huge. The field measurement, such as mud logging, is one of the most indispensable tools which provide mass of information and data for such non-conventional reservoir (shale gas). There are usually several hundreds of oil wells in a large-sized oilfield, and the mass of database generated by these wells are stored in the database. These data can be processes as resulting force forecast model of oil well, so as to evaluate the production performance of oil well. It is very important to judge and analyze main factors to production performance of reservoir according to the data generated by such several hundreds of wells, and this is related to the Company's operation cost.

8.3.2 Structural Characteristics of Drilling Engineering Data

The big data in the petroleum engineering possesses typical diversification characteristics. Its diversification means various types of generated, stored and analyzed data. The types of data recording equipment and sensor are different, so the generated data can possess different sizes and formats. The formats of generated data can be text, image, audio or video. The classification can be completed in a more technical way, such as structural, semi-structural and non-structural data. Meanwhile, there will be continuous real-time data flow, and these data comes from well head, drilling equipment (EDR, LWD, MWD and mud logging) and flow and drilling pressure sensing. Most of generated oil & gas data from ground and underground facilities, drilling data and production data refers to structural data. These data can be time series data, which is recorded through a certain time interval. Another source of structural data includes assets, risk and project management report. The structural data is usually processed with specific application procedure. These applications include management investigation, processing and imaging, exploration planning, oil reservoir modeling, production and other upstream activities, and there will be external structural data sources, such as market price and weather data, which can be used for forecast. The structural data types include PPDM, SEG-Y, WITSML, PRODML, RESML, etc. However, during the above-mentioned process, there will be mass of non-structural data, such as images, log curves, well log, maps, audio, video, etc. The non-structural data sources of oil & gas industry include drilling log, daily written report and CAD mapping of drilling, various types of data volumes, achievement files, maps and statements, instrument data, etc., and for non-structural data, the overall management is relatively weak; in the petroleum and natural gas industry, there are various kinds of experimental and computer simulation practices, which can generate data for further analysis. These data can be classified into semi-structural data.

The drilling database refers to multi-layer table, i.e. the well information table is regarded as root; through establishing step-by-step association of external key, the described object will be detailed constantly. In the drilling database, most of data tables refer to description of level progress of drilling tool combination, well cementing and drilling fluid performance. The differences only lie in breadth and depth. Based on purpose, the data in the drilling engineering can be divided into three categories, drilling site level, oilfield block level and group company level. For the data collected from multiple sources has the well-known complexity, usually, due to increasing of data amount collected at a high speed, the data analysis is to be conducted via efficient processing mechanism. It is very difficult to utilize these data effectively in traditional data processing way, while the big data technology is applicable to processing of these diversified data.

8.4 Application of Big Data Technology in Drilling Engineering

At present, most of data in the oil & gas industry is applied into the control and monitoring, not to optimize the assets performance. If the oil & gas industry can analyze and comprehend all data generated by itself, the operation efficiency can be enhanced by 20%. The research of Bain Capital shows that the bid data analysis can enhance the performances of oilfield and factor by 6–8% [18]. For now, the oil price and the economy are in downturn, and the profit gaining capability at the upstream will be challenged, which makes the completion among oil & gas companies more and more violent and cruel. How to reduce the operation cost of upstream enterprise, enhance production efficiency and avoid environmental risk? This is a subject that each oil & gas company must be confronted with.

The big data technology can be used to treat mass of isomeric data easily, thus providing valuable reference information for the oil & gas industry, so as to help the operation enterprise make a decision better. The digital technology can change the upstream business, and create extra profit from existing production capacity, thus making the industry more productive and flexible.

8.4.1 *Design of Drilling Fluid*

With regard to design of drilling fluid, spacer, cement slurry and fracturing fluid, in the traditional flow, the design process of fluid is shown as below generally: firstly, a preliminary fluid formula design will be proposed according to personal experience of the engineer; and then, the performance testing will be conducted in the laboratory, and after multiple times of repeated tests, an ideal fluid design will be obtained. This process relies on the prior knowledge of the Engineer. The trial and error process needs mass of time and cost; meanwhile, the mass of sample data generated by trial and error is also difficult to be processed for the experimental personnel.

Based on the machine learning algorithm, the artificial neural network can be adopted to make the process of fluid design intelligent, so as to save cost. The intelligence is mainly divided into four processes: (1) prepare data, and select dataset of data mining from historical data record; (2) conduct pre-processing of data, and enhance the model training efficiency; (3) conduct dataset calculation; (4) conduct mining verification, compare the original data set which hasn't been mined with the mining result, so as to judge whether the data mining can be forecasted; (5) explain the mining result, and utilize the mining result for forecast. Take raw materials of different drilling fluids as the data input set, and after training, the influence law of different factors on drilling fluid performance can be obtained.

8.4.2 Big Data and Non-conventional Drilling

The development of digital technology enables the energy development of non-conventional reservoir. Although the development of petroleum drilling technology has reduced the development cost of non-conventional reservoir, the oil & gas exploitation of non-conventional reservoir is still confronted with the risks of huge investment and easy occurrence of drilling accident.

The method for enhancing recovery rate [19] via hydraulic fracturing has been widely applied into increasing of both production and income. Only designing the well completion better in a scientific way and enhancing effectiveness of hydraulic fracturing strategies can the oil & gas income increasing be enhanced. For the hydraulic fracturing and drilling & well completion, it is necessary to get to know underground geologic conditions more deeply. The oil well production capacity cannot be enhanced unless the reservoir characteristics are improved, and the drilling, well completion and production capacity increasing methods are optimized by applying formation rock physics and geo-mechanical information.

It is very important to utilize a relatively small amount of capital to obtain higher production effect. The traditional mud logging method is featured by high cost and high risk, while a precise high-resolution geologic model can be established by utilizing existing industrial data and relying on big data technology. For the non-uniformity of non-conventional reservoir geology, the exploitation needs establishing reservoir model of higher precision and resolution. The conventional well completion methods and fracturing strategies are not practical or effective in these unforeseen heterogeneous environments. Along with the borehole, in case of meeting rocks which influence fracturing efficiency, the production effect of oil well will be poor.

The big data is a very favorable tool for drilling of non-conventional reservoirs and under non-conventional conditions [20]. After utilizing the big data to analyze various relevant oil well datasets, such as rock debris analysis, gas detection, tracing measurement and real-time pump pressure, the reason for difference between the actual oil well production capacity and predicated production capacity can be known in an all-round way. After applying the big data technology into the oil well production, the relationship among well completion, production increasing processing and the final oil well production capacity can be analyzed. Finally, according to the analysis result of big data, adjust the stage positions and dimensions, while optimizing the cluster position, so as to reduce the difference in closure stress among inter-stage formations to the greatest extent. Optimize the stimulated rock volume, enhance the percentage of production potential, and enhance the fracturing benefit; meanwhile, the formation fracturing pressure, reservoir ductility and geologic anomaly of formation can be known clearly. This makes operation personnel minimize the anomaly during the fracturing process, and make sure that the loading capacity of propping agent is maximized, thus finally guaranteeing the fracturing effect.

8.4.3 Fault Identification and Equipment Maintenance

Nowadays, major oil companies are mining big data through the micro-sensor installed on the production equipment and they hope to save the costs by avoiding shutdown of drilling machine, identifying potential safety hazards and improving the supply management [21]. For example, Chevron's Tengiz oil field located at Kazakhstan expects to include about one million sensors.

8.4.3.1 Fault Diagnosis of Drilling Mechanical Equipment

Due to the complexity and concealment of stratum, frequent relocations of the drilling machine as well as poor fieldwork conditions and other features, the fault diagnosis of the drilling machine is more difficult than the fault diagnosis of other enterprise mechanical equipment. As for the down-hole drilling tools, it cannot make direct monitoring on the state of the equipment or make use of the data collected from the pump and well, such as drilling pressure, pumping pressure, pump flow, rotation speed, drilling speed and torque parameters to predict the operation conditions of the tools. That is, during the drilling process, there are many complex and uncertain influencing factors. It is very difficult to utilize precise modeling to establish the mathematical model suitable for the practical drilling process and then make predictions to the faults and equipment maintenance during the drilling process.

The development of the data mining and artificial intelligence theory allows people to make use of the actual input and output data of the drilling system and the rich knowledge and experience of related experts to establish the model not seriously relying on the internal mechanism of the drilling system. Big data has been used for the intelligent drilling platform and pipeline infrastructure to realize the problem prediction and fault prevention [22].

In this aspect, the most commonly used big data algorithm is the decision tree algorithm and neural network model. Take the data records at the time of historical overhaul or fault of the drilling equipment as the feature parameters and take the major factors influencing the fault and overhaul of the drilling equipment as the input factors to make the adaptive learning and establish the model of fault feature, fault and maintenance method. Make use of such model to get the maintenance program and then realize the fault judgment on the automation of the drilling process.

8.4.3.2 Preventive Equipment Maintenance

Generally speaking, the downtime of the oil gas industry is as high as 10%, equivalent to three times of the average value of American industry. The equipment fault occurred in the oil gas industry leads to great economic losses. Big data occurred in the oil gas industry forms the "Industrial Internet", which combines the machine learning and extraction of intelligent data in nature to improve the quality of the preventive asset maintenance.

In the oil & gas industry, the preventive maintenance of the equipment is not a new concept. The traditional method is to record the non-fault history of the machine as the reference sample to monitor the pressure, volume, temperature and other features of the operation of the new machine, and then compare it with the historical data to predict the potential fault. As for the upstream of the oil & gas industry, the equipment maintenance is also very important, especially whose professional equipment with expensive costs.

The development of the well site digitization makes the sensor applied into each position of the drilling equipment. The sensor can collect the data from the pump, well and other equipment. It has higher frequency and lower cost than manually collecting the same data. Transmit the monitoring data of the equipment to the big data processing software to make the processing and analysis. It also can be enriched with other data stream related with the weather, seismic activity and social media sentiment to more completely describe the conditions occurred on the site. Understand how the equipment is influenced by the pressure, volume, impact, vibration and other variables during the maintenance interval by the algorithm parsing of the large data set in Hadoop to adjust the maintenance plan, prevent the occurrence of the fault or make fault prediction, get the prediction results of the equipment operation conditions, compare it with the actual results of the equipment, and then judge whether the fault time is earlier than the expected time. When it predicts any fault, judge the time of the maintenance detection according to the on-site operation state, optimize the on-site dispatching and make the team arrange the on-site equipment maintenance more effectively. Guide the workers to maintain or replace the machine as per the results of the fault analysis to avoid the losses due to the equipment fault.

8.4.4 Optimization of Drilling

As for the upstream enterprise, the expenditure of the drilling costs per hour is very huge. Shortening the drilling time can cut down much cost for the emprise. Big data technology can excavate the potential of the cost optimization in the drilling operation; thus, it is very important.

In the traditional sense, the engineer makes the analysis for the daily drilling report generally, and then adjusts the drilling performance. Oil companies need to make real-time and continuous monitoring on each link to make use of 2D, 3D and 4D seismic technologies to make the geologic modeling simulation for the drilling operation and acquire accurate seismic data or the operation conditions of the equipment asset. This is usually completed by plenty of sensors distributed in the underground boreholes and surface facilities. Therefore, it needs thousands of sensors distributed in the whole drilling process. The underground sensor or the sensor on the ground can record the information of the drilling process continuously in real time. The drilling operation process will generate lots of data. These data is of complex form, thus its explanation and analysis are very difficult. The daily data flow of Chevrolet exceeds 1.5 TB.

The big data technology can avoid the disadvantages of the traditional analysis method, realize the real-time analysis of the drilling monitoring data and then optimize the drilling operation performance, improve the drilling efficiency, evade the drilling risk, shorten the drilling time and reduce the costs. Integrate the ground logging data and report information (such as daily drilling report and bit report) together and measure the lithology, trajectory and other key data during the drilling to create a complete set of drilling performance analysis.

The automation [23, 24] of drilling process management is the trend for the drilling industry. The automation of drilling process needs to make in-depth analysis for the data in nearly real time and then make the adjustment for various drilling parameters. The predictive analysis on the data generated during the real-time drilling process is very important. The big data technology must generate lots of data during the real-time process and make the real-time handling. Once it is lagged behind, the control on the drilling will be delayed and bring uncontrollable dangers. Therefore, the researches on the application of the big data technology in the real-time drilling are very special.

The mechanical drilling rate is an important indicator for the drilling engineering. The drilling rate of the bore is affected by the layer, lithology, rock compressive strength and other geological factors and meanwhile influenced by the bit type, drilling parameter, drilling fluid density and other engineering factors. Make use of the neural network algorithm and take the major factors influencing the drilling rate as the neuron. As shown in Fig. 8.4, it may be influenced by the drilling rate prediction and other factors, mainly including the geological factors and engineering factors, wherein the geological factors include the layer, lithology, rock compressive strength and other objective factors, and the engineering factors include bit type, drilling parameter, drilling fluid density and other factors subject to human

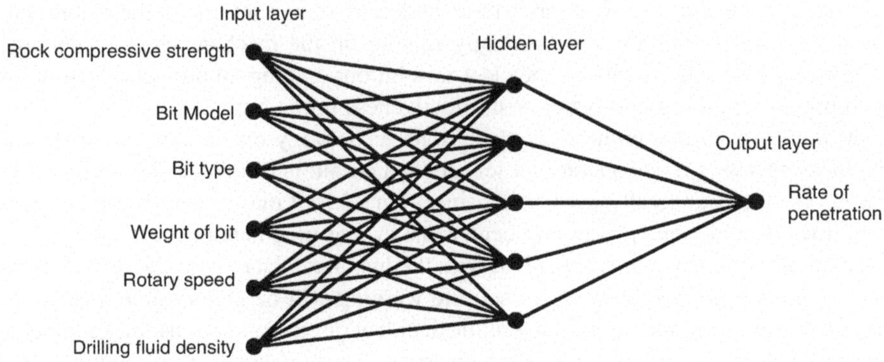

Fig. 8.4 Drilling rate prediction model based on neural network

interventions. Take the major factors affecting the drilling rate as the neuron, including the rock compressive strength, bit size and bit type (roller bit, PDC bit and TSP bit), pump pressure, rotary speed and drilling fluid density. Make use of the neural network algorithm to establish the prediction method of the bore mechanical drilling rate.

8.4.5 Drilling Safety

The demands on the drilling safety are very critical. Mexican oil spill event [25] may remain fresh in our memory till now. The Secretary General of China Chemical Safety Association (CCSA) Lu Nianming stated that: "In recent years, the Chinese government pays high attention to the problem of safety in the oil refining and petrochemical industry." Although the total number of the accidents happened in Chinese oil refining, petrochemical and other chemical industries is decreased greatly, the total number is still great. Chinese society has low tolerance on the worker safety. Lu Nianming stated that: "Therefore, although the profits have declined, the petrochemical industry still has to increase the investment on the safety measures." "The monitoring of the government is improved. When the accident happens, the facility will be immediately ordered to halt production."

60%–80% of the industrial accidents are caused by human errors. The digital management system can effectively reduce the occurrence of the accidents. The advantages of the digital and comprehensive data management of the oil & gas system on the safety production are obvious. However, it also will generate new dangers requiring management. Especially the network security problem, using the sensor and monitoring technology to improve the safety is an important research topic.

The digital technology will pay a more and more important role in the natural gas industry. Even in the areas traditionally relying on the mechanical and hydraulic safety may increasingly rely on the electric solutions and digital protection solutions with higher efficiency and better cost effectiveness in the future.

Making use of the application of big data technology on the field of oil & gas engineering safety relies on various sensors and monitors installed. These sensors in special positions generally work unceasingly for 24 h. It may generate one constant data flow. The big data processing center makes real-time handling and analysis on the data of the sensor and compares it with the historical data. Once the data is usual and deviates from the safety range set, the warning will be given immediately. No matter it is the equipment fault, downhole accident or other situations, make the data analysis on the reasons of these problems to get the influencing factors with certain features. These features may indirectly cause the accidents, or are directly related with the accident. The big data analysis could get the safety threshold of these parameters. Set the safety threshold into the monitoring system for the real-time processing system of big data as the fixed point. Once the value monitored by the sensor deviates from the fixed point, give the warning and the engineer will adopt corresponding measures according to the situations monitored. For example, the monitoring equipment and sensors installed at the wellhead can collect the downhole information in real time. The real-time capacity of the big data technology can identify the abnormalities during the drilling process, monitor the downhole abnormalities, and make timely processing to prevent the well leakage and blowout. Make the prediction and analysis to reduce the non-production time, save the equipment losses during the drilling and well completion process and optimize the asset performance and production. The big data technology can make the drilling and well completion process more intelligent. Make use of the big data technology to analyze the problems possibly occurred during the drilling and well completion process to make the decision and processing plan in advance [26]. Monitor the drilling process and record any other change due to the gas leakage, water channel, pressure change and seismic activity.

Take the real-time oil well maintenance as an example and calibrate the safe values or ideal values of the pump pressure, rotary speed, flow rate and temperature of the oil well under normal production. These values are usually obtained by the mining analysis of the historical data. The big data processing system Hadoop makes the real-time analysis and warning on these feature values. Once these set values deviate from the predefined scope, corrective measures shall be adopted.

The application of the big data in the safety field of petroleum engineering is in multiple aspects. Get the tome nodes and environmental conditions that are easy to be damaged for the drilling tool by mining the historical drilling data and then prepare for the maintenance and accident handling; Devon Energy integrates Hadoop, SAS and the text data, confirm the reasons for the non-production time and then reduce the non-production time by 30%. By reducing the affected non-production time, petroleum and petrochemical companies can save USD 500,000–1,000,000 per day in average.

8.5 Big Data in Oil & Gas Production

In addition, the big data analysis can improve the oil production by 6–8% [18]. In fact, the estimations of Chevron Corporation show that, the completely optimized digital oil filed can improve 8% of the productivity and 6% of the gross recovery.

ShellGas Corporation cooperates with Hewlett-Packard and Amazon WebServices and makes use of the big data to reduce the oil exploitation costs. They utilizes the data stream collected and transmitted by optical fiber as well as the big data technology to predict the geographic space data, petroleum and natural gas report and make use of these water to acquire to accurate downhole information. It can evaluate the layer, develop new prospect and offer advantages for the competitive bidding. It also can improve the recovery ratio of the current oil well by analyzing the earthquake, drilling and production data. Such information is helpful for confirming when to change the oil deposit and when to change the oil extraction method. While the could computing and big data are also helpful for more accurately predicting the oil production. The big data technology is used to analyze and predict the oil & gas price trend to guide the production, help the oil & gas companies to realize the optimization of the resource allocation and then acquire the optimal cost control and profit value. The world's largest listed oil service group Schlumberger released new software named del-fi. It improves the production of the whole oilfield to the largest extent by improving the design of the oil well and the coordination of the well drilling. Only this new system expects to reduce 40% of the production costs in the future 10 years.

As for the oil & gas production, selecting the storage layer and production area with high productivity is of great necessity. Big data can help analyze the oil& gas reservoir, detect its non-optimal area and select the production scheme and prediction result by. It can make the predictive analysis on the changes on the storage layer of the oil deposit, offer intelligent analysis for the oil deposit engineer by integrating and analyzing the seismic, drilling and production data and then improve the recovery ratio. Meanwhile, predict the performance of the storage layer on the basis of the historical data, identify the production area inferior to the related standards, mark the aged well not satisfying the expected output threshold and then make the repairing immediately.

70% of global oil & gas comes from mature oil wells. The improvement in the analysis technology of sensor big data has made significant progress in enhancing the recovery ratio. However, there is still great potential in this area. Even if adding 1% of the output on currently active oil & gas factories, it also can increase 2 years for the global oil & gas supplies.

At present, as the discovery of the new oilfield becomes more and more difficult, how to ensure the output of the current oil wells is a problem that oil companies must be faced with. Decline curve analysis (DCA) [27, 28] makes use of the past output of the oil well to estimate the future output. The decline curve most commonly used to represent or extrapolate the production data are members of a hyperbolic family defined by the following differential equation [28].

$$\frac{d}{dt}\left(\frac{q}{dq/dt}\right) = -b \tag{8.11}$$

where: q is the oil production rate. b is the reservoir Factor.

This method has many advantages. The decline of the oil well when its service life finishes follows the nonlinear model. Generally the oil & gas production declines faster with the time. How to optimize the production parameters is the intelligent parameter (such as pressure, flow rate and thermal properties of the injection fluid mixture) management of prolonging the service life of the oil well to the largest extent. The big data can be utilized into the exploration and development process of the oil & gas and then help the managers and experts to make the decision on the drilling strategies. Make use of the big data technology to analyze the historical drilling data and production data, analyze the current drilling process, establish and evaluate the drilling model on the basis of the historical drilling data, and adjust and optimize the parameters of the drilling machine during the drilling process in real time.

In the later development stage of the oil deposit, making the directional windowing at the original well position to make the sidetracking to the oil layer is an effective oil increase method. Generally it needs the engineer to determine to utilize which well to make the sidetracking according to its experiences and related data. This traditional method has a big workload and strong subjectivity and it is unfavorable for the promotion and utilization. Take the oil well-related data implemented the sidetracking as the data source to train the neural network and it can predict the sidetracking effect. The data technology also can be applied in product optimization. For example, if the company integrates the enterprise data and production data, it will be convenient for the workers at the well field to offer the prediction of the layer information and convenient for the real-time decision-making. With the help of the big data analysis, optimize the oilfield depletion planning and de-risk the drilling to maximize the investment returns.

8.6 Big Data in Downstream Oil & Gas Industry

After the exploitation of the oil & gas, it needs to be transmitted by the oil & gas pipeline to the market. The oil & gas leakage during the transportation will bring lots of economic loses and safety risks. The midstream companies apply the big data technology into the leakage detection, warning and emergency stop, etc. The Internet of Things (IOT) is thoroughly changing the midstream pipeline operation through the environmental monitoring and infrastructure management application based on SCADA and by the monitoring and control on the operation of the pipeline infrastructure. The big data is applied into the real-time oil condition monitoring (OCM) technology [29]. It is helpful to product the midstream compressor. A reduction of

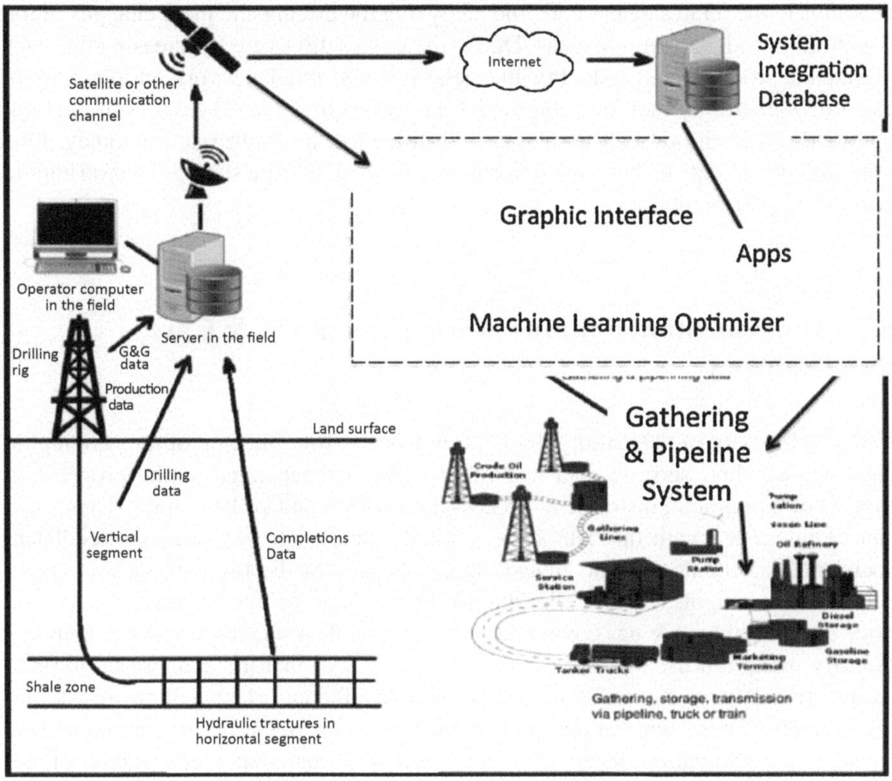

Fig. 8.5 The internet of things provides the nervous system for the PALM brain to provide near instantaneous cause-and-effect responses to problems detected in the field, and optimizing operations [30]

0.5% in the efficiency of the compressor may cause the compression operation of the natural gas lose USD 180,000 per year.

Recently, a kind of analysis system based on machine learning is named petroleum analysis learning machine (PALM) [30], as shown in Fig. 8.5. It is used to simultaneously analyze hundreds of properties of the Internet of Things of hundreds of horizontal wells in real time, including thousands of hydro fracture stages.

The data analysis may help monitor the pipeline and equipment and offer more predictable and accurate maintenance method. For example, the sensor may indicate when the equipment suffers usual pressure and it allows the operator to implement the preventive shutdown or intervention to avoid any accident or leakage. For example, the leading compressor manufacturer develops the customized sensor model and uses the predictive analysis software to actively monitor the readings offered by these sensors. It is helpful for the preventive maintenance of the equipment arranged for the midstream customers.

Using big data analysis in the processing, logistics and sales (downstream) expects to bring breakthroughs. The big data is helpful for predicting the oil product demands in the retail sale network and analyzing the pricing and price changes of the competitors and competitive areas. Due to the modes discovered, increasing the sales of related products and reducing the delays in the retail network (through more accurate logistics by fuel transport vehicle) may be attractive. The prices of the bulk commodities in the oil & gas market will influence the profit rate obviously, thus utilizing the big data technology to formulate the expenditure strategy measuring the costs is of great importance.

8.7 Implementation Status of Big Data in Oil & Gas Industry

The development of the drilling digitization leads to fold increase of the geographic location, weather, seismic data and other new data generated in the progress in aspects of equipment, instrument, process automation and collaboration. These data can be integrated with the "human generated" data, such as the market feedback, social media, email, text and image, and make use of the big data technology to acquire new insights. Meanwhile, due to the decrease of the oil price, the low oil price hampers the oil & gas companies investing on data scientists who can help this industry make full use of the big data; others consider that the collapse in oil price really creates more powerful demands for lots of data-oriented innovations and investments. These innovations and investments can improve the operation efficiency and automation level. It expects that the status quo and future of the exploration and production for the oil & gas industry will heavily rely on the big data and Internet of Things [21]. According to the trend investigation of Accenture and Microsoft digital energy in 2016, 56% of the executives in the oil industry consider that, in future 3–5 years, the utilization of big data analysis will be the key component of their business strategies.

The big data seems to have bright prospects in the oil & gas industry. However, currently the big data lacks the business support and the cognition on the big data within the industry is not deep. The data quality and the understanding on the problem complexity are also parts of the challenges that the big data application is faced with. Related analysis shows that, the application of big data analysis in the oil & gas industry is still in the experiment level, this industry lags behind in the aspect of adopting the big data and its application in the oil & gas companies is still in the starting stage. In addition, as the research hotspot in current Internet field, the technical accumulation and reserve of big data exist in large Internet companies, such as IBM, Google, Oracle, Microsoft and other oil & gas service companies Schlumberger, Halliburton,etc.

1. Chevron applies Hadoop to make the seismic data processing (IBM Big Insights) [31]. At present, the internal flow generated by Chevron per day exceeds 1.5 MB.
2. Royal Dutch Shell implements a pilot program applying Hadoop for the data collection of seismic sensor and puts the optical cable sensor into the well to

measure the data. The collected data is used to analyze the working mode of the oil well and the quantity of the remaining oil/gas. Mass data from the sensor is located in Amazon private cloud (Amazon VPC). Shell plans to deploy the optical cable within 10,000 oil wells in the future.

3. Газпромнефть analyzes the oil well operations applying Teradata ASTER system within the framework of big data pilot program completed in 2015 [32]. Such analysis uses the voltage reconnection logging data in about 200 million logging data and accident logs recorded by 1649 wells in 2014. The analysis result shows that, the utilization of big data can acquire a new cognition, and the interdependence relation previously unknown occurs in the pumping equipment.
4. Intertek is cooperating with British Robert Gordon University for a research program, which is aimed at helping the oil companies make use of the big data, improve the asset performance, raise the efficiency, reduce the operation cost and ensure safe operation.
5. Halliburton Company is applying the big data technology to optimize the seismic space, drilling space and oil well planning.
6. Schlumberger insists in applying the digital technology for creation and accelerating the continuous improvement. Schlumberger released DELFI cognitive E&P environment in September 2017. Schlumberger Petrel E&P software platform and Intersect high resolution oil deposit simulator are integrated into the Google Cloud Platform in Delfi environment. It is a safe, cognitive and cloud-based coordination environment, which makes use of the data analysis, machine learning and high performance computing and integrates the data, workflow and digital technology (such as artificial intelligence and machine learning). Offering the seamless integration technology among physical geography, geology, reservoir engineering, drilling and production area provides a new working mode for the asset team and realizes the coordination among teams.
7. Saudi Aramco adopts the method of big data analysis to develop the automatic system monitoring the well kick. This system adopts 5 kinds of big data machine learning algorithms to monitor the well kick, including Decision Tree, K-Nearest Neighbor (KNN), Sequential Minimum Optimization (SMO), Artificial Neural Network (ANN) and Bayesian Network. These models use the ground, drilling and logging parameters to predict the well kick, including pressure, flow, hanging load, drilling time, torque, pump speed, and bit pressure, etc. Mark "with well kick" or "without well kick" on each parameter set of the sample set collected, then adopt five models to learn these samples separately and finally make the anticipatory computing for each parameter set of the test set. It confirms the best monitoring model is the decision tree and K-nearest neighbor methods after assessment, and the prediction accuracies of these two ones both reach over 90%.

It must be definite that all data has a value. In addition, once the data is utilized, the seemingly un-intuitive pattern relationships in the data can be realized. As for an industry established on the basis of the data, selecting high-quality data from low-quality data is a difficult and time-consuming challenge for a long time. Possessing an ecosystem supporting all tool processing data volumes is also crucial.

The supplier solutions in the big data field of oil & gas industry are still in the early stage.

At present, Schlumberger is the leader in the big data technology of the drilling industry. Schlumberger cooperates with Microsoft, Google and other Internet companies and makes use of the host computers and cloud infrastructure of these companies to bring the expansibility and safety of the data processing. Schlumberger has over 250 software engineers working in American, Norwegian and Chinese drilling software development centers. The technical experts and data scientists of this company concentrate on various software technologies, including high-performance computing, cloud, big data, analytics, Internet of things (IOT), Industrial Internet, visualization and user experience, and offer digital supports for the drilling services of Schlumberger.

8.8 Main Challenge of Big Data Strategy

The development time of the big data in the drilling engineering field is not long. The integration of big data and traditional IT system is another problem for the big data in the field of drilling engineering. The big data infrastructure is established in isolation with the traditional IT system. The data moves from the master warehouse to HDF for analysis, and then moves from HDF to the master warehouse. This isolation causes the increase in the operation cost and the low efficiency and waste of the resources. In view of this, some researchers state that, large-scale development of the big data may not happen in the gas & oil industry. Although the oil & gas industry already understands the value of the data, how to make correct data management for mass data generated in current oil & gas industry is still the problem that the related personnel must solve. The correct data management directly affects the business efficiency as well as the success and risk of the new program. Establishing the big data development strategy of the oil & gas industry requires considering the problems existed in the data use field.

One of the main challenges of the big data in any industry including the oil & gas industry is the costs related with the management of the data records, storage and analysis. The data collection of the oil & gas industry costs lots of funds, but the data is not maintained. Most companies have no structured data management method. In addition, most enterprises in the oil & gas industry usually take the data as the descriptions and records of their assets or investment process, but not consider the data as the asset and just ignore the inner value of the data. Generally, there are lots of unprocessed old data archived in the oil & gas industry. With the big data technology, they can be transferred into more valuable knowledge. However, only the big data leaders (social network providers) of this industry are aware of the significance of taking the data as the underlying asset. The data shall be taken as the asset and incorporated into the operation philosophy of the organizations in the oil & gas industry, define which data is valuable and apply the overall data evaluation method. The data management shall be more highly regarded.

The oil & gas industry has mass data, though these data belongs to different fields. The value of the data extracted from the original data field (marketing, finance, manufacturing, etc.) is of great importance and difficulty. Although there are considerable researches, it has not achieved any actual effect.

In addition, integrating the digital tools and equipment into single system, adopting the comprehensive method for capital-intensive new programs for data collection and improving the application of the intelligent tools during production process are all problems that the big data in the oil & gas industry must be faced with. The biggest challenge for utilizing the big data in the oil & gas industry is lacking the consciousness and business supports. Other challenges discovered in the investigation include the data-related decision, lacing proficient personnel and the cost of the big data infrastructure. Therefore, let the staff and executive personnel be familiar with this technology and its application will greatly accelerate the implementation of the big data in the oil & gas industry.

8.9 Looking in the Future

Breakthroughs are happening in the oil & gas industry. This is not only the progress in aspects of the new downhole tools, designs of the drilling machine and operation procedures, but it is the revolutionary beginning of the overall trend in the oil & gas industry. The rapid progress in the big data management, digital connection and high-performance computing will change most costs and risks in the upstream programs concerning drilling and oil well construction and then inevitably change the game rules in overall health, safety, environment, efficiency and the financial performance of oil well. The application of big data is called "the dawn of the new era for the oil & gas industry" and "the fourth industrial revolution". Although the application of the big data technology in the oil & gas industry is still in the starting phase, it displays extraordinary potentials.

8.9.1 Big Data Awareness

As previously mentioned, the big data seems to have bright prospects in the oil & gas industry. One investigation made by IDC Energy finds out that, currently the big data lacks the business support, the cognition of the big data within the industry is not profound and the investigation in the oil & gas industry still lags behind. Among the professionals accepting the investigation in the oil & gas industry, only 36% of them plan to make major or moderate investment on the big data and analysis this year. The oil & gas companies must understand the concept of "big data" and its potential application and clearly know which technologies need to be updated. The main problem here is defining which data is valuable and analyzing it to acquire the market advantages and keep the competitive power.

Currently the oil & gas industry lacks the investment and construction on the big data infrastructure, and meanwhile the big data practitioners and the big data engineers in the oil & gas industry are also insufficient. Therefore, let the staff and executive personnel be familiar with this technology and its application will greatly accelerate the implementation of the big data in the oil & gas industry. In addition, although some oil & gas companies have paid attention to the big data technology, but the investments and researches in this aspect are still not enough. The facts prove that, the oil & gas industry is cost-oriented. If the big data technology can prove its capacity in reducing the operation cost, the oil & gas industry will increase its attention on it greatly. The oil & gas companies shall lay emphasis on the patent layout and technology layout of the big data technology, and remain invincible in the future ware of big data.

8.9.2 Data Processing of Sensor

With the development of the digital oilfield, more and more data appear in the oil & gas industry and various sensors and recording equipment generate millions of data per day. One key challenge that the digital oilfield is faced with is transmitting the data from the oilfield to the data processing facility according to the data type, data volume and data protocol. Integrate the digital oilfield plan with the big data analysis to realize better comprehensive operation and working process. This type of progress in the big data aspect expects to lead to lower operation cost, lower descending risk and less uncertainty and shorten the non-production time.

In fact, no matter it is the oil & gas industry or other industries, how to reduce the costs of managing the data records, reservoir and analysis is a challenge that the big data technology must be faced with. With the recent technical progress, fog computing, cloud computing and Internet of Things have applied to solve the problems related with data storage and computing. The cloud computing is connecting the off-site high speed computer with the on-site computing system to process the on-site situations. Amazon WebServices's Oil &Gas Director of Global Business Development of Arno van den Haak stated: "With the beginning of the cloud operation, we completely see the innovation ability of our industry now, not several years or several decades, but several weeks." An operator using this "innovate fast, fail fast" approach is, according to van den Haak, Australia's Woodside Energy. The company has fully embraced cloud computing capabilities in its daily operations.

The big data in the oil & gas industry, especially in the upstream field, relies on the sensor arranged in the well site. The precision of the big data processing is limited by the frequency and the data quality of the data recording sensor. However, during the process of collecting mass data, the capacity of trusting its effectiveness and source becomes more and more difficult. The oil engineering expert shall cooperate with the data scientist and correctly apply the big data tools to solve various problems in the oil engineering field. Each company shall develop its special big data tools, including data recording and storage facilities as well as the data analysis tools. This may reduce the software ownership cost and optimize the value of recording the data.

8.9.3 Interdisciplinary Collaboration

The process and decision related with the oil & gas exploration, development and production generate mass data. At present, merely in America, there are about 1,000,000 oil wells producing oil and/or gas, and there are more oil well monitoring performances. The data volume increase every day. The oil & gas industry realizes that, using the data in a more intelligent and convenient way may find out strong power and impending breakthroughs in these data. The oil & gas industry is a traditional industry, while the big data technology is an emerging Internet technology. The big data analysis adopts the statistical analysis, data mining, machine learning, simulation model, optimization method, data visualization, data aggregation and integration and other methods. Bridging the gap in vocabulary and culture between the data scientists and oil technical professionals is a problem required to be solved in the oil big data field. This needs interdisciplinary collaboration among various fields. In the oil & gas industry, the big data can be used for seismic analysis, reservoir modeling, drilling services and production reporting. Integrate the big data method and physics-oriented data analysis, establish the interdisciplinary team composed of the computer scientists and oil engineers, take the results as the user-friendly interface, take the demands as the orientation, solve the relation between the problem and the overall situation and thus make the integration of the big data technology in the oil engineering field closer. The oil industry is considered as the "No Man's Land" for the entrepreneurs in the new era, while the main technology provider costs billions of dollars to attempt to enter into such industry (such as General Electric, IBM and Microsoft). The big data analysis adopts the statistical analysis, data mining, machine learning, simulation model, optimization method, data visualization, data aggregation and integration and other methods.

8.9.4 Data Integration System

Various different types of data generated during the drilling and its processing are problems that must be faced with during the drilling. As for the drilling process and the features of dispersed data sources, data integration of the drilling data is the precondition for making the analysis and processing of the drilling big data. Related researchers commit themselves to establish the big data integration in the drilling engineering, namely the analysis system expects to make the real-time processing and utilization for the data during the drilling process, and offer decision-assistant data support for improving the drilling technology and promoting the drilling efficiency.

Forcibly transferring the data to single data model may cause a series of problems, such as reducing the performance and improving the integration complexity. However, the integrated data system supporting multiple data storages has begun to appear, and the data integration system will become a key component for the researches of big data.

The researchers propose the frame model for the integration system of the big data and propose a group of qualitative evaluation standards: heterogeneity, autonomy, transparency, flexibility and optimality; they analyze and propose the important direction of future integration system, especially establishing the principles of equivalence and inclusion for cross-model queries, valid data conversion (including the migration among data storages), performance monitoring and automatic load balancing as well as distributed locking and transaction management.

8.10 Conclusion

With the development of the drilling intelligentization and digitalization, the automation degree of the drilling system is continuously improved. The dehumanization of the production facility and the intelligent management system during the technology, energy, transmission and industrial process is also continuously developing. The data volume generated in the oil industry is continuously improved. In the future, the data utilization will increase exponentially. Who has bigger data utilization, who will take the superiority in the competition?

Oil is the commodity with the biggest transaction volume in the world. Google, Facebook and Amazon advanced data analysis now is more and more applied into the energy industry. The timeliness and the importance of this research are obvious. The upstream, midstream and downstream of the oil industry is an integral whole, inseparable. Driven by the upstream big data, the midstream and down stream's attentions on the big data technology also will be improved.

The big data technology will generate disruptive influences on the whole oil & gas industry. Better utilize the big data, improve the petroleum and petrochemical industries and take it as the global leading industry to take the preemptive opportunities on the tide of economic development.

The functions of the big data in the production are embodied in following aspects:

1. Evaluate the stratum, develop new prospects and offer advantages for the competitive bidding;
2. Predict and analyze the changes in the oil deposit layer. Offer intelligent analysis for the oil deposit engineers by integrating and analyzing the seismic, drilling and production data and then improve the recovery ratio;
3. Make use of the big data technology to analyze the oil & gas price trends to guide the production, help the oil & gas companies to realize the optimization of resource allocation and then acquire the optimal cost control and profit value;

4. Optimize the production parameters to slow down the attenuation of the oil & gas well;
5. The enterprise will utilize the big data to remold the production and supply chain and then realize the maximization of the business value.

Not merely the big data technology will generate influences on the oil & gas industry. The progresses obtained in computing technology, Internet of Things, cloud computing, mobile communication technology, robot technology and artificial intelligence bring new innovations for the oil & gas industry. Integrating the traditional production mode in the oil & gas industry with the rapidly developing Internet industry will definitely make the oil & gas industry glow the new vitality.

References

1. Economist A. Special report on managing information: data, data everywhere. Economist [J]. 2010.
2. Mark, "Gartner says solving 'big data' challenge involves more than just managing volumes of data," Gartner. Archived from the original on Jul. 10, 2011, Retrieved Jul. 13, 2011.
3. Mohammadpoor M, Torabi F. Big data analytics in oil & gas industry: an emerging trend [J]. Petroleum, 2018.
4. Feblowitz J. The big deal about big data in upstream oil &gas[J]. IDC Energy Insights, 2012: 1–11.
5. Baaziz A, Quoniam L. How to use Big Data technologies to optimize operations in Upstream Petroleum Industry[J]. arXiv preprint arXiv:1412.0755, 2014.
6. Jin W, Li ZJ, Wei LS, et al. The improvements of BP neural network learning algorithm[C]// WCC 2000-ICSP 2000. 2000 5th international conference on signal processing proceedings. 16th world computer congress 2000. IEEE, 2000, 3: 1647–1649.
7. Ben-Haim Y. Tom-Tov E. A streaming parallel decision tree algorithm[J]. J Mach Learn Res. 2010;11(Feb):849–72.
8. Joachims T. Text categorization with support vector machines: learning with many relevant features[C]//European conference on machine learning. Berlin: Springer; 1998. p. 137–42.
9. Manikandan SG, Ravi S. Big data analysis using Apache Hadoop[C]//2014 international conference on IT convergence and security (ICITCS). IEEE, 2014: 1–4.
10. Perrons RK, Jensen JW. Data as an asset: what the oil & gas sector can learn from other industries about "Big Data"[J]. Energy Policy. 2015;81:117–21.
11. Feblowitz J. Analytics in oil & gas: the big deal about big data[C]//SPE digital energy conference. Society of Petroleum Engineers, 2013.
12. Mohammadi AH, Anderson R, Tohidi B. Carbon monoxide clathrate hydrates: equilibrium data and thermodynamic modeling[J]. AICHE J. 2005;51(10):2825–33.
13. Xie H, Shanmugam AK, Issa RRA. Big data analysis for monitoring of kick formation in complex underwater drilling projects[J]. J Comput Civ Eng. 2018;32(5):04018030.
14. Qiang X, Cheng G. An overview of China intelligent digital oilfields development in 2017[C]// 2017 International Conference Advanced Engineering and Technology Research (AETR 2017). Atlantis Press, 2018.
15. Kragas TK, Williams BA, Myers GA. The optic oil field: deployment and application of permanent in-well fiber optic sensing systems for production and reservoir monitoring[C]// SPE Annual Technical Conference and Exhibition. Society of Petroleum Engineers, 2001.
16. Kersey AD. Optical fiber sensors for permanent downwell monitoring applications in the oil & gasindustry[J]. IEICE Trans Electron. 2000;83(3):400–4.

17. Aref SH, Latifi H, Zibaii MI, et al. Fiber optic Fabry–Perot pressure sensor with low sensitivity to temperature changes for downhole application[J]. Opt Commun. 2007;269(2):322–30.
18. Bertocco R, Padmanabhan V. Big data analytics in oil &gas[J]. Bain Brief, March, 2014.
19. Rubinstein JL, Mahani AB. Myths and facts on wastewater injection, hydraulic fracturing, enhanced oil recovery, and induced seismicity[J]. Seismol Res Lett. 2015;86(4):1060–7.
20. Noshi CI, Assem AI, Schubert JJ. The role of big data analytics in exploration and production: a review of benefits and applications[C]//SPE international heavy oil conference and exhibition. Society of Petroleum Engineers, 2018.
21. Hassani H, Silva ES. Big data: a big opportunity for the petroleum and petrochemical industry [J]. OPEC Energy Rev. 2018;42(1):74–89.
22. Baaziz A, Quoniam L. The information for the operational risk management in uncertain environments: Case of early kick detection while drilling of the oil or gas wells[J]. Int J Innov Appl Stud. 2013;4(1)
23. Thorogood J, Aldred WD, Florence F, et al. Drilling automation: technologies, terminology, and parallels with other industries[J]. SPE Drill Complet. 2010;25(04):419–25.
24. Macpherson JD, de Wardt JP, Florence F, et al. Drilling-systems automation: current state, initiatives, and potential impact[J]. SPE Drill Complet. 2013;28(04):296–308.
25. Michel J, Owens EH, Zengel S, et al. Extent and degree of shoreline oiling: deepwater horizon oil spill, Gulf of Mexico, USA[J]. PLoS One. 2013;8(6):e65087.
26. Hems A, Soofi A, Perez E. How innovative oil & gas companies are using big data to outmaneuver the competition[J]. Microsoft White Pap., 2013.
27. Baihly JD, Altman RM, Malpani R, et al. Shale gas production decline trend comparison over time and basins[C]//SPE annual technical conference and exhibition. Society of Petroleum Engineers, 2010.
28. Rahuma KM, Mohamed H, Hissein N, et al. Prediction of reservoir performance applying decline curve analysis[J]. Int J Chem Eng Appl. 2013;4(2):74.
29. Dong C, Mullins OC, Hegeman PS, et al. In-situ contamination monitoring and GOR measurement of formation fluid samples[C]//SPE Asia Pacific oil & gas conference and exhibition. Society of Petroleum Engineers, 2002.
30. Anderson RN. 'Petroleum analytics learning machine' for optimizing the internet of things of today's digital oil field-to-refinery petroleum system[C]//2017 IEEE international conference on big data (big data). IEEE, 2017: 4542–4545.
31. Zikopoulos P, Eaton C. Understanding big data: analytics for enterprise class hadoop and streaming data[M]. McGraw-Hill Osborne Media, 2011.
32. Bolen M, Crkvenjakov V, Converset J. The role of big data in operational excellence and real time Fleet performance management—the key to Deepwater thriving in a low-cost oil environment[C]//IADC/SPE drilling conference and exhibition. Society of Petroleum Engineers, 2018.

The manufacturer's authorised representative in the EU is Springer Nature Customer Service Centre GmbH, Europaplatz 3, 69115 Heidelberg, Germany. If you have any concerns regarding our products, please contact ProductSafety@springernature.com

Printed and bound by CPI Group (UK) Ltd, Croydon, CR0 4YY
22/09/2026
02221062-0016